Sustainable Plantation Forestry

Herman Hidayat

Sustainable Plantation Forestry

Problems, Challenges and Solutions

Springer

Herman Hidayat
Indonesian Institute of Sciences (LIPI)
Jakarta
Indonesia

ISBN 978-981-10-7652-7 ISBN 978-981-10-7653-4 (eBook)
https://doi.org/10.1007/978-981-10-7653-4

Library of Congress Control Number: 2017964497

Printed on acid-free paper

This Springer imprint is published by Springer Nature
The registered company is Springer Nature Singapore Pte Ltd.
The registered company address is: 152 Beach Road, #21-01/04 Gateway East, Singapore 189721, Singapore

Acknowledgements

The strong inspiration and motivation to write this book came from the wise words of the well-known Indonesian novelist Pramoedya Ananta Toer who has said that "the book is the eternal works for any scientist and should be transformed as knowledge for generation to generation by reading it." These words drive a strong desire for this author to collect my former research findings in Japan, ASEAN, and United States. In line with this, some papers have been published in academic journals in Indonesia, Thailand, Japan and so forth.

Relating to research findings, papers and academic journal collections—actually some papers have received comments from excellent reviewers and that they could eventually be published a book entitled: *A Sustainable Plantation Forestry: Problems, Challenges and Solutions*. For this reason, I would like to extend my grateful thanks to Springer Publishing Company for their assistance and cooperation in publishing and disseminating this book to readers. I would also like to thank the two reviewers who gave a positive indication from their critical comments; I as the author thank them and subsequently revised the manuscript based on their input.

While I carried out fieldwork entitled: "Development of Pulp and Paper in Japan" for the Institute of Developing Economies (IDE)-JETRO from August 2006 to March 2007 in Chiba, Japan, I am especially indebted and thankful to host scientist Dr. Yuri Sato, Prof. Masahisa Fujita as President of IDE-JETRO and others. My appreciation and thanks are also presented to Prof. Yasuyuki Kono as the host scientist when I received a scholarship from the Japan Society Promotion of Science (JSPS) Post Doctoral Fellowship in 2008–2010 and conducted field work entitled: "In Search of Sustainable Plantation Forestry in ASEAN countries" and Prof. Kosuke Mizuno, Director of Center for Southeast Asian Studies (CSEAS)-Kyoto University. I would also like to recognize colleagues in several institutions, such as Kasetsart University, Khon Kaeng University, Siam Cement Group (SCG) of Paper Industry officer in Khong Kaeng; Faculty of Forestry of the University of the Philippines in Los Banyos, CSDC's Wood Processing Mill Officer in Talacogon-Butuan City; Vietnam Forestry University, Ministry of Agriculture and Rural Development (MARD); The Department of Forestry officer in Hanoi,

Research Institute of Pulp and Paper Industry (RIPPI) officer, Vinapaco Paper and Bai Bang Company, The Forest Research of Center Office in Phu Ninh District, Phu Tho Province; Sinar Mas Group (SMG) Officer in Jakarta and Asia Pulp and Paper (APP) officer, Perawang-Riau Province for obtaining data collection through in depth interviews with informants and participatory observation in addition to obtaining interviews with academic communities for discussion on conceptual and theoretical review comments of papers; for instance, Prof. Ukan Abdullah (University of Padjadjaran-Bandung), Prof. Makoto Inoue (The University of Tokyo), Dr. Masatoshi Sasaoka (Hokkaido University), Andreas Neef (University of Hohenheim-Germany), Masaaki Okamoto (CSEAS), Masayuki Yanagisawa, Prof. Koji Tanaka, Prof. Wil de Jong, Prof. Mamoru Kanzaki, Kazuo Watanabe, Kazuya Watanabe, Fumikazu Ubukata, Osamu Kozan, Hideki Hayashida (Doshisha University) and so forth.

The chance arose to be Visiting Professor in November 2012–March 2013, while I carried out fieldwork out entitled: "Yakushima National Park and World Heritage" in Yakushima island under the sponsorship of Kagoshima University Research Center for the Pacific Islands (KURCPI). I would like to express my sincere thanks to Prof. Shunsuke Nagashima and Sota Yamamoto as host scientists. Other scientists who contributed to my data findings include: Prof. Kei Kawai, Prof. Shinichi Noda, Prof. Sueo Kuwahara, Prof. Tsuyoshi Yoneda, and Prof. Satoru Nishimura. Special gratitude to Dr. Carol J. Pierce Colfer as host scientist in Kahin Advance Research Center, Cornell University, Ithaca, New York, while I obtained AIFIS-Henry Luce Foundation Grants for 2014 (August–October) to conduct fieldwork entitled: "A Sustainable Forest Management in New York State". In data collection and discussion, I express my gratitude to Prof. Martin Hatch, Prof. Benedict Anderson, Dr. Audrey Kahin, Dr. Richard Dudley, Dr. Anto Mohsin, Dr. Thomas B. Pepinskey, Dr. Steven A. Wolf, Dr. Thamora Fishel, Prof. David Sonnenfield (State University of New York-College of Environmental Science and Forestry), and Prof. David H. Newman. Finally, I am pleased to address my gratitude and thanks to my family: my wife, Dr. Yani'ah Wardani M.A.; my daughters Ana Sabhana Azmy M.Si., Kholilah Saadah M.Sc., and my son M. Alfatih M.Sc., who have always accompanied me in times of happiness and sorrow. I do hope that the publication of this book will be a great contribution and sharing of knowledge for academia, government officers, politicians, businessmen, forest scientists and society at large.

Jakarta, Indonesia
December 2017

Contents

Part I
Introduction

Chapter 1
Introduction

Introduction

Currently, global society is aware that 'climate change' is having a negative environmental impact on such aspects of the environment as global warming and the tides of sea water. Climate change impacts 'forests' through the increased intensity and frequency of extreme weather events (*storms, floods, droughts*) and it may lead to an increase in forest fires and outbreaks of pests and diseases. In Japan, for example, 170 died and 54,216 suffered sickness in hospital during the summer months (July–August 2010) (*Gatra* 14–20 October 2010). Inn 2003, Europe suffered one of its hottest summers ever. Across the continent, temperatures soared; in France, the thermometer hovered around 40 °C for a fortnight. More than 52,000 Europeans died from heat related conditions in the summer of 2003.[1]

The International Conference on 'climate change', which was held in Bali on December 3–15, 2007; on December 24–30, 2008, in Poznan, Poland; and December 26–31, 2009 in Copenhagen, Denmark aimed to shape a new global agreement to fight 'global warming'. Based on the critical impact of climate change noted above, world leaders, including previous Prime Minister of Japan, Yukio Hatoyama, made a commitment to reduce down to 25% until 2020, factors responsible for climate change based on the agreement of Protocol Kyoto 1997 (*Gatra* 14–20 October 2010). In line with these conditions, scientists and environmentalists want mechanisms to reward the developing world for saving its forests incorporated into any such treaty. Forests and jungles absorb carbon dioxide, the main greenhouse gas, which is stored in trees. Cutting trees down releases CO_2

[1]For a better understanding about the impact of "Global Warming already kills 150,000 people every year," see Williams (2007). Further information see article by Janet Larsen, "Setting the Record Straight: More than 52,000 Europeans Died from Heat in summer 2003," July 28, 2006. See in http://www.earth-policy.org.

H. Hidayat, *Sustainable Plantation Forestry*,
https://doi.org/10.1007/978-981-10-7653-4_1

and triggers the emission of additional greenhouse gases from denuded soils. Forest loss and land degradation could be responsible for 20% of the planet warming gases attributable to human activities, some experts suggest (World Growth, December 4, 2008).

As is known, deforestation in tropical countries, forest fires, forest land conversion for agricultural farming (oil palm, coffee, cacao, etc.), and industrial activities in developing and advanced countries, as well as consuming oil, LNG, coal and other energy sources eventually produce gas emission (CO_2). One of the positive decisions of the United Nations regarding 'climate change' is to present a scheme for 'Reducing Emission from Deforestation and Forest Degradation', the so called R.E.D.D. Many Western governments and environmental activist groups like WWF, Conservation on Nature, and Friends of the Earth (FOE) have launched critiques of 'deforestation'. For instance, they are now proposing that all deforestation be halted in tropical zones—a move that would undoubtedly increase poverty in developing countries (World Growth, Media Released, December 2, 2008). In contrast, a growing number of experts in climate and development are calling for the expansion of 'sustainable forestry' (cultivating, maintaining, and developing forests), as a cost-effective way to reduce carbon emissions and combat the current global recession, while environmental activists are continuing to campaign for absolute restrictions. As the statement of Alan Oxley, Head of World Growth group says: "If forests are managed in sustainable ways, their cultivation can both raise living standards and reduce greenhouse gases."

Wulf Killmann, Director of FAO's forest products and Industries Division, says 'plantation forestry' is playing a useful role in counterbalancing deforestation and forest loss. They harbor less biodiversity than natural forests, but provide not only timber but they also protect water resources and soils. By sequestering carbon dioxide, plantation forests help mitigate climate change (http://www.metsabotnia.com). Many North American and European companies, including Finnish and Swedish, and Japanese companies have invested in 'forest plantations' and pulp and paper production in South America and Southeast Asia. From the global forest perspective, this development is beneficial.

It is widely recognized that pulp and paper industries in Japan and Association of Southeast Asian Nations (ASEAN) are very strategic. A previous research paper (2007) entitled: "Pulp and Paper Industries in Japan and Indonesia: From the Viewpoint of Political Ecology" revealed that thirteen private companies from these industries occupy a rank in the top 41 largest manufacturing companies in Japan, which contributed about 6.8 trillion Yen and absorbed 34,839 workers (Japan Pulp and Paper 2005). Also in Indonesia, as a member of ASEAN countries, currently pulp and paper industries are categorized as strategic industries which contributed about US$2,1 billion in 2005 and rapidly increased US$3,3 billion in 2007 and create huge employment, about 1,7 million workers in 2007.[2] This research highlights the role of 'stakeholders' such as the role of government, private sectors, local

[2]Statistic Indonesia: Economic Indicators, successive issues, 2007.

people, academics, and NGOs, domestic and international as well, such as Green activist groups like WWF, Greenpeace, and Friends of the Earth (FOEI). Japan and Indonesia could be categorized as among the 12 major countries producing paper and paperboard. For example, Japan produced 30,889 million tons (number three after USA and China) and Indonesia is number twelve with total production 7,678 million tons (Japan Pulp and Paper 2005). The demand for raw material such as wood is quite big for both countries. For example, Indonesia's wood demand reached 20 million m^3 in 2000 and rapidly increased to 28.9 million m^3 in 2005 and increased 37 million m^3 in 2008.[3] Japan also reached 87 million m^3 in 2003 and slightly increased 89 million m^3 in 2004, which is provided by domestic supply of about 16 million m^3 and imported supply of 71 million m^3.[4] As an illustration, in 1993, 38% of Japan's chip wood came from North America; 30% from Australia and New Zealand; 15% from Latin America; 8% from other Asian countries (Southeast Asia); and over 1% from Fiji and Papua New Guinea. Over the next decade, hardwood supplies from Southeast Asian plantations are likely to replace a significant portion of imports from more distant regions such as Chile or the southeastern US.[5]

Unfortunately, huge numbers of these productions are not inherently accompanied by properly planned 'plantation forestry'. As a result, the forestry industries are facing 'critical' problems in procuring timber and chip wood as raw materials in Japan and ASEAN countries.

Hence, 'plantation forestry and forest conservation' were one of the key issues at the 1992 Earth Summit. In other words, between profitability and environmental conservation regarding upland areas where competition arises in land use between agriculture and forestry.[6] Although many countries adopted Agenda 21, which called for actions to prevent 'deforestation', and the Forest Principles, the Earth Summit failed to conclude with the creation of a Forest Convention. After the Earth Summit, a number of international initiatives emerged, such as the *Intergovernmental Panel on Forests* (IPF), the *World Commission of Forest and Sustainable Development* (WCFSD), and others, in order to find ways to halt worldwide deforestation and degradation of all types of forestlands.

This paper aims to discuss the dynamic process of 'plantation forestry' from the viewpoint of a 'stakeholders' movement. Meanwhile, sustainability level analyses will be focused on economic: soft loan, technical innovation,[7] subsidiary, training

---

[3]Indonesian Pulp and Paper Association (APKI) for 1988–2000; Poyry (1998) for 2005–2010 projection.

[4]See Iwai (2002); Forestry Agency (2000).

[5]Carrere and Lohmann (1996).

[6]See Kono et al. (1994).

[7]Selecting of seeds and developing of new species of (*Acacia mangium*, *Eucalyptus*, etc.) for fast growing trees whether in planting and harvesting as well. See: The project of Humanosphere Professor Kawai, Shunichi, "Study on Nature-inspired technologies and institutions". Interviewed with Prof. Kono Yasuyuki, in CSEAS, Kyoto University, on December 29, 2008.

manpower; social: people's participation,[8] rules, networks, etc. and ecological issues: the protection biodiversity, water and soil.

Plantation Forestry

The initial title uses "in search of sustainable plantation forestry". This hypothesis means that plantation forestry in developing countries (**ASEAN**: *Indonesia, Malaysia, Philippines, Thailand, Vietnam*, etc.) are not yet sustainable from the viewpoint of economics: the lack of economic incentive, low production, etc.; social: lack of participation, land disputes; and ecological: plantation eventually affects ecological damage, such as loss of biodiversity, soil erosion, lack of water catchment, etc. To explain in terms of 'participation' in the plantation forestry is significant. In the context of development plans and programs, participation can be defined as "the process through which stakeholders influence and take part in decision-making in the planning, implementation, monitoring and evaluation of programs and projects". The issue of participation and its forms, potential and problems raises the question of the optimal level of involvement of *local people*. If participation was maximized, local people would have complete control over the natural resources. Thus, the aim of participation may not necessarily be to transfer the decision-making power completely to local communities, but rather to initiate a process of negotiation among stakeholders: integrating individual, communal and national interests in balanced way.[9]

From this description, therefore, the question is how to transform from not sustainable to sustainable. The appropriate answer to how to actively involve 'stakeholders' in planting trees in terms of economic, social and ecological issues are significant for carrying out sustainable plantation forestry.

Sustainable Plantation

The word 'sustainable' is not as new to the forestry profession, including forest economists, as it may be to some mainstream economists. The Faustmann Formula, one of the main pillars of conventional forest economics, is based on the idea of a sustained supply of timber for an infinite number of rotations. However, the recent concerns about 'sustainability', which were signaled by the publication of "The Limits to Growth" by Meadows et al. (1972) and "Our Common Future" by WECD (1978), are not limited to a specific product, but include all natural systems and human life. In simple words, 'sustainability' involves ensuring opportunities for a

[8]Inoue (2003).

[9]See Heyd and Neef (2006).

desirable 'quality of life' for all future generations as well as for the present one. Humans' quality of life includes not only the economic dimension but at least two others: the ecological and the social.[10] In line with this definition, according to Fikret Berkes and Carl Folke, 'sustainability', as used here, is a process and includes ecological, social and economic dimensions. The term *ecological system* (ecosystem) is used in the conventional ecological sense to refer to the natural environment. We hold the view that social and ecological systems are in fact linked. Therefore, when we wish to emphasize the integrated concept of humans-in- nature, we use the term *social-ecological system* and *social-ecological linkages* (Berkes and Folke 2000: 4). The implementation of a social-ecological system based on the opinion of (Norberg and Cumming 2008: 155) needs complex adaptive systems (CAS) approach that requires societies to: (1) to learn to live with change and uncertainty; (2) combine different types of knowledge throughout the learning process; (3) create opportunities for self-organization toward social-ecological resilience; and (4) Training the development of capacity for renewal and reorganization.[11]

When implementing the economic theory of 'sustainability', forest ecosystems can be of enormous use due to numerous reasons. First, forest ecosystems are important components of all the international agreements related to the sustainability-convention for climate change, biodiversity convention, and Agenda 21. Second, interactions between human systems and forest ecosystems can provide an experimental setting to study interactions between ecological, social, and economic dimensions of human welfare. Third, the concept of sustainability, even though in a limited sense (related *timber*), has existed for about 150 years in the thinking about forestry, including forest economics.[12]

It is widely accepted that 'forest resources' should be managed to meet the economic, ecological and social needs of present and future generations. Therefore 'plantation forestry' must respond to *economic, environmental and social* issues. This requires feedback of relevant information between the planning, implementation, control and impact of forest management.

The concept of criteria and indicators (C&I), which was designed to be used in assessing the sustainable 'plantation forestry' has been evolving since 1995 when this issue gained recognition by the Intergovernmental Panel on Forests (IPF). Several international institutions have developed guidelines and C&I for sustainable plantation forestry.[13] For example, ITTO launched criteria for assessment of sustainable plantation forestry; WWF and IUCN developed guidelines for timber plantation, environmental, social and cultural issues relating to commercial afforestation; *LEI* also developed C&I for Sustainable Forest Management (SFM) of natural and plantation forests in Indonesia. Centre International on Forestry

---

[10]Kant and Berry (2005).

[11]For more information 'Living in a Complex World', see Walker and Salt (2006).

[12]Ibid., p. 4.

[13]See Muhtaman, Siregar and Hopmans (2000).

Research (CIFOR) has developed a generic process for the development and evaluation of C&I for natural forests (Prabu et al. 1999).

Clearly a more holistic approach to plantation development and management is required with due consideration to not just wood production but also *environmental, social and economic* factors. Long-term sustainability will only be achieved by taking into account: the ecological capability of the site; intensity of management; soil, water and other environmental values, economic benefits; and social goals. In general, C&I can be formulated to serve at various levels of scale, namely global, regional, national and sub national, or at the Forest Management Unit (FMU) level.

The teams of CIFOR agreed that sustainable development of plantations must improve the socio-economic condition and well-being of the local community. The criteria proposed for the principle of human well-being addressed the following issues:

(a) security of land tenure and land use;
(b) participation in forest management;
(c) sharing of social and economic benefits;
(d) industrial relations and responsibilities of stakeholders.[14]

Economic Benefits and Rural Development

Plantation forestry in the tropics can significantly aid economic development, especially through earnings, foreign exchange from exports of forest products or import substitution. The highly successful use of government incentives in Chile and Brazil to encourage plantations[15] has made these countries become leading exporters of wood pulp. While governments may support plantations for financial benefits alone, more often it is for broader economic reasons and to generate social and environmental benefits. For instance, employment opportunities through developing new forest industries, watershed protection, enhanced landscape amenity values, recreational opportunities, and land rehabilitations are some of the justifications for government involvement in plantation development, either through state, forestry companies, or the provision of a range of incentives.

Meanwhile for ecological protection, industrial plantations of *A. mangium* are managed primarily for wood production, but the intensity of management varies considerably between companies responsible for the various concession areas. Long-term sustainability of this plantation resource requires management to take into account not only wood production but also ecological and environmental values (e.g. *stabilizing soil, prevention of erosion, controlling water runoff in catchment areas, biodiversity*, etc.) as well as socio-economic issues. It is therefore

[14]Ibid., p. 12.

[15]Evans and Turnbull (2004).

important to evaluate the impact of plantation development on the structure, function and resilience of the entire ecosystem of the FMU.

On the other hand, when elaborating on economic issues, Inoue Makoto pointed out some guiding principles such as: the role of central and local government is very significant to contribute soft loans, subsidiary (seedling, fertilizer, etc.), benefit sharing of advantages, training manpower (to maintain plantation forestry), and technical innovation (to provide seeds to bolster production)[16] to private sector and local community. Furthermore, on social issues, Inoue highlights participation, rule, and network. Meanwhile, when it comes to ecological issues, the role of government is necessary to protect and evaluate bio-diversity, soil quality, water catchment, flood and drought.[17]

Why Plantations?

It must be highlighted that overall there is a worldwide 'shortage' of timber. Several studies have addressed global supply and demand for wood. Sedjo and Lyon (1996) indicated that average annual demand for industrial roundwood would increase from 1700 million m^3 in 1995 to about 2300 million m^3 in 2045. Sohngen et al. (1997) assumed regional supply indicated from existing forest resources and the potential role of plantations were expanded and FAO (1998) has estimated consumption, production and trade in forest products globally to 2010 (Table 1.1). By 2010, Asia, enhanced by increasing population growth and economic development, Europe and the former Soviet Union are expected to increase their share of global industrial roundwood consumption while the North and Central American region is expected to decrease its share (FAO 1998). Therefore, consumption of pulpwood for wood-based panels and paper is expected to approximately double to 1330 million m^3 in 2045. In 1996, 70% of industrial roundwood was used in developed countries (FAO 1999) (Evans and Turnbull 2004: 13).

Actually, there were some studies—FAO (2001), Evans and Turnbull (2004), Jiang and Zhang (2003), Guizol and Aruan (2006), and Forest Inventory and Planning Institute, Vietnam (2005), etc.—that highlighted the necessity of expanding plantation forestry, especially *Eucalyptus* and *Acacia* trees, as raw materials to overcome timber shortages for pulp and paper industries. FAO (2001) reported that plantation forestry has increased in the last two decades (1980–2000). Globally, FAO resource inventory data suggests that plantation estates have increased from 17.8 million hectares in 1980 and 43.6 million hectares in 1990 to

[16]Interview with INOUE Makoto, Professor from Department of Global Agricultural Sciences, Graduate School of Agricultural and Life Sciences, The University of Tokyo, on January 16, 2009 in Tokyo.

[17]Interview with Abdullah, Oekan S, Visiting Scholar in CSEAS, on January 25, 2009 in Shugakuin Koryu Kaikan, Kyoto.

Table 1.1 Current and future forecast global forest production and estimate consumption by products, 1996–2010

Products	Production 1996 (mil. m^3)	Estimate consumption 2010 (mil. m^3)	Growth 96–2000 (%)
Industrial roundwood	1490	1872	26
Sawn wood	430	501	17
Wood-based panels	149	180	20
Wood pulp	179	208	16
Paper and paperboard	248	394	39
Fuelwood and charcoal	1860	2210	27

Source FAO (1998, 1999), Evans and Turnbull (2004). *mil* Million

187 million hectares in 2000 (Evans and Turnbull 2004: 30) and rapidly developed to become 271 million hectares in the year 2005.[18]

Plantation in Other Countries

In line with the FAO (2001) report, the findings are that in China, plantation forestry also appeared as a significant policy taken by the government. Therefore, protecting natural forests and expanding the plantation forests has become a national priority in China.[19] There are two rationales behind this scenario: Firstly, the environmental issues that have been raised from ecological disasters in recent years (e.g. flooding, soil erosion, decrease of biological species, shortage of water resources etc.), resulting in hundreds of billions of dollars in losses and unmeasurable long-term environmental, social and economic impacts. Secondly, the increasing economic demand, as world population and consumption of wood products increased. The official statistical data shows that China imported 1.12 billion USD of wood and wood products in 1981, 2.6 billion USD in 1992, and 5.3 billion USD in 1996, rapidly growing to become 6.3 billion USD in 1998. To overcome this 'shortage of timber', the government took the policy of expanding plantation forestry. Today, China has the world's largest plantation resources. The National Forest Resources Survey shows that plantation areas in China have reached 46.67 million hectares in 1999, which accounts for about 20% of the entire world forest plantations. It is estimated that China's plantation forests can annually

[18]See "Wood from planted forest: a global outlook 2005–2030" (http://www.entrepreneur.com/tradejournals/article/191954507.html).

[19]See Jiang and Zhang (2003).

provide 130 million m^3 by 2015, which can meet domestic needs. Thus, the imbalance between demand and supply in this country can be resolved (Jiang and Zhang 2003).

Indonesia launched a policy to boost plantation forestry which mostly highlighted conservation and economic aims. The forestry industries (plywood, sawmill pulp and paper mills) needed a substantial amount of timber in this country, which eventually led to large scale extraction in primary forests, causing massive deforestation. Current annual log supply shortage is 35 million m^3, i.e. between log supply from primary and plantation Forests 15 million m^3 and log demand is 50 million m^3.[20] Therefore, to mitigate the negative impacts of timber exploitation and generate alternative timber supplies, the government set up a reforestation fund and promoted large scale industrial timber and plantations of fast growing plants (*Acacia and Eucalyptus trees*), which actively engage private companies, cooperative, forest state enterprise (Inhutani) and farmers. The official aim of promoting HTIs (timber plantation forestry) was to create wood resources on unproductive forest lands located in 'productive forests'. However, in practice, because of a lack of control and collusion, it triggered, in many places, clear-cutting of rich natural forests. During *Repelita* V (Five Year Planning) (1990–1994), 900,000 ha of large scale plantations were planted, amounting to 60% of the 1.5 million hectares target. Overall, it was a period of rapid HTI expansion, although the annual target of 300,000 ha planted was only met in one year that was 1993.[21]

The main target of the plantation forestry which started in the mid 1980s–2000s and actively engages stakeholders in primary forests (*protection, conservation forests and production forests*) was economic aims and conservation in order to recover forest coverage. In line with these aims, the Minister of Forestry launched five priority programs for 'forestry development' during the period 2001–2004, namely: (i) combating illegal logging; (ii) controlling forest fires; (iii) restructuring the forestry sector; (iv) establishment of plantation forests and reforestation and (v) decentralization of forestry activities (Ibid). Therefore, in the case of item 4, which mostly emphasizes conducting plantation forestry and reforestation, is aimed at providing forest conservation and timber for forest industries. In 1990, plantation forestry registered 3.7 million hectares and increased to become 9.8 million hectares in 2000 (FAO 2001). The main species is *Acacias* spp. and *Eucalyptus* plants, composing 80% of the pulp plantation inputs.

In the case of Vietnam, there is more emphasis on economic and conservation aims for plantation forestry. As registered, there were 600,000 ha total plantation forestry in 1990 that increased to become 1.7 million hectares in 2000 (FAO 2001). The driving factors on rapid plantation are: (1) Economic reform is in force in Vietnam shifting centrally planned systems to a market oriented economy. The main portrait in the forestry sector has been the application of a policy of 'land and forest allocation' to persons and entities effectively involved in forest land farming

---

[20]See Aruan (2005).

[21]For further information see Guizol and Aruan (2006).

and plantations; and (2) the government has also encouraged domestic and foreign investors to become involved in a wide range of plantation activities. This policy has attracted investors, cooperative sectors, farmers, and NGOs to tree planting in Vietnam. Today *Eucalyptus* planting has reached a climax rate; however, the State and many scientists, express caution and are of the opinion that *Eucalyptus* plantation programs will not be effective unless there is a suitable choice of species, sites and appropriate management of concentrated plantations, are adopted.[22]

From an intra regional illustration, what does the Thailand case mean at a macro level in terms of plantation forestry? In the case of Thailand, how will the country overcome timber shortages and extend the plantation forestry? As a consequence of the rapid growth of pulp and paper, sawmill plywood and furniture industries at the end of 1990s to the beginning of the 2000s, wood demand reached 17.2 million m^3 in 1999 and rapidly developed to become 20.1 million m^3 in 2006. There are two ways in which the Thai government is overcoming the shortage of timber. Firstly, the Thai government invites *stakeholders*, such as private companies, academics and local farmers to be actively involved in planting and developing 'plantation forestry' (*Eucalyptus camaldulensis*) for commercial trees. Secondly, the government launched 'economic incentive' policies: providing accessibility to credit, tax relief on machinery imports and other goods for paper factories and *infrastructure* (*highways and port facilities*). These policies taken by the government eventually led to widely expanded plantation estates throughout the whole country and highlighted how significant *plantation forestry* could be for providing the raw material for pulp and paper industries in the future. Thailand and her people welcomed plantation forestry, especially *Eucalyptus* trees. According to some studies, Thaiutsa et al. (2003), Forestry Research Center of Kasetsart University (1989), Thailand Development Research Institute (1991), and Forest Research Center for Royal Forest Department's paper (2008), it was recommended that *E. camaldulensis* was a suitable plant according to soil and climate conditions and promoted as 'commercial trees' for Thai people. These findings have raised questions as to what the 'specialties' are of Thailand compared with other countries? For instance, China's specialty is that plantation forestry primarily highlighted the goal to boost greening areas, because of highly polluted air. In the case of Vietnam and Indonesia, the emphasis has been on the balance of economic and environmental issues, aiming to recover natural forest and to boost wood production for domestic wood demand and to expand 'forest coverage'. These conditions lead to preventing ecological disasters such as floods, soil erosion and drought.

There were two arguments to support Thailand's specialty. Firstly, the Thai government adopted and legitimized plantation forestry for 'economic' benefit rather than environmental issues. Therefore, the focus on *Eucalyptus* trees was adopted for their commercial value for farmers and private companies, aiming to boost production, provide income generation and jobs for rural communities. Secondly, the government encouraged easier procedures to obtain timber

[22]For further information see Tran (2005).

concession areas, gave subsidies to farmers, access to credit, tax holidays and provided infrastructure to private companies and farmers who wished to plant and develop plantations. The government supported private companies creating synergic cooperation with local farmers under the scheme of 'contract farming'. This contract requires the company to give seedlings, fertilizers, accessibility to credit from banks and a guarantee of market at harvest time for timber products from the farmers. The schemes provide a win-win solution for both parties (company and local farmers). As a result, plantation forestry in Thailand has rapidly grown from 560,000 ha in 1990 to become 4.9 million hectares in 2000 (FAO 2001). Therefore, in order to deal with these policies, it is significant to understand general policy and its impact of Thai government on plantation forestry.

Most problems in plantation forest development are *policy*-related and institutional in nature—caused by human actions. Relevant *stakeholders* critically examine the following problems such as:

a. Lack of updated statistical data and information about existing types of plantation forest resources in Indonesia and other ASEAN countries covering: (i) amount; (ii) age; (iii) species; (iv) location; (v) site quality; (vi) schemes; (vii) present protection conditions; (viii) existence and nature of claims/conflicts, community interest and involvement. These important basic data are needed for better long-term policy analysis in the forestry sector including management plans.
b. Inadequate capability at decentralization level is apparent, among others, in (i) the poor state of forest management; (ii) low status of human resources development, and (iii) lack of planning capability. Moreover, changing over to a decentralized management system is a challenge and simultaneously an opportunity for potential improvement.
c. Other constraining factors are: (i) lack of adequate access to financial sources for potential plantation investors and revenue sources for local governments; (ii) lack of dialogue between relevant stakeholders, i.e. local communities, local and central governments, private sector, NGOs, etc. More than 20 years after the IX World Forestry Congress 1978 in Jakarta, this "forest for people" philosophy is finally being promoted (1999) in the decentralization context.[23]

Those problems have particular possible solutions, such as practical supports and development of regional innovative mechanisms and relevant legislative frameworks. Then, it is very clear, that plantation forest development is one of the central agendas for degraded production forest areas. This urgently needs applicable incentives.

[23]See Aulia (2009).

Purpose

The study focuses on 'plantation forestry' and the scope of period is limited to other ASEAN countries from the 1980s until 2000s, because they began to promote forest rehabilitation program and plantation forestry in these periods.

To fulfill these scopes, ***research purposes*** are mentioned below:

(1) to clarify the role of stakeholders especially government and private sectors in the implementation of sustainable "plantation forestry" programs;
(2) to examine local people "participation and subsidiary" in plantation forestry under the program of social forestry in the region;
(3) to discuss the critiques of NGOs, academics and local people toward implication on sustainable of plantation forestry;
(4) to explore the implication of plantation forestry program management on environmental spheres.

Methodology

Theoretical Framework

The study uses "political ecology" as an analytical framework. We must define political ecology. Many scientists (Bryant 1997; Blaikie and Brookfield 1987; Abe 2003) define it differently. Paterson (2000) notes that, "political ecology as an approach that combines the concerns of ecology and political economy to represent an ever-changing dynamic tension between ecological and human change, and between diverse groups within society at scales from the local individual to transnational as a whole." Other scientists define it as, "political ecology" a framework to understand the complex interrelations between local people, national and global political economies, and ecosystems (Blaikie and Brookfield 1987). The concept has been adapted in a variety of ways, such as Third-world political ecology, where (Bryant 1997) notes that: "political ecology may be defined as the attempt to understand the political sources, conditions and ramifications of environmental change." Most current political ecology tends to overlook ecological dynamics and focus upon the structure of human systems (Rocheleau et al. 1996). Abe (2003) defines political ecology, as "a collective name for all intellectual efforts to critically analyze the problems of natural resource appropriation and political economic origins of resource degradation, be they for the purpose of academic study or practical applications".[24] In other words, political ecology is concerned with the political dimensions of natural resource use and subtleties of those politics. Apparently, the scope of political ecology has been referred to as 'a method of

[24]Abe et al. (2003).

analysis', rather than a unified scientific discipline or sub-discipline, which is usually characterized by a set of related ideas, premises, and theories.

Meanwhile, commented: political ecology is similar to a method applied by human ecologists analyzing policy-relevant environmental questions that is 'progressive contextualization'. This approach starts with actors, in this case, direct resource users, and considers the contexts within which they act, or do not act, in a particular way towards a resource. This approach also intends to explain why people use the environment in particular ways, sometimes causing resource decline or degradation detrimental to their own and others' uses of the resources (Peluso 1992).

From the above definitions, apparently, Bryant's definition, which emphasizes 'putting politics first' on the political ecology of sustainable development aspects is more operational on 'In Search of Sustainable Plantation Forestry: Political Ecology Analyses on Stakeholders'. There are three reasons for this condition: First, that 'demand for plantation forestry' to supply as raw materials for pulp and paper industries in Japan and ASEAN countries have rapidly increased from the early 2000s. Second, the implementation of sustainable plantation forestry to bolster trees production could be evaluated from viewpoint of regulation, economic incentive, people's participation, technical innovation and linkage of institutional interaction. Third, is the impact on environmental conditions, such as flood, soil erosion, forest fire, depletion of biological species, etc.

'Political ecology' is a framework to approach the subjects mentioned. It is a generic term used for this research connecting two levels of study. This study highlights the viewpoint of politics in the study of ecological conditions. It includes a small-scale study centered on how local communities carry out plantation forestry with other stakeholders in their villages (e.g., using the approaches of *cultural anthropology, and applied anthropology*). Besides, focusing on a large-scale study from local, national, and regional standpoints, such as plantation forestry products distribution in ASEAN countries and Japan (e.g., political economy) and eventually utilization of the products as raw material of pulp and paper industries. After reviewing existing studies, I decided to adopt the framework focused on the movement and *logic of stakeholders* (actor analysis) among other frameworks of political ecology in this study (Figs. 1.1 and 1.2).

Actors' Movements

The paper would like to concentrate on actors' movements for two issues: namely identification of actors and the role of actors. Observing the above description, there are two critical reasons to identify actors' movements in the plantation forestry.

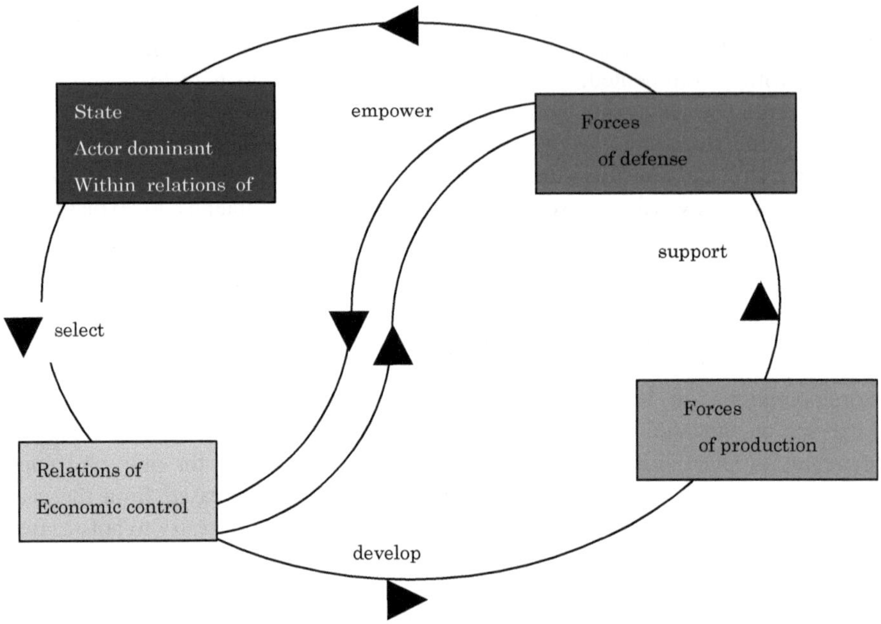

Fig. 1.1 An interrelational model that modified by Alan Carter (1993)

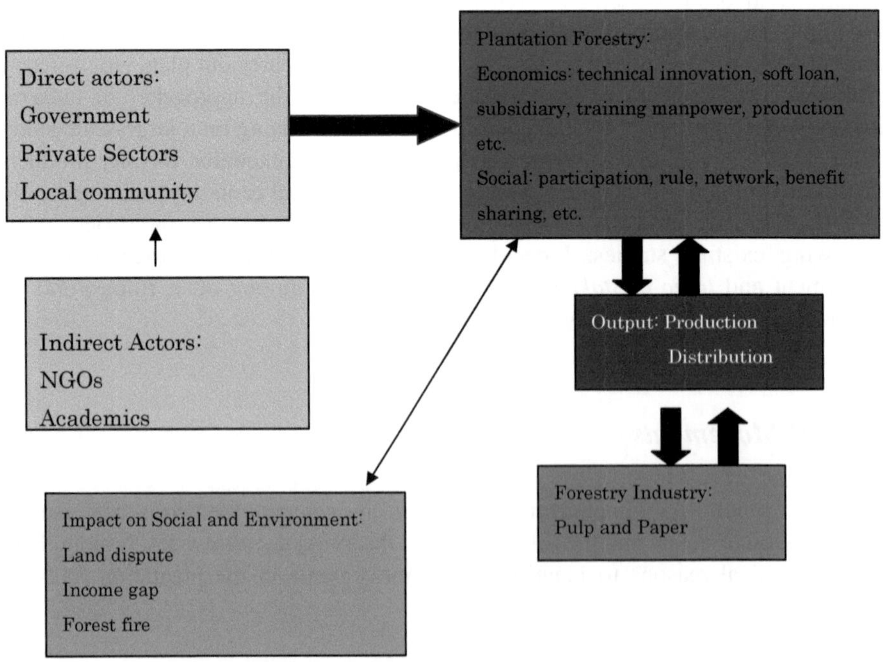

Fig. 1.2 Conceptual framework of political ecology of plantation forestry

Table 1.2 Political and economic forces and relations

Issues	Relations	Forces
Political	Relations of political control	Forces of defence
Economic	Relations of economic control	Forces of production

Source Carter (1993)

Firstly, is to clarify that direct actors can be categorized as the state,[25] (Table 1.2; Fig. 1.1), private sector, and local community. Secondly, is to review critiques from indirect actors, such as grass root movements, local government, academics, and NGOs.

The government (jointly performed between executive and legislative boards) as an actor on "plantation forestry" policy is represented by introducing and passing legislation and issuing permissions, monitoring and giving sanctions to the plantation forestry industries. On the other hand, businessmen represent the private sector and also local people independently own—which have legal concessions with the time limit of 7–8 years on operating plantation forestry, industrial timber plantation (HTI), and implement the replanting of trees in the state or private *production forests*, based on the law, regulations, economic incentives, participation, technical innovation, and institutional cooperation. The operational mechanisms carry out on the logging, timber, plantation forestry production and the paying of taxes for government income in the Department of Forestry. In contrast, the governments, through the Department of Forestry apparatus are responsible for controlling all of these operations and sanctions.

Seemingly, indeed there is a relationship between both roles, whether direct or indirect actors on plantation forestry industries. If both parties truly cooperated and created symbiotic relations in the near future based on a sustainable plantation forestry, the effects of deforestation will be prevented and reforestation (plantation forestry) programs to provide raw materials for forestry industries, such as plywood, sawmill, pulp and paper industries, etc., will continuously recur in ASEAN countries.

[25]I might now be in a better position to describe the actual role of the modern state. It can be argued that those who enjoy a dominant position within the relations of political control (ordinarily state actors) select or stabilize relations of economic control that are in their interests. For further information, see Carter (1993). There are new 'interdisciplinary' approaches necessitated by environmental and related issues for social sciences (political and economic aspects). See Dickens (1992).

Method for Data Collection

The main part of data and information are available through three techniques:

(1) Literature reviews of many books, journals and newspapers;
(2) Making 'guided interviews' and conducting 'in depth interviews' with *stakeholders* of plantation forestry programs, such as government officers, private sectors officers, academics, NGOs and local people;
(3) Field observation of plantation forestry programs in ASEAN countries (Indonesia, Philippines, Thailand and Vietnam) from the perspective of economic, social and ecological issues.

Research Sites

The focus of the study is highlighting the role of stakeholders on plantation forestry from the viewpoint of 'political ecology' analyses. To achieve this the research sites appealing case study on plantation forestry carried out in ASEAN countries (Indonesia, Philippines, Thailand and Vietnam) in the fiscal years 2009–2010. The initial field research in summer 2009 in July was carried out in Thailand: Bangkok, Khon Kaen province, Mancha Kiri sub-district and in August 2009 in the Philippines: Manila, Caraga Region (Mindanao), Butuan City, Talacogon-Agusan Del Sur. And in fiscal year 2010 in March, field work was carried out in Vietnam: Hanoi, Phu Tho province and Phu Ninh district. Meanwhile, field research in Indonesia was conducted in April and May 2010 in Jakarta and Riau Province. The rationale for choosing those countries, is based on two arguments. First, currently ASEAN forestry, which is focused on Indonesia, Thailand, Philippines and Vietnam are facing forest degradation and deforestation in many areas. Second, the rapid development of pulp and paper industries in ASEAN countries urgently needs huge timber supply to maintain operational industries. To overcome the critical condition, the government and other stakeholders, such as private companies, state forest enterprises, local farmers, etc. are carrying out planting trees in production forest to provide raw material of timber for pulp and paper industries. On the other hand, government, international agencies (World Bank, ADB, JICA, SIDA, CIAR), etc. and NGOs launched affirmative actions by carrying out reforestation and rehabilitation forest in protected and conservation forests to prevent soil erosion and flooding in respective countries.

Expected Results

This research will make a great academic contribution to the field of forest policy and conservation, because there has been no research to explore the development of *plantation forestry* in ASEAN countries comparatively in terms of *political ecology* analyses. The recommendations for these actors will also make a certain contribution to affect their future activities and forest policy making process.

Part II
Southeast Asia

Chapter 2
Thailand

Discourse on Thailand's Plantation Forestry (*Eucalyptus camaldulensis*)

Introduction

The initial question rises in terms of plantation forestry. In the case of Thailand, how does the country overcome timber shortages and extend plantation forestry? As a consequence of the rapid growth of pulp and paper, sawmill, plywood and furniture industries at the end of the 1990s to the beginning of the 2000s, wood demand reached 17.2 million m^3 in 1999 and rapidly developed to become 20.1 million m^3 in 2006. There are two ways in which the Thai government is overcoming the shortage of timber. Firstly, the Thai government invites *stakeholders,* such as private companies, academics and local farmers to be actively involved in planting and developing 'plantation forestry' (*Eucalyptus camaldulensis*) for commercial trees. Secondly, the government launched 'economic incentive' policies, providing accessibility to credit, tax relief on machineries imports and other goods for paper factories and *infrastructure* (*highways and port facilities*). These policies taken by the government eventually led to widely expanded plantation estates throughout the whole country and highlighted how significant *plantation forestry* could be for providing the raw material for pulp and paper industries in the future. Thailand and her people subsequently welcomed plantation forestry, especially *Eucalyptus* trees. According to some studies, Thaiutsa et al. (2003), Forestry Research Center of Kasetsart University (1989), Thailand Development Research Institute (1991), and Forest Research Center for Royal Forest Department's Paper (2008), it was recommended that *E. camaldulensis* was a suitable plant according to soil and climate conditions and promoted as 'commercial trees' for Thai people. These findings have raised questions as to what the 'specialties' are of Thailand compared with other countries? For instance, China's specialty that plantation forestry highlighted the aim of boosting greening

H. Hidayat, *Sustainable Plantation Forestry*,
https://doi.org/10.1007/978-981-10-7653-4_2

areas, because of highly polluted air. In the case of Vietnam, the Philippines and Indonesia, the emphasis has mostly been on the balance of economic and environmental issues, aiming to recover natural forest and to boost wood production in the production forest for domestic wood demand and to expand 'forest coverage' by launching reforestation and forest rehabilitation programs in protected and conservation forest These conditions lead to preventing ecological disasters such as floods, soil erosion and drought.

There were two arguments to support Thailand's specialty. Firstly, the Thai government adopted and legitimized plantation forestry for 'economic' benefit rather than environmental issues. Therefore, focusing on *Eucalyptus* trees was adopted for their commercial value for farmers and private companies, aiming to boost production, provide income generation and jobs for rural communities. Secondly, the government encouraged easier procedures to obtain timber concession areas, gave subsidies to farmers, access to credit, tax holidays and provided infrastructure to private companies and farmers who wished to plant and develop plantations. The government supported private companies creating synergic cooperation with local farmers under the scheme of 'contract farming'. This contract requires the companies to give seedlings, fertilizers, accessibility to credit from banks and a guarantee of market at harvest time for timber products from the farmers. The schemes provide a win-win solution for both parties (company and local farmers). As a result, plantation forestry in Thailand rapidly grew from 560,000 ha in 1990 to become 4.9 million hectares in 2000 (FAO 2001). Therefore, in order to deal with these policies, it is significant to understand general policy and the impact of the Thai government on plantation forestry.

Hence, the objective of this paper is to examine three research objectives as follows: (1) to examine the impact of the *Royal Forest Department* (*RFD's*) policy on plantation forestry on private companies and local farmers; (2) to discuss how private companies and local farmers responded to forest plantations and the economic incentives provided by the government; and (3) to investigate the debates from academic and NGOs on ecological damages.

Study Sites

Study Sites

The study sites were located in Khon Kaen Province, Hua Naklang village, Tusala sub-district, Mancha Khiri district, 50 km from Khon Kaen city and in Samchan village, about 10 km from Khon Kaen city (Fig. 2.1, Map of Khon Kaen). Geographically, Khon Kaen province is located in Northeast Thailand and about 449 km from Bangkok (Fig. 2.2). There are two rationales for selecting this province. First, the Khon Kaen province is mostly covered in forest plantations (*E. camaldulensis*) planted by the local farmers and private companies, based on field surveys that indicated that soil and climate conditions were appropriate for

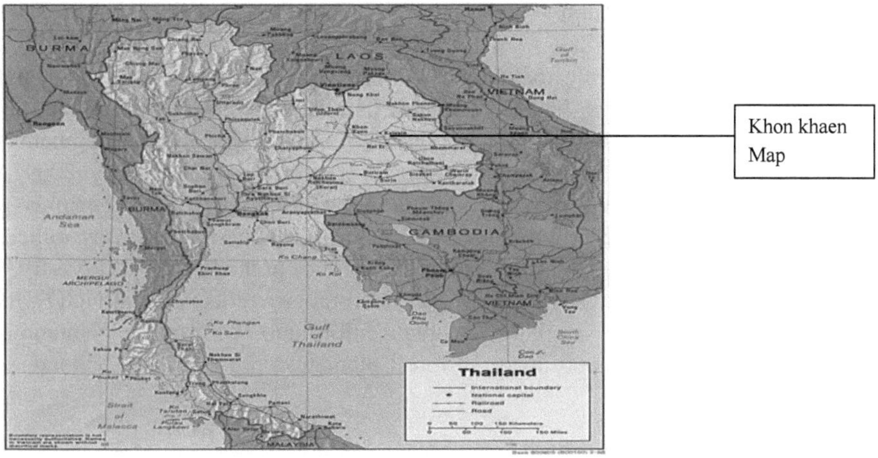

Fig. 2.1 Map of Khon Khaen area among Thailand. *Source* From Google, Khon Khaen in Thailand, 2008

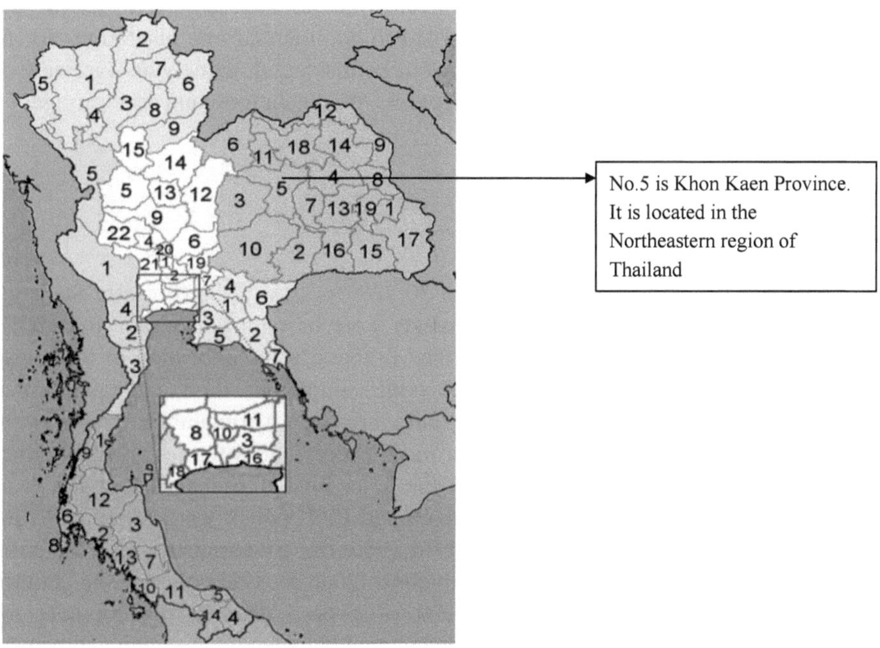

Fig. 2.2 Khon Kaen province is located in the Northeast of Thailand (number 5). *Source* Google

these plantation trees. Second, the villages that were selected could be categorized as dependent farmers, or so called 'contract farmers' with Siam Cement Group (SCG) Company, for instance, in Hua Naklang village, Tusala and Mancha Khiri.

Method for Data Collection

The majority of data and information was collected through three techniques: First, literature reviews of books, forestry statistics of Thailand (2007), the internet and journals about Thailand's forest policy, plantation forestry and the development of pulp and paper industries. Second, fieldwork was carried out in July 2009 in Bangkok and Khon Kaen province. Using guided interview questionnaires to conduct in-depth interviews with stakeholders (*Royal Forest Department* officer; Forestry Research and Development officer; academics in the Faculty of Forestry of Kasetsart and Khon Kaen Universities; private company officers, local NGOs and local farmers—purposive sampling) to obtain information on forestry plantations. Third, field observations were made of forest plantation programs in Thailand.

Royal Forest Department Policies

The Royal Forest Department (RFD) plays a significant role in policy development for plantation forestry. As mentioned in the first research objective, it was the RFD's policy on plantation forestry that led to commercial tree plantations for *E. camaldulensis*. There are four main issues to be discussed, namely an overview of forest policy, plantation forestry, the progress of plantations and foreign agency support.

Overview of Forestry Policy

Thailand was covered by 158,652 km^2 or 30.92% forest in 2006 (Forestry Statistics of Thailand 2007). In the first phase, forests were brought under state ownership and management. The RFD was authorized to classify the allocation of forest use and issued concessions. As part of the policy implementation, the institutional structure was built up and adjusted, and various government programs were implemented. The legal status of the permanent forest areas was established either as protected areas or forest reserves. In the latter phase, 'timber concession' was given to private companies and the state-owned FIO[1] which were eventually cancelled when the logging ban was issued. In 1985, the government issued adoption of National Forest Policy in an attempt to consolidate sectoral policy in the country and to place forestry within the context of overall national development for instance: (1) promoting shared forest management between government and private

[1]*Forestry Industry Organization* (*FIO*) is a Government Forestry Enterprise Company. It was originally established on July 25, 1956 by RFD (*Royal Forest Department*). The main task of FIO is to carry out tree plantation forestry in concession areas, logging in non-concession of the state's project areas (*dam and reservoir sites*), and the use or sale of confiscated wood either illegally cut or illegally imported into Thailand.

sector; (2) specifying the target forest areas at 40% of the country's area (15 conservation forests and 25% commercial forests); (3) reducing forest destruction by improving agricultural technology; (4) intensification of private forest plantations to meet the needs of forest industries; (5) creating incentives for private forest plantations, etc. (ITTO Objective 2000: 81).

While the government issued a logging ban on natural forests in 1989 to prevent deforestation and forest degradation that eventually affect soil erosion and flood, reforestation and afforestation programs were highlighted as significant strategies to supply wood demand and protect forests and land. As a result, a reforestation program was encouraged by the government (RFD) in 1991 aimed at encouraging the private sector, FIO and farmers, to engage in tree planting. The tree planting identified the need for partnerships between the public and (*commercial*) private sectors. Forest laws and regulations were reviewed and revised. For instance, in 1992, the government issued the Forest Master Plan with a focus on 'rural development' and 'community forestry' (Ibid: 81). Hence, RFD was directed to encourage local communities to participate in tree planting and cooperate with the private sector.

Plantation Forestry

Since logging in natural forests was banned in 1989, timber production in Thailand has shifted from *natural forests to planted forests*, particularly teak and rubber wood and non-forest sources supplemented by imports. Tree plantations are an integral part of the reforestation program. RFD has the authority to issue 'timber concession' areas to government agencies (FIO) and private companies. The government's farm program (1994–2001) was a response to the deteriorating wood supply situation with a target area of 1.28 million hectares. The program subsidized the private sector and farmers to plant trees (ITTO Objective 2000: 8). The program encouraged the private sector and farmers to plant specified economic tree species on their lands at 1250 seedlings/ha. The government gave seedlings and fertilizer to farmers. The aim was to make use of all unutilized marginal farmland and areas for environmental benefit, and to reduce rural poverty. Planting was subsidized by the government with US$469/ha and the farmers were free to harvest the trees at the beginning of year six. The program absorbed about 80,167 farmers and the planted areas covered 169,400 ha, mostly located in the northeast and central regions. Unfortunately, the program only reached 13% of the target, with the main reason provided being that more attractive subsidies were offered for rubber, but the inherent obstacles to investing in tree crops below have obviously been important as well. Because Thailand's farmers have been accustomed to attaining a lucrative business from rubber plantations, the risks from trying *Eucalyptus* trees were too great (ITTO Objective 2000: 41–42).

Unfortunately, the planting of timber species has progressed slowly due to a series of constraints. Initially farmers did not respond well to the timber plantations

Table 2.1 Plantation area by species in 2000

Species	1000 ha
Rubber	2019
Teak	836
Eucalyptus spp.	443
A. mangium and other A. spp.	148
Other broadleaved species	541
Pinus merkusii and other P. spp.	689
Other conifers	148
Total	4824

Source FAO (2001)

because of the following reasons: (1) no proper credit facility was available; (2) no guaranteed market at harvest; (3) forestry species were competing with other cash crops such as cassava, sugar cane, etc., which were also being actively promoted and (4) the registration problem of reserved species at the time of harvesting represents a source of uncertainty (Ibid: 40) (Interview, July 11, 2009). But, these constraints could be overcome when the RFD (*Royal Forestry Department*) officers invited private companies to conduct synergic cooperation with local farmers. The impact of the policy indicates that in the early 2000s the pulp and paper in Thailand increased profitability and engaged many stakeholders. As a result of the cooperation among parties, plantations are growing and the total extent of planted forests in 2000 (Table 2.1) was estimated at 2.81 million hectares of rubber and teak, while other plantations, such as *Acacia mangium, E. camaldulensis, Pinus mercussii* and other *conifers* reached 2.0 million hectares (Ibid: 8).

Estimates of the actual area of commercial tree plantations vary considerably. According to RFD statistics, by 1981, 2.6 million rai had been reforested, with another 1.125 million rai reforested between 1987 and 1990. Some scholars predicted reforestation as follows: Hurst (1990) put the area of timber plantation planted by the RFD by 1982 at 2.5 million rai, by the FIO at 312,000 rai, and by the Thai Plywood company at 18,750 rai, but they estimated the survival rates at a third to 10%, which would give an actual area under state sector tree plantations of 280,000–930,000 rai. Puntasen et al. (1992) gave an estimate of 500,000 ha of *Eucalyptus* plantations nationwide for 1987 (Carrere and Lohmann 1996: 91), while FAO (2001) estimated around 443,000 ha (ITTO 2000). Notwithstanding these discrepancies, it was clear that the area under *Eucalyptus* plantation had to be increased significantly.

Therefore, to achieve this aim, a close system of collaboration between the Thai government and multinational conglomerates emerged and an ambitious state programme of reforestation were developed. The Fifth National Economic and Social Development Plan (1982–1987) also included the reforestation target of 300,000 rai per year by the private sector. Accordingly, the 1985 National Forest Policy not only represents a general shift in policy but also a specifically new role for the RFD. In recognition that the future of administering logging was limited, the RFD was then charged with organizing the setting up of large-scale plantations. First, in Article 4 of the document, 25% of the country's total area is set aside as production forest, with 15% as protected forest. Given the degraded state of most areas, this could only mean building up industrial tree plantations. Secondly, the document specifically prioritizes the pulp and paper industry in Article 13: "the state shall encourage integrate wood using and pulp and paper industries to realize the whole-tree utilization concept." Thirdly, articles 5, 12, and 19 emphasize the role of private plantations (Ibid: 92).

In order to advocate plantation forestry among locals and the private sector, the RFD set up a special Office for the Promotion of Private Forestry Plantations, which formulated the long-term goal of 30,000 km^2 of private plantations. The National Forestry Policy resolution also set up a National Forestry Policy Committee, which served as a joint think-tank of the RFD and private forestry enterprises. This committee set the goal of 41,600 km^2 plantations by corporations and an additional 20,000 km^2 to be planted by villagers. To this aim, various state programs were initiated to create public support for reforestation. These included the USAID-supported Village Woodlot Programm, initiated in 1981, which hoped to set up plantations in 1560 villages between 1987 and 1991. A Thai-Japanese joint venture, the Forest Plantation and Wood Industry, won a contract to supply 10 million *Eucalyptus* seedlings to this project. *Eucalyptus* plantations were awarded similar promotion as pulp production (Oliver 2005: 72) (Table 2.2).

It was claimed that by the end of 1989 100 forest protection units and 46 forest villages were established, with 500,000 rai of forest plantations planted and additional areas under fruit tree and rubber plantations. Most forest plantations were in private hands, with Rom Klao Oil, Philpittiwat, Treethip, Phoenix Pulp and Paper and the FIA among the companies reported to have especially close links with the projects.

Table 2.2 BOI-promoted *Eucalyptus* plantation

Name of company	Year approved	Area (rai)	Export (%)	Investment (mil. baht)
Kian Chunprasert	1988	1500	0	55.0
Suan Siam Kitti Reforestation	1988	20,000	80	120.0
N. R. *Eucalyptus* Agr, Co., Ltd.	1987	15,000	0	110.0
Siam Forestry Industrial Ltd.	1986	3000	0	23.15
Siam Vana Forestry Industrial	1986	3000		
Suan Siam Kitti Reforestation	1986	10,000	20	70
Rich Forest Co., Ltd.	1984	10,000	0	200.0
Siam Agrotex Co., Ltd.	1984	5200	0	97.0
Vanapan Co., Ltd.	1984	3000	0	46.0
Vanapan Forestry Co., Ltd.	na	3000	0	46.0
Total		73,700		767.15

Source Oliver (2005)

Progress of Plantations

The government functionalized research centers to produce the best alternative trees to be planted for commercial aims. Two surveys were published in 1989, one by the *Forestry Research Center of Kasetsart University* and submitted exclusively to South-East Pulp Co. Ltd., the other conducted by the *Thailand Development Research Institute* and entitled "Potential of Commercial Fast Growing Tree Plantations in Thailand", followed by Forest Research Center for Royal Forest Department's paper (2008). These centers' findings recommended large-scale *Eucalyptus* plantations as the best solution to the industry's raw material problem (Oliver 2005: 91). These valuable findings were adopted by the RFD, where they then issued the National Forest Policy that declared the need for 'commercial tree' plantations of fast-growing species, such as *E. camaldulensis*, to supply the raw material for the pulp and paper industry. Thai forestry academics believed that tree plantations would revive and improve the negative impact on existing forests from cultivation and illegal logging, as well as "the village people's encroachment on the forest for farming and the shifting cultivation by the hill tribes." They also believed that tree plantations would contribute to economic production and generate local and national income and employment (Kuaycharoen 2004: 9).

Due to developing wood demand over the last decade, fast growing tree plantations have been promoted. There are about 26 tree species, such as *pine, persian lilac, eucalyptus, leucaena, casuarina acacia, duabanga, acrocarpus and mahogany.*[2] *Eucalyptus* is a genus of tree originally derived from Australia and neighboring islands and was introduced to Thailand years ago as an ornamental tree. This genus consists of almost six hundred species, but only *E. camaldulensis* shows promising growth performance in commercial plantations and plays a significant role in the wood chip and wood pulp industries in Thailand. The domestic demand for logs for the wood chip industry, estimated at 6,395,000 tons/year over 90% (Fig. 2.3) of the total demand for logs, is provided by *E. camaldulensis.*[3] According to a study of the possibilities of changing land use to *Eucalyptus* plantations in five (5) provinces in Thailand, in Khon Kaen, *Eucalyptus* could constitute a total area of about 295,915 ha, in Chauyapkum, 341,663 ha, in Nakorn Racha Srima, 1,498,667 ha, in Buri Rum, 815,750 ha and in Maha Sarakham, 3,159,942 ha (Table 2.3). The findings recorded that the suitability of changing from existing land uses to *Eucalyptus* plantations was categorized into three levels: high suitability, moderate suitability and low suitability. Based on such land use factors, it was found that *986,480 ha are highly suitable for growing Eucalyptus,* including the areas of existing *Eucalyptus* plantations, cassava fields and degraded forests, and the most suitable areas were in Khon Kaen province.[4] The areas considered as

---

[2]For further information see Arbhabhirama et al. (1998).

[3]See Bunvong et al. (2003).

[4]Interview with informant on July 3, 2009 in Bangkok. Khon Kaen area is widely used for farm holding land, and *Eucalyptus* trees could grow well there.

Fig. 2.3 *Eucalyptus camaldulensis* trees in Northeast region, Khon Kaen. *Source* Individual photo collection, July, 2009

Table 2.3 The possibilities for changing existing land use to *Eucalyptus* planting

Province	Total rai	Total hectares (ha)
Khon Kaen	1,849,470	295,915
Chayapkum	2,135,391	341,663
Nakorn Racha Srima	9,366,665	1,498,667
Buri Rum	5,098,437	815,750
Maha Sarakham	19,749,633	3,159,942

Source Thaiutsa et al. (2003)
Thailand's account for land use is widely called "rai". 1 ha = 6.25 rai or 1 rai = 0.16 ha

moderately suitable are 201,108 ha, currently covered by sugar cane and maize, while 1,575,100 ha of land covered by paddy rice and secondary forests were considered as low suitability.[5]

Private companies, such as Advance Agro and Siam Cement Company, are actively engaged in *Eucalyptus* plantations. The first group, Advance Agro Alliance Company, planted 40,000 ha and farmer contracted 60,000 ha, mostly located in the central and eastern regions of Thailand. Siam Cement Company, in paper and packaging, planted 25,600 ha by the contract farming system, mostly located in the northeast and through non-members residing within a radius of 150 km around the

[5]Bunvong Thaiutsa. *Loc Cit.*

Table 2.4 *Eucalyptus* plantation in 1994–1996 (1000 ha)

Actor	Area (ha)
Planted by the government	1088
Planted by the private sector	
Advance Agro (AA)	100
Siam Cement Group (SCG)	756
Total	2844

Source Journal of Forest Management (2008)

factory more than 50,000 ha (Table 2.4). The pulpwood, 3–5 years old, is harvested and sent to the factories with an average price of approximately US$30 per ton depending on log diameter class (Laemsak 2008: 118). In 2004–2006, the Ministry of Agriculture and Cooperatives promoted farmers intensively planting *Eucalyptus* by agro forestry systems.

Foreign Agency Support

The planting of *Eucalyptus* eventually encouraged foreign donor countries to extend funds such as Japan, Canada, the USA and Finland to provide investment in this activity. They benefit, either from sales of machinery and consultancies and the cheaper import costs of Thai furniture, particle boards and veneer. In 1981, the *Japan International Cooperation Agency* (*JICA*) set up a trial *Eucalyptus* plantation in northeast Thailand to foster research and training in the field, and in the early 1990s funded **a nursery project**, growing 20 million trees seedlings per year. Japanese taxpayers' money has also been channeled through the Overseas Economic Cooperation Fund to support farmers' participation in the Thai-Japan Reforestation and Wood Industry Co., which was designed to supply raw materials to a consortium of Japanese papermakers. *CIDA,* the Canadian government aid body, has helped finance the Canadian consultant firm H. A. Simons' work with Soon Hua Seng as well as plantation research by the Thai Development Research Institute. Meanwhile, the Finnish government has pumped millions of dollars of its taxpayers' money not only into export credits for Finnish forestry and pulp and paper equipment, but also into a political lobbying exercise known as the Thai Forestry Sector Master Plan, which was conducted by the Jaakko Poyry consulting firm between 1990 and 1994 (Carrere and Lohmann 1999: 233–234).

Private Sector Responses to Forest Plantations

The private sector, including private companies (SCG, Advance Agro, Phoenix etc.), FIO (*Forestry Industry Organization*), and wood traders, launched strategies to get raw material (timber), to boost production and access to the market, cooperating with individual farmers and members of cooperatives.

SCG Company Responses to the Government's Policy

The significant of Siam Cement Group (SCG) as respondent due to this private company as a main actor of pulp and paper in Thailand. The significant role of SCG is the single actor who actively engages on contract farming with farmers in Khon Kaen district.

SCG responded positively toward the RFD's (*Royal Forest Department*) policy of plantation forestry concession because the investment atmosphere was good for investors and SCG, in particular. The study focuses on SCG. SCG staff highlighted response to the policy, such as lower taxes, credit provided for the company and local farmers, simple procedures for obtaining land tenure, timber concession areas and the provision of good *infrastructure* (road and port facilities) as their reasons. Credit availability from the bank for the private company depends on rationality of the feasibility study. In the case of SCG, it partnered with The Farmers Bank[6] for tree plantation and factory mill establishment. Because of this close relationship between The Farmers Bank and SCG, the company could obtain a hundred million US$ dollar credit to invest in plantation forestry, pulp and paper mills and others.[7] The company searches for raw materials (timber) with other parties. For example, the Company looks for cooperation with government agency such as FIO (*Forest Industry Organization*) for planting trees in the district of Khon Khaen and surroundings and with local farmers to procure timber through *MOUs* (*Memorandum of Understanding*) with the scheme of "Contract Farming". The paper products of SCG are 65% sold in the domestic market and the remaining 35% is for overseas markets in Southeast Asia, Hong Kong, Japan, USA and Europe (Interview, on July 9, 2009).

Private Company (SCG) with Government Enterprises (FIO)

SCG conducted a joint venture in 2003 with FIO, one of the state enterprise forestry businesses belonging to the RFD, planting *Eucalyptus* trees on 12,000 rai of land in Khon Kaen area. The SCG provided the capital, in the form of seeds and maintenance until harvest, while FIO provided the land, counseling, guidance and inspection of plants until harvest. SCG provided capital from land clearing, weeding, planting, maintaining to harvesting the plantation per rai 4000 Baht $\times$ 12,000 rai = 48,000,000 Baht. The harvest after six years of one rai

---

[6]The Farmers Bank is categorized among the fifth biggest financial conglomerates in Thailand, such as Bangkok Bank, Bank of Ayudhya, Bangkok Metropolitan Bank and Bank of Asia. The major shareholder of The Farmers Bank is the Royal Family Group. For further information see Suehiro (1989).

[7]The close relationship between SCG and The Farmers Bank occurred because the majority shareholder of SCG is the Thai King (Bumibol). The data is based on an interview with Kriangsak Salee, a Promotion Department Manager of SCG (*Siam Cement Group*) on July 9, 2009 in Khon Kaen.

Table 2.5 Production of pulp and paper in Thailand year 2006–2008 (1000 tons)

Product	Domestic production			SCG Prod in 2008	% of share in Domestic Prod
	2006	2007	2008		
Pulp	1129	1169	1106	414	37
Paper	4308	4322	4235	1687	40
–Printing, writing	1221	1108	1014	390	38
–Paper and board	239	348	369	147	40
–Kraft paper	2600	2625	2604	1150	44
–Others (newsprint & sanitary paper)	248	241	248	–	–
Total	5437	5491	5341	2101	39

Source Sustainable Report of SCG Company (2008)

produced 14 ton × 12,000 rai = 168,000 ton. The price in 2009 reached 900 Baht/ton × 168,000 ton = 151,200,000 Baht.

The profit after six years reached 151,200,000–48,000,000 Baht = 103,200,000 Baht. Then the profit each year reached 103,200,000: (12,000 × 6) = 1433.3 Baht per rai. The profit at harvesting time (after six years) for SCG received 60% (0.6 × 103,200,000) = 61,920,000 Baht. Meanwhile FIO received 40% (after six years) (0.4 × 103,200,000) = 41,280,000 Baht.

The profit at harvest time, after tax, was distributed 60/40 to SCG and FIO respectively. According to Salee, an SCG staff member, this cooperation provided a profit for both parties. For instance, FIO could develop management and networking. On the other hand, SCG as the investor, could obtain bank credit and conduct contract farming with local farmers, thereby creating jobs and increasing the socio-economic strength of the rural areas.

The target of SCG for pulp production was about 414,000 tons (37% of the total national target) and for paper about 1.7 million m^3 (40% of the total national target) in 2008 (Table 2.5).[8] This target was reached by the company in cooperation with other actors (*FIO, Oji Paper, and Local farmers*) for providing raw materials (*Eucalyptus trees*). For instance, in 2010, the SCG cooperated with *Oji Paper* from Japan to carry out plantation concessions of around 22,000 ha in Laos, because forest land in Thailand is very limited. So far, fundraising and technical preparation have been conducted by both parties (Oji and SCG Company), while the Laos government has provided state forests and approved plantation concessions to plant *E. camaldulensis.*

[8]For further information, see *Sustainable Report of SCG in 2008*, pp. 26.

Private Company (SCG) with Local Farmers

SCG conducts strategic 'partnerships' with local farmers for securing raw material. The guarantee of providing raw material (*E. camaldulensis*) is significant to the company. Therefore, the company holds contracts with local farmers in order to ensure the provision of raw material. Within these contracts, SCG is obliged to provide seedlings of *Eucalyptus*, fertilizer and guarantee a market at harvesting time for farmers and the farmers are required to sell their product to SCG at the agreed price. Independent local farmers not contracted by SCG can sell to other traders if the price is agreeable between the two parties.

In this context, SCG likes to support local communities where it operates by sustaining their economic growth and increasing income generation. Farmers as suppliers of timber are considered the first priority for purchasing their products. The SCG policy supports *job creation* in rural areas that decreases migration to urban areas, such as Bangkok, by villagers.

Major examples in 2007–2008 programs of contract farming through local hiring are as follows: (1) Promotion of converting cash crop to *Eucalyptus* planting for farmers at the village level and establishing centers for seed distribution in the provinces of Kanchanaburi and Khon Kaen[9] in order to produce raw material for pulp production. The advantages include increasing income for farmers and providing alternative sources of raw material for SCG Paper. As an illustration, raw material produced from the company's plantation is around 75 and 25% from contract farming (Table 2.5).

(2) Establishment of a *Community Occupational Training Center* which has programs such as traditional Thai Massage, and production of handicrafts produced by housewives and the unemployed in the nearby community of Phoenix Pulp and Paper Mill[10] and SCG. In addition, the Company provides the market for the communities by inviting them to sell their products and services in SCG Paper exhibitions held in every mill as well as purchasing their products to be souvenirs for special guests and customers.

SCG established some **nursery centers** in the northeast, such as in Chum Phae, Ban Phai, Mancha Khiri, Nong Song Hong and Nong Ruea. The aims of these centers were as follows: (1) to provide best seedlings for farmers; (2) to provide an opportunity for SCG staff to explain how to intensify *Eucalyptus* planting production as follows:

---

[9]The field survey was carried out on July 7–18, 2009 in Khon Khaen area. The province of Khon Kaen such as Mancha Khiri, Tusala, Shamchan, etc., are located in the northeast of Thailand. Some areas could produce paddy rice based on rainfall, irrigation system is very little, and the land is infertile. But, some of the land areas are very suitable for *E. plantation* because these trees do not require much the water. This information is based on interviews with scholars in Khon Kaen University, July 9, 2009.

[10]Actually some shares of Phoenix Pulp and Paper had been made an acquisition by SCG in 2001. Then a part of management is controlled by SCG.

(1) to explain what is suitable land for planting;
(2) to teach people how to maintain plants from planting until harvest;
(3) to teach people how to realize first cutting of plants (after 4–5 years);
(4) to show people how to manage stumps, in order to grow just 3 plants, no more.

In the Mancha Khiri sub-district, where a research site is located, about 80% of local farmers who own land are categorized as members of contract farming with SCG. The remaining 20% of local farmers are categorized as independent farmers.

The *benefits* of contract farming for farmers are: (1) they received free fertilizer from SCG in the first two years (2004–2005); (2) they received discounts of almost 40–50% for seedlings in 2006; (3) they have access to credit from the local bank as a member of the village cooperative; (4) they are guaranteed a market of logs at harvesting time (with spot price). In contrast, the *losses* for farmers are in terms of the 'spot price' at harvesting time, which is often 'lower' with the company compared with independent traders. For example, in 2009, SCG determined on a price of 900 Baht per ton, while the general market value was 1000 Baht per ton. From this point of view, farmers under 'contract farming' lost 100 Baht per ton in selling at harvest time. Also, the provision of people to assist in maintaining and harvesting timber (*Eucalyptus*) in rural areas since the beginning of the 2000s has been difficult, because they prefer to work in the big cities such as Bangkok, Chiang Mae and Khon Kaen, where the salaries provided are greater than in rural areas (interview with Bualai, July 7, 2009).

There are many local farmers actively engaged in plantation forestry, irrespective of whether they are individual contract farmers, independent farmers or members of village cooperatives. As a member of a cooperative, they receive access to credit from agricultural banks and free seedlings from the local government. Below is a picture of an individual contract farmer called Bualai Yenchai. Bualai is 59 years old and has three children, while her 63 year old husband is called Thon Puan.[11] Bualai owns a gasoline station and shop for daily necessities such as rice, cakes and drinking water for local people. She owns about 110 rai of land, some inherited from her parents and most bought from other local farmers. The remaining 10 rai is reserved for paddy fields (Fig. 2.4).

Bualai initially knew about SCG from company staff who visited her home explaining about *Eucalyptus* as an alternative commodity tree. She became an individual contract farmer in 2004. Previously she planted cassava and sugar cane, but these crops did not attain much profit and required a great deal of effort and time to maintain. She explains that it is currently difficult to find labor in rural areas, because most people go to the cities for greater amounts of money, such as Bangkok, Chiang Mae and Khon-Kaen. Therefore, she chose *Eucalyptus* trees to plant for the following benefits: (1) the *Eucalyptus* is fast growing and produces greater profit; (2) not much time is required to manage the crop; (3) no extra labor is required for maintenance and harvest; (4) an ensured price at harvest time; (5) free

[11]The interview with Ms. Bualai Yenchai was carried out on July 9, 2009 in Manca Khiri district, about 50 km from Khon-Kaen city.

Fig. 2.4 This woman is Bualai, in her shop in Mancha Khiri sub-district, speaking with the SCG officer, acting as guide and translator. *Source* Individual photo document, July, 2009

fertilizer from the company in 2004–2005 as a promotion; (6) cheaper prices for seedlings and fertilizer and (7) a guaranteed market at harvest time.

Farmers' strategies to boost production

Production depends on many factors, such as quality of land, quality of seedlings, maintenance, fertilizers and sufficiency of capital for a successful harvest. Although SCG provides seedlings and organic fertilizers, the company also must improve seedlings and fertilizers in order for farmers to boost production. Currently, 250–270 trees with a spacing of 3×3 m and 4 bags (50 kg/bag) of fertilizer (200 kg) are required for one rai. This produces 10–14 tons per rai after 6 years (*first harvest*). In contrast, if farmers do not use fertilizer, the production declines to just around 7–8 tons per rai. The price per ton was 1200 Baht in 2007, which declined to 900 Baht per ton in 2009.

Mostly farmers who own land in Mancha Khiri sub-district are individual contract farmers and members of cooperatives. The benefits they obtain as members are: discount prices for fertilizer and seedlings, credit from the agricultural bank and life insurance. They pay an annual fee of 2000 Baht. If a member dies, the family receives 150,000 Baht from the life insurance.

Bualai informed us that 3200 Baht[12] per rai is needed for clearing land, weeding, planting, maintaining to harvesting the plantation. She planted *Eucalyptus* trees in 2004 and will harvest them in 2010. Bualai owns 100 rai, therefore needing to invest 320,000 Baht. As an individual contract farmer and member of the cooperative, she borrowed 300,000 Baht from a local agricultural bank, using her land as collateral. She must repay about 30,000 Baht annually over the next 10 years and is so far able to make repayments each month. The estimated harvest of one rai should produce 10 tons, and at a price of 1000 Baht (estimated price in 2010), total profit should be 1000 tons × 1000 Baht per ton = 1,000,000 Baht, or 1,000,000–320,000 Baht = 680,000 Baht (after six years). The profit in each year should be about 680,000 Baht: 600 (100 rai × 6) = 1133 Baht per rai. She would like to use this profit to expand her shop, buy more land and pay education fees for her children and possible large medical expenses.

Private Company (SCG) with Member Cooperative Sector

Most farmers who live in Tusala sub-district are involved in contract farming and village cooperatives. I interviewed members of a cooperative: Showaeng, 57 years old, and Kaseem, 51 years old. Kaseem owns 50 rai and Showaeng owns 100 rai of land, mostly bought from other farmers. Both farmers live in Hua Naklang village, Tusala sub-district.[13] Showaeng has two daughters and one son and also has a shop that sells daily necessities for local people (Fig. 2.5).

These two farmers initially learned of SCG when SCG staff visited their homes to explain the benefits of *Eucalyptus* tree plantations and becoming a member of the cooperative. They were interested in becoming contract farmers because: (1) they would learn how to intensively plant *Eucalyptus* commercially, (2) a guaranteed market at harvest time and (3) they would receive discount prices on quality seedlings and fertilizer. For instance, as members, they could pay 3 Baht per seedling, whilst non members paid 4 Baht, whereas they received fertilizer for free from the company in 2004–2005 and normally they would pay 200 Baht for one bag (50 kg), while a non-member would pay 250 Baht.

To optimize production

The *Eucalyptus* trees planted from seedling require 6 years before the first harvest. Showaeng planted *Eucalyptus* trees in 2005, and has been fertilizing, maintaining and weeding the plantation since.

Although he could have applied for credit, Showaeng did not obtain credit from the agricultural bank and instead planted in parts: 50 rai in the first year and 50 rai

[12]The breakdown cost per rai composes many activities such as buying seedling, fertilizer, paying workers for clearing land and planting seedling and distributing fertilizer, and for cutting trees at harvesting time (see at Showaeng's cost per rai).

[13]The interview with informant was carried out on July 11, 2009 in Hua Naklang village.

Fig. 2.5 The farmer Showaeng (middle) with his wife (Thonsa) and grandchild in Hua Naklang village, Mancha Khiri sub district. The guide and translator is on the left. *Source* Individual Photo document, July, 2009

in the second year, using his own savings of about 220,000 Baht. He used this money to buy seedlings, fertilizer and maintain the plantation, cleaning weeds and harvesting, with only himself and his family laboring on the plantation. The investment needed is 3218 Baht per rai from planting until harvesting: 6 years. The total money could be spent as follows:

(1) for buying seedlings: 256 × 3 Baht/ seedling = 768 Baht;[14]
(2) for buying fertilizer: 4 bags (50 kg/bag) × 200 Baht = 800 Baht;
(3) for paying workers land clearing and weeding: 150 Baht per day × 1 worker × 3 days = 450 Baht;
(4) for paying workers for planting seedlings and distributing fertilizer: 150 Baht × 2 workers × 2 days = 600 Baht;
(5) for cutting trees at harvesting time: 150 Baht × 2 workers × 2 days = 600 Baht; Then Showaeng needs capital for 50 rai × 3218 Baht = 160,900 Baht.

Currently, his *Eucalyptus* trees are about 4 years old and planned to be harvested in 2011 (Fig. 2.6). Harvesting will produce 11 tons/rai and with 50 rai, the estimated total will be 550 tons in 2011. At a price of 1000 Baht/ton in 2011, he

[14] I ha usually need 1600 seedlings. 1 ha = 6.25 rai; 1 rai = 0.16 ha (1600 m). The respondent owns land 50 rai means 50 rai × 0.16 ha = 8 ha. 1 rai = 0.16 × 1600 seedling = 256 seedlings.

Fig. 2.6 4 year-old *Eucalyptus* trees in Hua Naklang Village. *Source* Individual photo collection, July, 2009

will receive 550 tons × 1000 Baht/ton = 550,000 Baht, leaving him with a profit of 550,000–160,900 Baht = 389,100 Baht. Then, annual profit of 389,100: 300 (50 rai × 6) = 1297 Baht per rai.

He plans to use the profits to increase his savings, buy more land, pay educational fees for his children, buy life insurance and open a shop and plant *Eucalyptus* trees again.

Critiques of Academics and NGOs

In this Sect. 1 highlight what some academics and NGO feel about changes in the political ecology in the region. I would like to focus on to what extent academics and NGOs criticize the ecological damage impact throughout environmental issues.

Environmental Impact

Actually, there are some activist NGOs, such as Kuaycharoen, Khemchalerm and scientists Ubukata, Colchester, Sawaeng, and Lohman who actively launched critiques on the ecological damage of *Eucalyptus* plantation.

Kuaycharoen is a famous advocate of NGOs; he reported that local people had to bear the brunt of the social and environmental costs when their cultivation lands and community way of life were damaged by the large stands of monoculture tree plantations and the operations of pulp and paper industry. For instance, communities living around the Phoenix Pulp and Paper mill located near the Phong River in Khon Kaen province suffered the impacts of the *toxic pollution* of the river and farmlands while farmers lost large areas of farmlands and village commons to the Suan Kitti plantation company as it aggressively expanded its *Eucalyptus* plantations in the eastern region (Kuaycharoen 2004: 13).

Besides, huge protests occurred in 1995 and village networks in the Northern Province attempted to eliminate *Eucalyptus* from the areas altogether, forcing the RFD to suspend its *Eucalyptus* operations over a wide area. Responsibility for existing plantations, meanwhile, was passed to other authorities, whom villagers pressured to cut the *Eucalyptus* and distribute the profits locally. Throughout their intensive campaigns, northeastern villagers and their NGO allies researched and publicized multi-purpose native alternatives to *Eucalyptus* which are responsive to the diversity of food, construction, medicinal and ecological needs of different localities; launched supplementary plantings of native trees on degraded sites and posted new areas as community forests (Carrere and Lohmann 1996: 238).

Wiboon Khemchalerm, a former senator, NGOs activist and well-known organic farmer from Chachoengsao Province, explained that "after the existence of *Eucalyptus* plantations in nearby areas, underground water has dried up. Small-scale farmers could not survive by planting *Eucalyptus*. Obviously, there was no guarantee that those who had left the land would not encroach on other forests and cause further problems". In addition, there was no guarantee that a *monoculture* tree plantation covering a huge area would not inflict unexpected impacts on the *environment* in the long run (Kuaycharoen 2004: 17).

In line with Wibon's statement that *Eucalyptus* trees absorb much water and negatively affect surrounding cash crops, such as cassava (with the distance 3–10 m), causing dry and infertile land; this is in accordance with my field observations with Dr. Sawaeng Ruaysoongnern, lecturer from Faculty Agriculture, Khon Kaen University, Thailand in Samchan village, about 10 km from Khon Kaen (Fig. 2.7). Sawaeng proposed that local farmers should try other options by carrying out planting of local tree species (shift from *monoculture to multicultural* trees) in their forest commons and farmland. The aim of planting local tree species is to ensure conservation, to prevent soil erosion and recovery of soil fertility in the near future (an interview, on July 8, 2009).

Ubukata has said that "the negative aspects of *Eucalyptus* production to ordinary villagers may be exacerbated by their persisting perceptions of agro-ecological impacts." In fact, planting *Eucalyptus* did not mean that the villagers came to hold positive perceptions on these aspects. As mentioned in the section on anti-movements, many villagers felt that *Eucalyptus* damages water and nutrients in the soil, despite the state's continuous efforts to emphasize its harmless nature (Ubukata 2009: 23).

Fig. 2.7 Dr. Sawaeng next to *Eucalyptus* and Cassava trees. The picture shows that *Eucalyptus camaldulensis* absorbs much water and its surrounding cassava trees subsequently become dry and infertile

Colchester and Lohmann contended that *Eucalyptus* plantations alone threaten to displace millions of peasants, as the RFD, the military and other bureaucratic and business interests have promoted plans to lease as much as 40,000 Km^2 of so-called *degraded forest* (consisting mainly of farmland, pasture and community woodlands) to commercial planters to feed foreign and domestic wood-chip and paper-pulp demand while supposedly reforesting the country. Even so, *Eucalyptus* operations have received loans from the *Asian Development Bank* (ADB), *United Nations Development Programme* (UNDP), the Japanese, Australian, Canadian and Finnish aid agencies, Britain's Commonwealth Development Corporation as well the army-initiated Green North-East Programme (Colchester and Lohmann 1995: 210–211).

Concluding Remarks

Plantation forestry in Thailand had a positive impact in terms of providing cash flow of money, raw material for pulp and paper industry and strengthening the socio-economic capacity of farmers in rural areas. The rationale for this success is explained as follows: The Thai government adopted and legitimized policies on plantation forestry of 'commercial trees', highlighting economic benefits rather than the environmental impact of plantation forestry. The economic facilities provided by the government include easy accessibility to timber concession areas, credit, tax breaks on import of machineries and other goods for pulp and paper factories.

The Royal Forest Department (RFD) actively invited research centers in universities and the government to carry out studies on trees that are most suitable for plantation forestry. According to some studies such as that of the Forestry Research Center of Kasetsart University (1989), Thailand Development Research Institute (1991) and the Forest Research Center for Royal Forest Department (2008), *E. camaldulensis* is the most suitable plant based on soil and climate condition and it was promoted as a 'commercial tree' for Thailand.

The policies adopted by the government subsequently led to a wide expansion of plantation estates by private companies, FIO (*Forestry Industry Organization*) and farmers in provinces and districts and a significant growth of plantation forestry of *E. camaldulensis* as a raw material for pulp and paper industries. The response from private companies, such as SCG (*Siam Cement Group*) and Advance Agro were positive and they conducted a joint venture with FIO in 2003 by planting about 12,000 rai of *Eucalyptus* trees in Khon Kaen area. The company provided the capital, seedlings, fertilizer and maintenance cost of plants until harvesting. Meanwhile, the FIO provided land, counseling, guidance and conducting inspections of plants. SCG also made an agreement to expand plantation forestry from about 22,000 ha in 2010 in Laos with OJI Paper from Japan. The Laos government agreed to provide state forest land for a three parties' joint venture. In order to ensure the provision of raw material for the future, the private companies such as SCG Company and Advance Agro carried out 'contract farming' with local farmers. The companies have obligations to provide seedlings, fertilizer and to guarantee a market at the time of harvest. In line with contract farming scheme, farmers can easily access credit from the local agricultural bank and guarantee their products in the market. This type of agreement between companies and farmers is called a 'strategic partnership' of the spirit, providing win-win results and eventually leading to job creation and economic and social growth in rural areas. As a result, the stakeholders actively participated in plantation forestry (including *E. camaldulensis*) and it was subsequently expanded in Thailand from 284,000 ha in 1996 to 4.8 million hectares in 2000, consisting of about 2 million hectares of rubber trees, 836,000 ha of teak trees and the remainder a mix of several types of tree species, such as *Eucalyptus* spp., *A. mangium, Pinus mercussi* and Conifers. On the other hand, academics and NGOs criticized the ecological damage caused by monoculture plantations of *E. camaldulensis,* which absorbs more water than native trees, affecting cash crops surrounding the plantations. From this perspective, academics and NGOs made recommendations to reduce the monoculture setup of plantations and adopt mixed plantations with local tree species in order to improve conservation and soil fertility recovery. In this case. local farmers who own common forest land and farmland should plant local tree species. The villagers get more economic benefit from mixed plantations in the longer term as they will still attain the economic benefits of their plantations, but also reduce the strain on their land, allowing for continuous use of their land for a longer time.

Review on Emerging Pulp and Paper Industry

Introduction

Thailand is the second largest producer (4.5 million m^3 ton) after Indonesia (7.6 million m^3 tons)[15] of paper production in ASEAN countries as of 2007. The last two decades (1980s–2000s) have seen an acceleration of emerging pulp and paper industries in Thailand. There are two main factors driving these emerging industries: Firstly, the government readily welcomed *foreign direct investment* (*FDI*) in many fields, including forest industries (pulp and paper) and approved joint ventures and even 100% ownership by foreign companies of domestic ventures. Secondly, the Thai government actively supported the development of the industry through easy access to credit, *infrastructure* (port and highway) construction, and subsidies to plantation owners, tax relief and favorable import duties on machinery.

These policies encouraged a positive response from domestic and foreign private companies to invest their capital in establishing pulp and paper factories and plantations for the raw material in Thailand in the 1980s. Among those investing were some famous companies, such as Siam Cement Group (SCG), who established Siam Pulp and Paper; Soon Hua Seng Group (SHS) with Advance Agro (AA); The Phoenix Pulp and Paper Company, owned by the European Overseas Development Corporation (EODC); Ballarpus industries, an Indian Industrial conglomerate; Suan Kitti Corporation, Hi-tech Paper; Shin Ho Paper Co; from South Korean conglomerate, Thai Cane Paper Co, etc. (Soonenfeld 1996; Carrere and Lohmann 1996).

As a result, the pulp and paper, plywood, veneer, sawmill and furniture industries expanded and demand for timber significantly increased. The question that then arose was how to maintain a sustainable supply of wood? In order to answer this, the government and private sector carried out two separate activities: First, the government invited other stakeholders [private companies, local farmers, FIO (*Forest Industry Organizations*)] to manage forest plantations. About 4.8 million hectares of plantations, or about 38% of total forest cover (16.2 million hectares), were managed in Thailand in 2000, consisting of rubber (2 million hectares), *Eucalyptus* (1.9 million hectares) and others (FAO 2001). Second, they carried out importing of log and sawn timber from overseas. The positive impact of plantation forestry by the private sector and under contract farming between private sector companies and local farmers contributed to job creation in rural communities and empowered local farmers socio-economically. Although some academicians and NGOs criticized the impact of plantations on the ecology, through soil erosion and too much water absorption that subsequently adversely affects surrounding crops, generally farmers and private companies were accepted and rapidly developed '*Eucalyptus camaldulensis*' as commercial trees for farmers in Thailand.

---

[15]*The Japanese Pulp and Paper Industry in Charts and Figs.* (2005), published by Japan Pulp and Paper Co., Ltd, pp. 45.

Hence, this paper aims to clarify the dynamic process of the emerging pulp and paper industry in Thailand by examining government's policy on economic development (*macroeconomics policy*), the impacts of policy toward development of pulp and paper industry, using two companies Advance Agro (AA) and Siam Cement Group (SCG) as case studies in how they developed and managed their companies, and concluding as noted major findings.

Government's Policy on Economic Development

Capital Accumulation

The attitude of the Thai government toward *Foreign Direct Investment* (FDI) has been generally positive, though it has varied somewhat over time. For instance, in the early 1960s, the government welcomed foreign direct investment in the manufacturing industry and approved even 100% ownership in the import substitution industry. However, usually the government encouraged foreign investors to enter into joint ventures with domestic partners; FDI also became part of this process (Pasuk and Baker 2000: 19). In the 1980s, Thailand also invited more investments from East Asian countries such as Japan, Taiwan and Hong Kong. Taiwanese companies, mostly producing labor-intensive products, invested in Thailand for reasons similar to that of the Japanese, whereas Hong Kong companies were there because of uncertainty regarding the future of the territory. The then political instability of the country had been overcome, by the early 1980s. The unstable democratic period was ended by the military coup of October 1976 and a system of government established, in which the military and politicians shared power, which lasted until the late 1980s, although the balance of power was in favor of the former.[16]

During the 1980s, the Thai economy experienced massive expansion and restructuring that profoundly impacted on the strategies and influence of the leading strategic groups. By the end of the 1970s, agro-industry-led economic growth began to enhance products, partly because of substantial drops in the world market prices of such goods due to overproduction. After a slump in the Thai economy in 1984, a period of unprecedented boom based on the export of *manufactured goods* occurred. Between 1985 and 1995, GDP rose from 1191 to 2912 billion Baht, an annual growth rate of 9.4%. The proportion of manufactured exports, already at 39% in 1984, rose to 67% in 1987 and to 84% of total exports by 1995. If in 1980, the big export earners were rice and cassava, by the late 1980s they were textiles and cheap-labor manufactures, and by the early 1990s, computer parts and electronics. Hence, until the early 1990s, there were two main sources of capital formation: the Thai commercial banks and foreign direct investment.

---

[16]See Foreign Direct Investment (FDI) in Thailand, in Yoshihara (1999).

The expansion of the *agro-industry* had already led to a massive accumulation of capital and to the formation of powerful corporate conglomerates that were increasingly economically independent of the state. This concentration can be seen by the fact that in 1985, the 641 companies with over 200 employees (representing 1.6% of the total number of companies) employed 41% of the workforce and held 54% of fixed assets. Ninety percent of capital of the top seventy Thai financial firms was held by sixteen conglomerates, which controlled fifty of these firms (Suehiro 1989: 218–219). This trend increased after 1985. Gross Capital Formation rose from 346 to 1215 billion Baht, that is, from 29 to 42% of GDP between 1985 and 1995. The Thai economy became dominated by groups of companies, which incorporated finance capital, agro-industry and manufacturing.

The following discussion focuses on the agro industry, which consists of plantation forestry and led to the pulp and paper industry, driving factors of emerging factors of pulp and paper industries in Thailand, and how the government's policy to encourage and accessibility other facilities for investors.

Development of Pulp and Paper Industry

The forestry sector was associated with agro industry and mainly covered plantation forestry such as *Eucalyptus, A. mangium* and *local tree species*. This sector grew in the 1980s and was predominantly occupied by government agency (FIO: *Forestry Industry Organization*) and big paper companies such as Siam Cement Group (SCG), Advance Agro (AA), Phoenix, etc. In forestry, the old state-dominated *timber industries* declined as forest resources were depleted. During the period from 1980 to 1984, export revenue from logs and sawn wood dropped to an annual average of 22 million Baht whereas imports rose to over 2 billion Baht a year (Oliver 2005: 81–82). This trend was finally completed by the end of the decade with the general ban of logging in 1989. However, a new corporate group within forestry emerged in the form of conglomerates interested in *pulp and paper* production. The increase of manufacturing and the export boom led to a huge increase in the demand for paper for communications, printing, writings, administration and packaging.

The discussion focuses on the development of pulp and paper companies in Thailand by exploring case studies in terms of strategy to obtain capital, procurement of raw material, production and marketing of two emerging companies: Advance Agro and Siam Pulp and Paper. The Thai pulp and paper industries can be looked at in terms of the economic and political analyses of power relationships, capital movement and institutional structures and regulations, stressing the dynamic interaction between states and markets on national, regional, and global levels.[17] In this sense, Siam Cement Group (SCG) is closely linked with the royal family, political elites and high-level banking officers. Also, within Advance Agro, the chairman of the company was the former advisor to the Bangkok Bank,

[17]For further information, see Shiraishi (2005).

Virabongsa, and had close ties with the Prime Minister, General Chavalit Yongchaiyudh. The company has received loans to a total of US$850 million from the government bank.

The Impact of Government's Engagement

The Thai government actively supports the development of industry, through credit, subsidies, pro-cash crop and plantation policies, tax relief and favorable import duties on machinery imports. The *Royal Forest Department* (RFD) issued a policy on the provision of subsidies for seedlings and fertilizers to tree plantation farmers. In the period from 1981 to 1984, the government invited local farmers to plant *Eucalyptus* on their land, providing subsidies of approximately 1500–2000 baht per rai.[18] The response from farmers was positive, however, the unstable political conditions in the country led to inconstancy in the provision of subsidies during the 1990s. Hence, in order to continue sustainable plantation forestry and to empower the socio-economic strength of farmers, the government requested large companies such as Advance Agro, Siam Cement Group (SCG), Phoenix, Panjapol, etc., to create contracts with farmers under the scheme "Contract farming", thus providing the seedlings, fertilizers and a guaranteed market through private companies that would profit from this scheme themselves at harvest time. This program eventually categorized success achievements to provide 'raw material' (timber supply) to the company. Besides farmers could earn income to lift up their socio-economic condition and provide job creation for rural communities.

Domestic Capacity of Pulp

Total production capacity for short fiber pulp increased by 86% from 331,000 to 615,000 dry tons per year in 1996. The additional capacities were from two new pulp producers: Advance Agro Company Ltd, with 175,000 tons per year and Panjapol Paper Company, with 99,000 tons per year. Meanwhile, Phoenix Pulp and Paper Ltd increased to its full production capacity of 210,000 tons. The existing three pulp mills, Siam Pulp and Paper, Siam Cellulose Co and Bang Pain Pulp and Paper Mill still maintained their capacities of 68,000, 60,000 and 3000 tons per year, respectively (Table 2.6).

In contrast, from the first to the third quarter of 1996, Thailand imported 282,000 tons of short fiber and long fiber pulp, mainly from USA, Canada, Chile, Brazil, New Zealand, Sweden and Indonesia. At the same time, Thailand also imported 464,000 tons of wastepaper during the same period from USA, Singapore, Germany, the Netherlands, and New Zealand. The major kinds of imported wastepaper were old corrugated containers (OCC), representing 65% of the total

[18]An interview with Veirapol Suthiponpalangkul, Royal Forest Department (RFD Officer) on July 3, 2009 in Bangkok.

Table 2.6 Pulp companies' production capacity (1000 tons)

Company	Year 1995	Year 1996
Phoenix Pulp and Paper	200	210
Advance Agro	–	175
Panjapol Pulp Industry	–	99
Siam Pulp and Paper	68	68
Siam Cellulose	60	60
Bang Pa-in Pulp and Paper	3	3
Total	331	615

Source FAO (2004)

imported amount. But in the same period, 88,000 tons of pulps were exported. The main destinations were mostly in Asia, in particular, India, China, South Korea, Indonesia, Taiwan, Japan and Italy (FAO 2004).

Domestic Capacity of Paper

In 1996 there were 47 paper mills registered, with a combined annual capacity of 2,842,000 tons, a 16.3% increase compared to the 2,444,500 tons in 1995, including the new pulp and paper mill, namely, Advance Agro Company Ltd, with 217,000 tons per year. The total paper capacities in 1996 could be categorized into 1,625,000 tons for kraft paper, 631,000 tons for printings and writings paper, 264,000 tons for paperboard, 110,000 tons for newsprint paper, 132,000 tons for household and sanitary paper and 80,000 tons for gypsum plaster board liner (other papers) (Table 2.8).

When the economic crisis hit Asia in 1997, there was a reduction in domestic consumption of paper and board in Thailand. Consumption fell from 2,042,000 tons in 1997, to 1,604,000 tons in 1998. Domestic pulp consumption also fell from 802,000 tons to 644,000 tons (FAO 1998). Meanwhile imports of pulp and paper in 1998 decreased to 43% of 1997 levels. At the same time, production capacity increased. Exports increased, partly because of the surplus created by reduced domestic demand, but also because companies were desperate to earn hard currency in order to repay foreign loans after the collapse of the value of the Baht. In 1997, around 525,000 tons of pulp and paper was exported, while the 1998 figure almost doubled to 971,000 tons.[19] Advance Agro and Phoenix currently have ambitious plans to increase capacity, driven by their needs to repay debt and by the demand for cheap pulp price internationally (Table 2.7).

[19]For more information, see Paperloop (2000).

Table 2.7 Paper production capacity (1000 tons)

Various paper	Year 1995	Year 1996	Percentage %
Kraft paper	1416	1625	57
Printing & writing paper	465	631	22
Paperboard	264	264	9
Household and sanitary	110	132	5
Newsprint paper	110	110	4
Other papers	80	80	3
Total	2445	2842	100

Source FAO (1999)

Table 2.8 Production, export, and import of pulp and paper 1979–1994 (1000 tons)

Year	Production		Export		Import	
	Paper	Pulp	Paper	Pulp	Paper	Pulp
1975–79	275.2	30.8	8	–	115.2	94.2
1980–84	374.8	45.2	8.2	0.8	171.6	96.0
1985–89	574.4	131.8	44.2	17	234.4	118.1
1990–94	1221.0	202.2	94.6	30	470.8	289.8

Sources TDRI Thailand Development Research Institute (1989), RFD (1986, 1990), *TPPIA* Thai Pulp and Paper Industries Association (1987, 1997)

Other Pulp and Paper Companies

Six companies with production capacities of over 100,000 tons controlled another 37% of production. These were: Phoenix Pulp and Paper Company, owned by the European Overseas Development Corporation (EODC); Ballarpur Industries, an Indian Industrial conglomerate, the largest producer of bleached virgin pulp; Soon Hua Seng (SHS) Group, the largest rice-exporting and agro-industry conglomerate in the country, which moved into the industry in the late 1980s with its plantation company Suan Kitti Corporation and Hi-tech Paper Co. Ltd. (33,000 tons of printing and writing paper) and Advance Agro (175,000 tons of pulp and 217,000 tons of paper) in 1994, with major shareholders being the Japanese multi-national New Oji paper and Shin Ho Paper Co. Ltd. from a Korean conglomerate (100,000 tons newsprint); Thai Cane paper Co., Ltd. (100,000 tons kraft paper) and United Paper Co., Ltd. (99,000 tons kraft paper) (Sonnenfeld 1996; Suehiro 1989; Carrere and Lohmann 1996). From this description, the growth of the pulp and paper industry can be seen in Table 2.9.

The development of paper production was increased by an annual 10% between 1976 and 1984, and pulp production by an annual 27%. Between 1980 and 1984, pulp production rose by a remarkable 45%. Throughout the 1980s, production expanded, with paper production in the early 1990s over three times the volume of the early 1980s, and pulp production over four times larger. However, this increase

Table 2.9 BOI-promoted pulp manufactures by 1988

Company	Year approved	Capacity (tons)	Investment (ml. baht)
Kimberley-Clark	1984	5952	18.00
Panjapol Pulp & Paper	1978	50,000	235.00
Pech Paisal Industry	1984	5952	50.00
Phoenix Pulp & Paper	1982	50,000	735.00
Siam Pulp & Paper	1979	na	500.00

Source Charit (1989), Oliver (2005: 88). *BOI* Board of Investment

did not satisfy domestic demand. In the early 1980s, paper imports were at around 46% of production and pulp imports were more than double domestic production.

During the 1980s, a concerted effort was made to increase supply. As Sonnenfeld (1996) puts it: "expanded pulp and paper production was a national development objective, to lessen dependency on outside suppliers, decrease imports, and provide intermediate goods to growing, export-oriented industries". The stated aim was to gain self-sufficiency but also to expand into export markets and this was apparent in the policies for the paper and pulp industry. Although 70% of the Thai industry's fiber requirements are supplied by *wastepaper* (in particular, for products like kraft paper for export packaging), virgin pulp is a necessary ingredient and high pulp price increases in the 1980s made dependence on imports especially painful. Therefore, a growing international market for pulp and wood-chips (for pulp production) made it attractive for Thai conglomerates to produce pulp for export (Table 2.4). Multi-national firms also started taking an interest in Thailand as a possible producer of *Eucalyptus* pulp.

The government promoted the nascent Thai pulp industry through the following means:

(1) Increasing the import duty for pulp from 1 to 10% in 1982;
(2) Placing four annual levies of an extra surcharge of 20% for imported pulp during 1982–1986;
(3) Providing exemptions from corporate income tax for three to eight years after a company starts earning income. Losses occurring during the exemption period could be deducted from revenue for five years after the exemption period (TDRI 1989) (Oliver 2005: 88).

Several companies were given preferential treatment by the BOI (*Board of Investment*) (item 3) (Table 2.10).

Government policy sought to protect the infant pulp industry in order to establish a competitive domestic basis from which it could then begin to export as well as in the promotion of wood chip production. Projections were made by the pulp and paper industry itself in 1989 showing confidence in an increasing demand and capacity (Table 2.12). Domestic demand of pulp was expected to reach 426,000 tons in 1993, 487,000 in 1995, 548,000 in 1996 and estimated to reach 712,000 in 2004, and slightly developed to become 780,000 in 2006. Production capacity was

Table 2.10 Firms with BOI privileges for wood chips production

Name of Company	Year appr.	Capacity (tons)	Export (%)	Investment (ml. baht)
Siam Vana Forestry Co.	1986	510,000	100	70.02
Siam Agrotex Co., Ltd.	1986	561,000	100	77.94
Thai Wittawat Corp.	1988	65,280	100	65.00
N.R. *Eucalyptus* Agr. Co.	1987	65,000	80	50.00
Thai United Chip wood	1988	100,000	100	53.26
C. Nakon International	1986	150,000	100	29.25
Siam Forestry Industrial	1986	255,000	100	40.02
Thai Wood and Pulp	1988	46,000	100	15.44
Kitti Garden Co.	na	65,000	na	70.00
V. P. *Eucalyptus* Chip wood	1987	240,000	na	185.00
Total		2,057,280		655.93

Source Charit (1989), Oliver (2005: 89). *appr* approved

expected to reach 152,000 tons in 1990 and rapidly developed to become 717,000 tons in 1996, 1180 million tons in 2003 and 1169 million in 2007 (Table 2.13). Meanwhile, domestic production of paper reached 2698 million tons in 2001, 3795 million tons in 2004 and rapidly developed to 4516 million tons in 2007.[20] To fulfill the demand for wood-chips, about 482,000 ha of plantations was required in 1990, rapidly growing to 4920 million hectares in 2000 (FAO 2001). However, the projected rise in production capacity could only be met if a sufficient supply of raw material for wood chips and pulp was made available. For this reason, the industry stepped up its efforts to achieve a substantial increase in the area of fast-growing tree plantations (Table 2.11).

Two Emerging Paper Companies

The discussion focuses on two emerging paper companies namely *Advance Agro and Siam Cement Group* (*SCG*). There are two main rationales for both companies to eventually progress. Firstly, both company's officers actively engage with political elites in the government by making "political collusion" in exchange for accessibility to facilities such as credit and loan. Secondly, the Thai's government launched 'affirmative policy' for attracting investors, such as tax holidays, tax relief for machinery imports, provide good infrastructure (port and high way), easy access of getting raw material (timber) for company, etc.

[20]Regarding total paper production comprises such as newsprint, printing and writing paper, kraft paper, paper board and packaging paper, and household and sanitary paper see Forestry Statistics of Thailand, 2007, pp. 28.

Table 2.11 Projected development of pulp demand, production capacity, and plantation area required in Thailand (1000 tons)

Year	Domestic pulp demand (TDRI)	Domestic pulp demand (TPPIA)	Pulp capacity (TPPIA)	Plantation area required (TDRI) 1000 rai
1988	–	185.50	110.50	389.25
1989	–	196.30	140.50	437.69
1990	118.00	268.00	152.00	482.84
1991	133.60	303.00	153.00	533.62
1992	149.20	340.00	203.00	590.02
1993	164.80	426.00	253.00	652.05
1994	180.40	480.00	552.00	719.69
1995	196.00	487.00	717.00	792.96
1996	214.40	548.00	717.00	879.11
1997	232.80	–	–	978.15
1998	351.20	–	–	1090.06
1999	269.60	–	–	1214.86
2000	288.00	–	–	1352.54

Sources TDRI (1989), TPPIA (1989, 1990a, b, 1991), TPPIA (1987), TPPIA (1990)

Table 2.12 Domestic production of paper and pulp (1000 ton)

Year	Paper	Pulp
2001	2698	919
2002	2517	953
2003	3598	1180
2004	3795	1000
2005	4308	1061
2006	4308	1129
2007	4516	1169

Source Forestry Statistics of Thailand (2007)

Profile of Advance Agro

Advance Agro (AA) in the branch of Soon Hua Seng Group (SHS). The company is the biggest fully integrated pulp and paper manufacturer in Thailand. Advance Agro's mill opened in 1996, and the second mill in 1998, at Tha Toom in Pachinburi province, 120 km from Bangkok. The Company has two pulp mills with a total production capacity of 580,000 tons a year and three paper plants with a total capacity of 600,000 tons a year. All short-fiber wood comes from *Eucalyptus* trees grown in a 150-km radius of Tha Toom mills. **Soon Hua Seng (SHS)** Company established a laboratory for carrying out research into *E. camaldulensis* and other fast-growing tree in 1982. Today, SHS Group's 67 ha complex of greenhouse and **nursery plots** churns out 65 million hybrid clone seedlings every year. The company distributes pulp and printing and writing paper in Thailand and exports to

Table 2.13 Total production of Thai Pulp and Paper Industries year 2007–2008 (Unit: 1000 tons)

Products	2007	2008	SCG Paper Prod in 2008	% of share in Domestic Prod
Pulp	1169	1106	414	37
Paper	4322	4235	1687	40
Printing writing & paper	1108	1014	390	38
Paperboard	348	369	147	40
Kraft Paper	2625	2604	1150	44
Others (newsprints & sanitary paper)	241	248	–	–
Total pulp and paper	5491	5341	2101	39

Source TPPIA The Thai Pulp and Paper Industries Association (2008)

Europe, Asia, USA, Africa and in the Middle East. 70% of the company's production of paper is exported overseas, the two largest markets being China and Japan (Woranuj 1998).

The company is part of the agribusiness **Soon Hua Seng (SHS) Group**, which was founded in the 1950s by members of the *Dumnernchanavit family*. SHS started growing *Eucalyptus on a commercial* scale in the late 1980s. Advance Agro's and SHS Group's management are well connected with Thailand's political elite. Virabongsa Rmangkura, chairman of the company, acted as an economic advisor to General Prem Tinsulanonda, Thailand's prime minister in 1980s. In the mid-1990s, he was finance minister during General Chatichai Choonhaven's administration and was deputy premier in charge of economic affairs under Prime Minister General Chavalit Yongchaiyudh (Lang 2002). Virabongsa is an advisor of Bangkok Bank, which has outstanding loans US$850 million with the SHS Group (AA www 1).[21] Advance Agro was incorporated in 1989, and listed on the *Stock Exchange* of Thailand in February 1995. The major shareholders are Soon Hua Seng Group (47.1%), officers and directors of Advance Agro (8.9%), Stora Enso, Finland (19.9%), Oji Paper (5.5%) and the Commonwealth Development Corporation (1.1%) (AA www 2).[22]

Hence, among the companies recently making large new investments in pulp and paper has been the *Soon Hua Seng Group*, one of Thailand's leading rice traders. In late 1995 the firm's subsidiary Advance Agro brought a new 217,000 tons per year Mitsubishi-Beloit paper machine on stream in east Thailand, with an integrated 175,000 ton-per-year Defibrator *Eucalyptus* kraft pulp mill following shortly thereafter. The Siam Cement Group, a company associated with the Crown Property Bureau and which holds eleven companies producing paper and pulp, has recently boosted its production capacity to around one million tons per year of pulp,

[21]See AA (www 1).

[22]AA (www 2).

paper and converted products, with three new machines slated to come on line by 1997. Hiang Seng and Panjapol have also added new capacity.[23]

Advance Agro was hard hit by the *1997 economic crisis*, when the value of the Thai Baht collapsed. Advance was left with debts of 22.6 billion baht, only 7 billion of which was baht-denominated debt. Advance subsequently breached the financial covenants on its loans. In November 1997, the company became the first Thai company to issue high-yield bonds with a US$111.35 million bond offering on US markets. The company raised more money by selling 19.9% (US$82 million) of its shares to *ENSO (which merged with Stora in 1998 to become Stora Enso) and 5.5%* (US$22 million) *to Japan's Oji Paper*. Through the deal Oji Paper won the exclusive rights to sell Advance Agro paper in Japan. In May 1999, Masahiko Ohkuni of Oji Paper became a board member of Advance Agro (*Bangkok Post*, 7 May 1999). The company exports 60% of its paper products to Hong Kong, China, Australia, and Europe.[24] In 2003, Advance Agro used 50% of its pulp to produce paper. Five per cent was sold in Thailand and the remainder exported to Australia, China, South Korea, and Malaysia among others.

Response from Local People

Villagers living near Advance Agro's pulp mill complain of ash from factory chimneys being deposited in their gardens. Some villagers have developed itchy skin and the pulp mill often smells. In August 2000, black, stinking water from piles of wood and charcoal in Advance Agro's factory compound leaked into a neighboring canal killing a large number of fish. The cause was a collapsed dyke inside the factory compound. The company routinely pours its waste water from pulp mills between the rows of *Eucalyptus* trees. The water is filthy and green and lies in channels in the stony infertile soil. People also supposed that Advance Agro's plantations have been one factor leading to the 'deforestation' of large areas of eastern Thailand. Other factors include road building (partly built during the war in Indochina to link bases for US troops with the port as Chon Buri and also to access Cambodia's forests), and the promotion of large-scale industrial agriculture by the government with support from the World Bank. One of the beneficiaries of these policies was Soon Hua Seng, Advance Agro's parent company.[25]

The company responded to minimize water pollution by buying water treatment equipment and installing it in its factory operation. The company launched a policy on 'Eco-friendly products' by implementing sustainable forest management with attention to three elements: namely economics, social and ecological aspects in realizing timber plantation.

[23]Carrere and Lohmann (1999).

[24]Thailand's Advance Agro rating raised to 'B', after completion of debt restructuring", Thai Press Reports, 10 June 2004.

[25]Lang (2003).

Company's Procurement of Raw Material

Advance Agro does not actually own any of its own plantations, instead purchasing the wood from Agro Lines, another member of the SHS Group. Agro Lines obtains its wood from four sources: (1) from its own plantations (which in 1997 covered 31,000 ha); (2) from plantations owned by the SHS Group; (3) from local farmers under "contract farming" with Agro Lines (in 1887, 46,000 ha), and (4) from individual farmers.

As one of the largest pulp and paper industries, the company needs a sustainable supply of raw material (*Eucalyptus*). To provide this need, Advance Agro planted timber plantation in some districts and launched a policy to have 'contract farming' of about 46,000 ha also with local people. The concentration area is located in the Eastern region of Thailand, such as: Pacinburi, Sakei, Chunbury, Cucunsaw, etc. Besides, the company had plantation concessions around 40,000 ha from Royal Forest Department (RFD) to be planted *Eucalyptus* trees.[26] In harvesting time, the logs are transported to the mill either by farmers or by a truck of the company which is also part of the SHS Group, working under subcontract to Agro Lines.

When Advance Agro's subsidiary Agro Lines started establishing its *Eucalyptus* plantations, villagers found they could no longer grow rice in neighboring fields. The company bought villagers' farmland in Prachinburi to convert the land to plantations. Allegations of intimidation of villagers surrounded the company's plantation in eastern Thailand throughout the 1990s. Canadian academic Keith Barney notes that "Thai NGOs have organized in opposition to the land displacement resulting from *Eucalyptus* farming in eastern Thailand associated with the mill".[27] Obviously, because of rising huge protests from Thai's NGOs, one possible source to obtain of new raw material for Advance Agro namely is overseas. Laos and Cambodia are neighboring countries, one of the options for expanding *Eucalyptus* plantation in the near future.

Development of Siam Pulp and Paper

Siam Pulp and Paper (*SPP*) is one of the holding companies of the **Siam Cement Group (SCG)**. The SCG Company is very well-known in Thailand, which was established by Royal decree in 1913 and is today one of Thailand's largest industrial conglomerates. SCG includes more than 16 branches of companies. One

[26]The discussion about procurement of raw material of Advance Agro in eastern region of Thailand had been interviewed on July 3, 2009 with Dr. Pasuta Sunthornhao, lecturer of Department of Forest Management, Faculty of Forestry, Kasetsart University.

[27]Barney (2005).

of them is Siam Pulp and Paper. In the line of SCG, Siam Pulp and Paper is categorized the third largest division of Siam Cement, after Cement and construction materials. In 1993, Siam Pulp and Paper produced half of all the packaging paper sold in Thailand, and around 40% of all printing and writing paper (Lang 2002).

SCG is closely related to the *Kingdom family being* among the greatest part of shareholders.[28] Because of this environment, the company has great accessibility in obtaining credit from national bank, cash capital in stock exchange market in Bangkok and eventually could expand other businesses. Currently, SCG develops business such as pulp and paper, cement, construction, container for packaging, polyester, chemical, distribution, etc. Focusing on pulp and paper, this business sector consists of the Siam Forestry Co, pulp and printing & writing paper, containerboard, corrugated containers and paperboard. As an illustration, SCG paper in 2008 increased sales from 2007 by 7.3% to 47,110 million baht, while net profit after tax declined by 30.1% to 1454 million baht (Fig. 2.1) (The profile of SCG Group 2008). But, it is very optimistic about paper business development in the future as told by Promotion Department Manager, because paper is one the necessary items of daily life, and the people's consumption will be rapidly increased in the future. In spite of consumption amounts to an annual average of 58 kilograms per capita, and will be boosted double based on the economic development. Paper, consequently, may be considered an index of prosperity. Table 2.8 shows the contribution of SPP Company to total production of Thai pulp reached 414,000 tons (37%) and paper products 1.6 million MT in 2008 or equivalent with 40% of whole domestic production about 4.2 million MT.

Historically, SCG moved into pulp and paper almost accidentally. In the 1970s, Siam Kraft Paper produced sack kraft for Siam Cement in its 25,000 tons-per-year mill in Ratchaburi province. Siam kraft was Thailand's first modern pulp mill, established in the late 1960s. Raw material for the mill came from *baggase*, the residue from sugar case processing. Because Siam Cement is dependent on a regular supply of *cement sacks* products. Therefore, Siam Cement loaned money to Siam Kraft Paper to maintain the operation to produce cement sacks. In 1976, Siam Cement took over Siam Kraft Paper, which at that time had 50,000 tons a year mill at Ratchaburi. Siam Cement discovered that producing kraft paper from waste paper was cheaper than using *baggase*. Then the company continued to produce baggase pulp, but with a new bleaching line so that the pulp could be sold. *Yuen Foong Yu*, a Taiwanese Company has experience in producing and selling baggase pulp. So Siam Cement made cooperation with Yuen Foong Company to sell it. In 1982, Siam Cement, Siam Pulp and Paper and Yuen Foong Yu established a new

[28]The King Bhumibol Abdulyadej who currently is 84 years old is the world's longest-reigning monarch in Thailand. He is the central figure of beyond Thailand politics and widely respected by Thai people and political elite. Therefore, the role of the Kingdom is a very strategic position, SCG Company that most shareholders owned by the Kingdom family has great accessibility to obtain cash 'credit' from bank institutions and could expand easily her holding business in many leading sectors (interview with SCG Officer, July 9, 2009).

company, Thai Paper. In 1983, the three companies bought a 51% share in Thai Union Paper, one of Thailand's oldest paper makers (Lang 2002). Siam Pulp and Paper in November 1992 officially opened three new affiliated pulp and paper in Kanchanaburi: (1) Siam Cellulose capacity 50,000 tons of bleached pulp per year; (2) Thai Kraft Paper capacity 250,000 tons of kraft paper a year; and (3) Thai Union Paper capacity 70,000 tons coated paper and gypsum paper per year (*Bangkok Post* 17 November, 1992).

The rapid development of Siam Pulp and Paper happened in the mid-1990s; the Company has a capacity of more than one million tons a year of pulp, paper and converted products. Industrial paper accounted for 46% of 1999 revenues; printing and writing paper, 36%; packaging and printing, 18%, and a nominal amount from paper pulp.

Its Market

After the success in the domestic market, Siam Pulp and Paper expanded abroad and approached the Laos government about a possible US$250 million project to establish a 150,000 tons-per year pulp mill. The mill was to be supplied from 32,000 ha of plantations in Savannakhet province in central Laos (Carrere and Lohmann 1996: 241). In 1996, Siam Pulp and Paper also bought 30% of Nityasa Prima Company, a proposed US$520 million, 350,000-tons-a-year pulp plant in East Kalimantan, Indonesia. The Indonesian company Suryaraya Wahana, part of the *Astra International Group*, owns 60% of the venture, with local investors holding the remaining 10%. For this reason, Chumpol Nalamlieng, Siam Cement's president told the *Financial Times* his company had to buy pulp overseas because Thailand's supply was limited due to deforestation and government restrictions on tree farming. "Indonesia has huge wood resources—biggest than the (*Association of South East Asian Nations*) region combined," Mr. Chumpol said. "Clearly it is going to be cheaper to produce pulp in Indonesia for the foreseeable future" (Barnes 1996). Unfortunately, a year later, this project was also cancelled (Saragosa 1997).

The Asian economic crisis in 1997 was a disadvantage for Siam Cement. The company was caught with US$4.2 billion of foreign loans and Siam Cement suffered the biggest ever deficit by a Thai company-recording a loss of US$1.3 billion in 1997. The solution was only by increasing the valuation of its assets by 75% could the company avoid posting a negative net worth. The company responded by hiring Mc Kinsey Consulting, who drew up a plan to sell one-third of Siam Cement's assets (Barnes 1996). Apparently over the next two years, Siam Cement managed to increase its exports, convert some of its dollar to baht and properly managed to make a profit. The selloff was scaled down to a 13% asset-scale. Siam pulp and Paper exports around 50% of its production to more than 40 countries. The amount is one-third of its industrial and packaging products to export to China and Malaysia (Umesh Pandey 1999).

Strategy to Expand Company

The strategy to obtain 'cash capital' in January 2000, that Siam Cement was to issue up to US$1.34 billion of domestic bonds. Wongbuddhapitak, Siam Cement's chief financial officer, told Reuters "We have a foreign exposure of about US$1.8 billion, while our foreign exchange revenues are about US$1 billion. We need to lower this gap to lower the risk of volatility".

In October 2001, Siam Pulp and Paper bought 24.98% of Phoenix Pulp and Paper from Janpath Investment. In January 2002, Siam Pulp and Paper bought a 20% stake from Thai Farmers Bank raising its holding to 44.98%. After a tender offer, which finished early in January 2002, Siam Pulp and Paper owned more than 61% of the Phoenix. Siam Pulp and Paper plans to go ahead with a planned expansion of Phoenix, including a US$30 million upgrade of the pulp mill and US $60–80 million on new facilities.[29]

On the other hand, SCG has demonstrated leadership and commitment to uphold *sustainable forest management policy*, in keeping with the globally certified standards of *Forest Stewardship Council* (FSC). In this sense, SCG could be categorized to be the first Thai paper company to be recognized for the FSC standards, including pulp products that have been certified FSC Chain-of-Custody (Coc); tracking of FSC certified material through the production processes all the way to the store. Apparently, these certifications offer consumers new and better environment-friendly choices (Fig. 2.8).

Concluding Remarks

The major findings of this paper are summarized below:

(1) The spectacular impact of more than three decades (1970s–2000s) of economic reform in Thailand is impressive. This reform transformed Thailand's foreign economic relations through the implementation of a radical open-door strategy. This development affected an increase in foreign direct investment (FDI) in cooperation with domestic investors in joint ventures and even one hundred percent foreign-owned companies. Also, the Thai government established *infrastructure* (*ports and highways*) and other facilities for investors: easy accessibility to credit, access to the stock exchange, tax relief and favorable import duties on machinery and so forth. These activities allowed for a great development on many sectors and affected the emerging pulp and paper industries in Thailand. For instance, the paper industr expanded in the end of 1980s, with paper production in the 1990s–2000s over four and five times the volume of early 1980s. As an illustration, paper companies grew from the 21

[29]See Siam Pulp and Paper takes over Phoenix, Reuters, 4 January 2002, website: http://www.pponline.com/inside/stories/wk12_31_2001/r36,shtml. Accessed 11 January 2002.

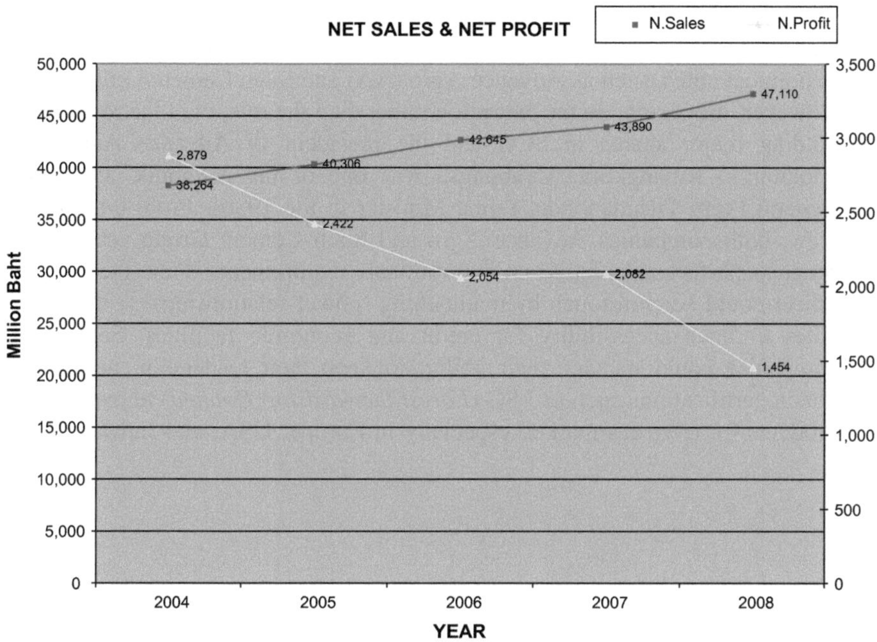

Fig. 2.8 Net sales and net profit after tax of SCG. *Source* The SCG company's profile in 2008, pp. 26. N net, NP net profit after tax

registered companies producing 574,000 tons in 1985–1989, and rapidly developed to become 47 units in 1996 (2.8 million tons) and 58 companies (4.3 million m^3 tons) in 2007.

(2) The rapid development of pulp and paper factories and plantation forestry (*Eucalyptus camaludulensis*) as 'commercial trees' was eventually accepted by local farmers and the private sector as a lucrative business to boost income generation and job creation in rural areas of Thailand. Apparently, better socio-economic incomes subsequently strengthened 'purchasing power' among local communities and people at large could affect domestic 'economic development' of Thailand in 1990s. However, the negative impact on the environment in the form of *toxic pollution* of rivers and *dried land* on cash crops surrounding *Eucalyptus* plantations raised critiques from academics and NGOs. At the end of the 1990s, NGOs and locals in the northeast of Thailand launched a huge protest to close factory operations, requesting water treatment facilities, because Phoenix Pulp and Paper mills disposed of toxic pollution into the Phong River in Khon Kaen province, killing the fish and shell species and contaminating down river agricultural farmland. These NGOs and northeastern locals refused the planting of monocultures and promoted the substitution of multi-tree plantations, with local tree species, providing for a greater bio-diversity, as well as proposing new areas for community forests.

(3) The strategies of pulp and paper companies to obtain credit, raw materials, management and marketing vary greatly. There are companies that have access to political elites, such as Advance Agro (AA) and Siam Cement Group (SCG). However, they are more the exception rather than the rule, with the royal family holding major shares in SCG and the president of Advance Agro being Virabongsa Rmangkura. Virabongsa was one of the economic advisors to General Prem Tinsulanonda, Prime Minister in the 1980s. From this point of view, both companies Advance Agro and Siam Cement Group achieved the progress performance based on two rationale requirements: First, the company officer could keep in touch by maintaining 'power relationships' with political elites to have accessibility for credit and economic facilities. Second, the companies could manage their ability to keep their leadership, networking, obtain certifications such as FSC (*Forest Stewardship Council*) to expand their products for overseas market, especially in Europe, USA, and Japan.

Chapter 3
Philippines

Review of Forestry Policy

Introduction

The Role of the Government in Reforestation

The *Forest Management Bureau* (FMB), a section under DERN (*Department of Environmental and Natural Resources*), is responsible as an institution for the management and monitoring of forest management in The Philippines. FMB issues plantation concessions for production forestry to the private sector, cooperatives and individuals under the IFMA and SIFMA schemes. Also, forest rehabilitation and reforestation programs in protected and conservation forests are under their authority. The 1990 Master Plan for Forestry Development estimated previous forest loss based on available information (DERN 1998) between 1934 and 1990 to an amount of 10.9 million hectares of forest cover for the entire country, equaling an average annual loss of 194,000 ha (Table 3.1) (Unna Chokkalingam et al. 2006, 11–12). Of this area, 10.37 million hectares (95%) was converted to other uses, while 0.52 million hectares was damaged from logging. From 1934 onwards, the loss rate increased dramatically, peaking at 300,000 ha/year over the decades between 1965 and 1975. Therefore, the rate of loss of forest cover gradually declined to 100,000 ha annually from 1985 to 1990.

The Philippines forests have degenerated over the years due to massive logging and conversion to agricultural land, including shifting cultivation. This is mostly due to population growth and the increasing need for more agricultural land.[1] This has reduced forest cover from about 21 million hectares (70% of the total land area)

[1]The Philippines population was estimated at 88,574,614 million in 2007, rapidly reaching 97,976,603 million in 2009 (*Wikipedia*). Rural communities still also carry out 'shifting cultivation' (*Kaingin)* in state forest land in the upper regions, which also affect soil erosion, causing flooding.

H. Hidayat, *Sustainable Plantation Forestry*,
https://doi.org/10.1007/978-981-10-7653-4_3

Table 3.1 Forest Cover Loss (in 1000 ha) from 1955 to 1990

Description	1955–1965	1965–1975	1975–1985	1985–1990	Average annual loss
Starting cover	13,900	11,600	8600	6600	
Less losses due to:					
[a]Conversion	2200	2835	1880	460	185
[b]Logging damage	100	165	120	40	9
Total losses	2300	3000	2000	500	194
Final cover	11,600	8600	6600	6100	

Source Based on secondary data interpretation by *the 1990 Master Plan for Forestry Development;* Cf: Unna Chokkalingam et al. 2006: 12
[a]Total forest covers loss
[b]Damage out of 5.3 million hectares logged

at the end of the 19th century (Garrity et al. 1993; Liu et al. 1993), to around 7.2 million hectares or about 23.9% of the total land area (Fig. 3.1).[2] Of the remaining forest cover, less than one million hectares can be considered primary forest (FMB, 2007). The remainder has been logged at least once or has suffered degradation through other activities (Pulhin et al. 2007: 866).

Based on FAO data (2006), the Philippines had the highest deforestation rate in all of South and Southeast Asia during the 1990s, recorded annually at about 2.8%. In comparison, Indonesia and Thailand recorded annual rates of 1.7 and 0.7%, respectively. In Vietnam, the forest area expanded by 2.3% annually. Between 2000 and 2005, the deforestation rate declined only marginally. Vietnam has certainly benefitted from massive investments in reforestation, while China and Thailand, like the Philippines, have very few forest areas left. The nationwide logging ban imposed in 1989 has curtailed uncontrolled forest conversion to some extent.

In the case of the Philippines, the main causes of deforestation have been attributed to broader structural forces, such as political patronage by political elites, poverty, inequitable access to forest resources, and corruption in the forestry sector (Porter and Ganapin 1988; Kummer 1992; Broad and Cavanagh 1993; Vitug 1993).

This paper will discuss decision makers in the Philippines over the last two decades and efforts to reverse the downward trend of forest degradation and address the mounting socioeconomic and environmental problems in the Philippine uplands. The paper focuses on land tenure and resource use in the Philippines, while also reviewing the two programs launched by the government to revitalize forest resources management: an affirmative program for *rehabilitation* of degraded forest lands and critical land covers with brush or grass and the *reforestation* of state forests in the upper regions, actively engaging local communities through

[2]Forest cover in the Philippines rapidly declined annually from 7.2 million hectares in 2004 to 5.5 million hectares in 2008, reaching the highest deforestation rate (2.8%) of the ASEAN region. This data is based on several discussions with informants in the Philippines on July 31–August 9, 2009.

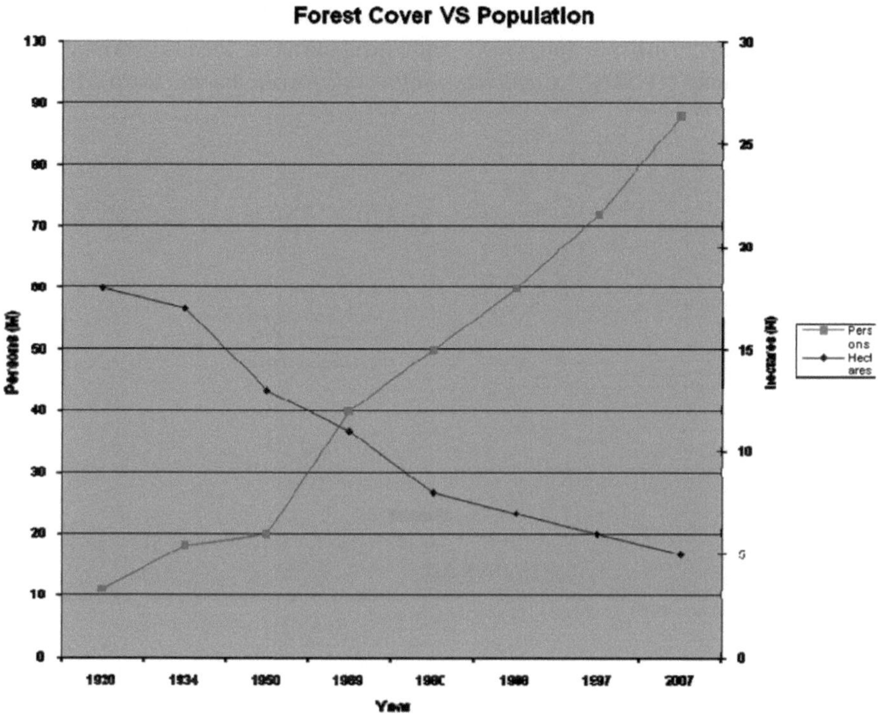

Fig. 3.1 The Philippines: forest covers development and population growth through the years. *Source* Modified data from Cf PCARRD, 2008, p. 1

CBFM (*Community-Based Forest Management*) in the 1980s. The CBFM program is well-known throughout ASEAN countries as 'community forestry'. On the other hand, two programs for timber concessions have been introduced: *IFMA* concessions (*Industrial Forest Management Agreement*), which actively invites the private sector to conduct logging, and *SIFMA* concessions (*Socialized Industrial Forest management Agreement*) for cooperatives, people organizations (PO) and individuals in state forests. The aim of these programs was to provide 'timber' as a raw material for forestry industries such as plywood, sawn mill, pulp and paper, furniture and construction for real estate and housing.

Study Sites

Interviews with stakeholders, such as government officers and academics were carried out in Manila and Philippines University (UP-Los Banos), Laguna. Meanwhile the study sites were located in the Caraga region, San Jose St, Butuan City; Talagon-Agusan Del Sur (Map). It takes one hour by airplane from Manila to

Butuan City. It was necessary to conduct field work to observe plantation forestry areas and conduct in-depth interviews with local farmers, local government and private companies (**CSDC**: *Casilayan Softwood Development Corp*) for date gathering.

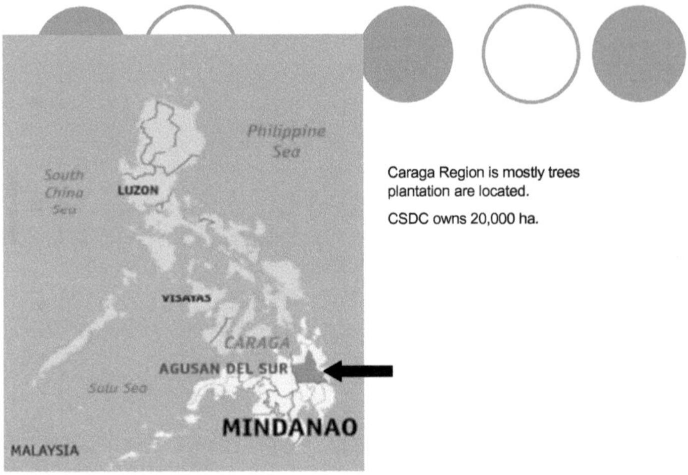

Land Tenure and Resource Use

Land ownership in the Philippines before the colonization by Spain was generally communal. Forests were accessible to anybody—whoever cleared and cultivated a piece of land first would own that portion. However, under the colonial rule of Spain, this was changed to ownership by the government.[3] The state forests were then owned by the national government after independence on 4 July 1946, followed by a radical change in land ownership during the Fidel Ramos regime in 1995, when the government recognized and issued thousands of hectares of 'customary rights forest land' for indigenous tribes. This policy encouraged the 'participation' of indigenous people (tribes) who commonly held customary forests in national economic development.[4] The government also legitimized the rights for allocating, classifying, regulating and managing forests and timberlands, reinforcing massive timber exploitation by the private sector and local communities,

[3]For a better understanding of land tenure under the Spanish until Philippine independence.
[4]Interview with informant in University of Philippines, College of Forestry, August 2, 2009.

through the conversion of lowland forests to pasturelands, agro-forestry and mining. Increased forest exploitation, therefore, marked the post-war period (Borlagdan et al. 2001).

To meet the demand of tropical timber, particularly in Japan, Europe and the USA, large-scale logging concessions were given to private sector companies (Boado 1988). As a consequence, logging grew, becoming very profitable and generating even greater revenues for the government (Borlagdan et al. 2001). The era of President Ferdinand Marcos was recognized as the zenith of logging. Marcos used *TLAs* (*Timber License Agreement)* to strengthen his political network, dispensing them as rewards to loyal cronies. Used as an arm for political patronage, the number of TLAs soared to as many as 400 during his regime (Vitug 1993; Inoue and Isozaki 2003) leading to the highest *deforestation* rate ever, amounting to as much as 300,000 ha/year. From 1971 to 1977, TLAs controlled one-third of the country's total land area of 30 million hectares, however by the late 1980s, TLA areas declined to the present 1.035 million hectares (FMB 1998).

The impacts of *deforestation* eventually grew too great to ignore, with environmental degradation causing flooding, soil erosion, landslides, and more, causing a great loss of agricultural production. Finally, a major milestone in the evolution of people-oriented forestry in the Philippines occurred in 1995 with the enactment of Executive Order No. 263, adopting community-based forest management (CBFM) as a national strategy for sustainable forestry and social equity. It unified and integrated the earlier community-oriented forestry program and projects, and provided a comprehensive mechanism for their implementation, thereby institutionalizing CBFM in the country.

Rehabilitation projects that have been carried out under the CBFM scheme include stakeholders such as DERN, NGOs, communities, private companies, etc. The CBFM program in the Philippines is considered progressive because of its land tenure and resource use rights features (Utting 2000; Pulhin 2007: 876). The issuance of various tenure instruments under CBFM promotes a "win-win" strategy for both the government and local communities, as granting of land tenure terminates the open access nature of forestlands, while devolving the responsibilities of management and protection to the local communities at minimal costs.

A closer analysis of the situation on the ground shows that the potential "win-win" outcome is often not being realized. Local communities continue to experience a strong sense of insecurity over their CBFM areas despite the issuance of rights. This is because of frequent government policy changes regarding timber utilization, such as when more than 1000 CBFM agreements were cancelled nationwide by the former DERN Secretary because of irregularities in some areas (Miyakawa et al. 2006; Pulhin 2006). Moreover, the associated bundles of rights have never been realized in most areas as a result of unstable policies exacerbated by excessive and tedious bureaucracy associated with timber utilization. Instead of providing rights to local people, different land tenure instruments such as CBFMs, have enhanced government control by limiting the devolution of responsibilities towards forest development and protection to local communities. The authority and rights to resources that local communities manage are often undermined, left unclear or even broken, which can leave people worse off instead of better off (Pulhin 2006).

Political Factors and Funding Availability

In the 1970s, the political landscape in the Philippines changed from forest extraction to a more friendly rehabilitation program. Major changes in political leadership affected both public and private efforts. Plantings usually peaked when new government administrations were implementing major forestry programs (Unna Chokkalingam et al. 2006: 23). For example, the declaration of Martial Law in 1972 was followed by the issuance of the Revised Forestry Code of the Philippines (*PD/Presidential Decree* 705) in 1975. This decree required nationwide 'reforestation' activities with private sector participation. P.D. 705 defined forest lands to be reforested as those with barren, grass or shrub cover, denuded areas within forest concessions, reserves and reservations, critical watersheds, national parks and other protected areas, areas covered by pasture leases needing immediate reforestation and miscellaneous areas such as river banks and roadways.

Forest Rehabilitation

The history of national rehabilitation initiatives is divided into three periods: (1) the colonial period (1910–1945), (2) post-war, government initiated projects (1946-mid 1970s), and (3) multi-sectoral efforts (mid 1970s-present).

By the end of the 1970s, the private sector, government agencies other than the Bureau of Forestry, local government units and citizens were actively involved in forest 'rehabilitation' efforts as a result of government proclamations issued since 1975. Generally, rehabilitation programs were the planting of local trees species in critical lands of watersheds, protected forests and conservation forests. The purpose of this program was to conserve water and prevent soil erosion and flooding from upstream to the lowlands. A total of 64,541 ha of trees were planted in 1981—33,834 ha or 52.42% was planted by groups besides the Forestry Bureau. Between 1973 and 1979 three 'people oriented forestry' programs were implemented, namely the *Family Approach to Reforestation* (FAR) Program, *Forest Occupancy Management* (FOM) Program, and the *Communal Tree Farming* (CFP) Program. In 1982, a major program known as *Integrated Social Forestry Program* (ISFP) was established through the issuance of *Presidential Letter of Instruction* 1260 (Pulhin 2007: 808) and the Community Forestry Program in 1987. ISFP consolidated the three earlier programs, while recognizing the vested interests of the forest occupants through the provision of a 25-year tenure security. This tenure security enabled the upland farmers to farm their land and enjoy the benefits of their labor without fear of being ejected from the government-owned forestlands.

Many of the projects were funded by foreign donors, such as the *Ford Foundation*, the *World Bank*, United States Agency for International Development

(*USAID*), Asian Development Bank (*ADB*), Japanese Bank for International Cooperation (*JBIC*) and the German Technical Cooperation (*GTZ*), and executed by, or in collaboration with, *DERN*. These projects tended to be small-scale *agro forestry* and *social forestry* projects targeted at meeting the livelihood needs of farmers and communities and addressing environmental degradation in the uplands.

After the end of the Marcos regime in 1986 and the new government of Qory Aquino, the Philippines regained its credibility with international funding institutions, enabling it to access huge financial assistance to conduct forest rehabilitation initiatives (Korten 1994). The major initiatives included the *Forestry Sector Projects* (FSP) I and II established in 1987 and 1995 respectively under the *National Forestation Program* (NFP). The NFP aimed to rehabilitate 1.4 million hectares nationwide from 1987 to 2000, or an average of 100,000 ha/year (Magno 2003). The long-term target was to reforest 6.5 million hectares of denuded lands, including 1.4 million hectares of critical watersheds needing immediate rehabilitation.

The *FSP* I method replaced traditional government-implemented reforestation with 'contract reforestation' involving families, local communities, NGOs, LGUs (*Local Government Units*) and the private sector. Under the contract reforestation scheme, contractors were paid a fee for reforesting and maintaining a particular area for three years with an expected survival rate of >80% and an average height of 0.8 m. After the contract period, the area was to be returned to DERN. FSP I was funded by a USD 120 million *Asian Development Bank (ADB)* loan with USD 120 million counterpart funding from the *Overseas Economic Cooperation Fund (OECP)* of Japan and USD 43 million from the *Government of the Philippines (GOP)*. FSP II was implemented through Community-Based Forest Management (CBFM). Under the scheme, communities were contracted to reforest and then were given tenure over the areas they developed. Executive Order (E.O.) 263 in 1995 adopted CBFM as the national strategy for sustainable forest management and social justice. Actually, the CBFM program unified all government people-oriented programs and projects, including those implemented in the early 1980s, entrusting local communities with the responsibility for forest rehabilitation, protection and conservation, with the promise of equitable access to forest benefits. FSP II was funded by a USD 39.7 million ADB loan, USD 55 million JBIC loan and USD 44.57 million GOP counterpart funding. Estimated costs of 20,410.06 pesos for establishing, protecting and maintaining a one-hectare plantation over three years were revised to 43,146 pesos per hectare under the Loan II component funded by *JBIC*. The most recent data from the FMB indicates that from 1987 to 2001, a total of 4927 million pesos (USD 98.54 million)[5] was spent under the comprehensive site development component of the FSP with 299,000 ha of trees planted.

---

[5]This excludes the costs of community organizing, which is a separate contract under CBFM normally granted to NGOs or assisting professionals to provide technical and social assistance before a comprehensive site development contract is awarded.

Reforestation

Reforestation programs were usually carried out in upper lands and previous production forests belonging to logging concession holders which predominantly occupied critical lands. FMB officers cooperated with local farmers in conducting *reforestation* programs in state forests. The *reforestation* scheme eventually became a win-win solution for both parties: the local farmers could improve their 'income generation' by planting, maintaining and harvesting trees, while the government could ensure that critical lands in the uplands and former previous logging concessions could recover with proper management. Reforestation efforts in the Philippines reached their peak in the 2000s, when a total of 191,663 ha of land were reforested (Fig. 3.2). The government played a major role in planting 80% of the area and the remaining 20% was replanted by the non-government sector. In 2004,

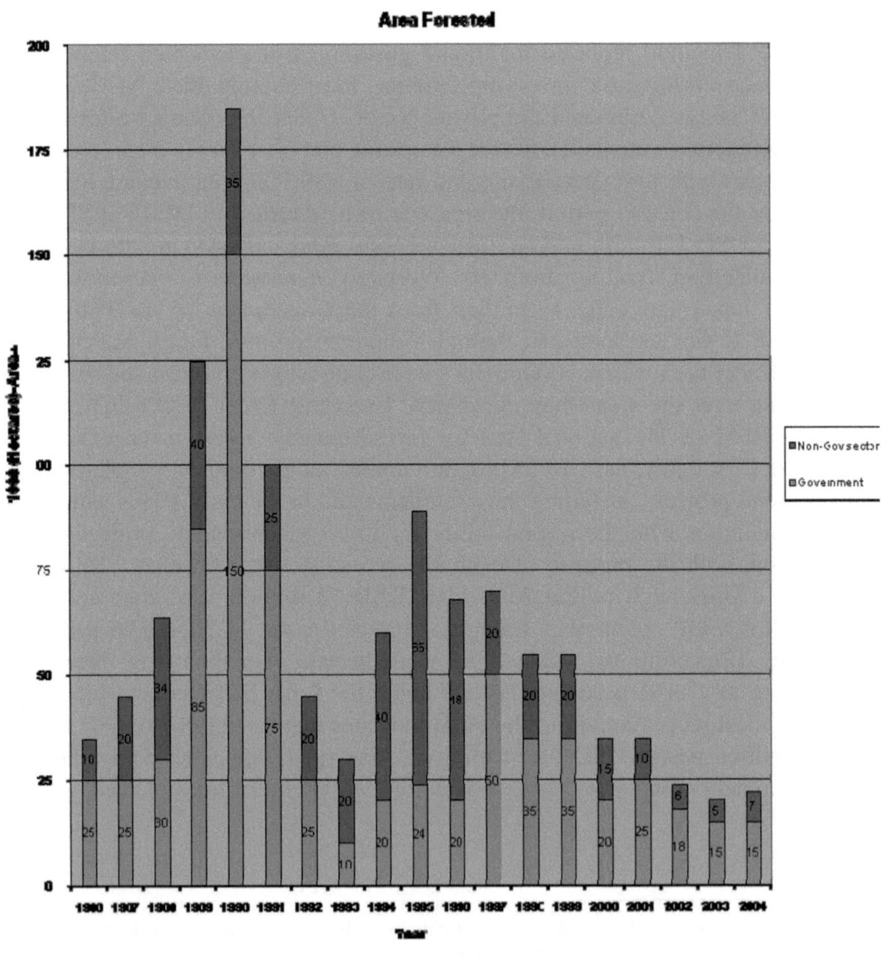

Fig. 3.2 Area forested by government and non-government sector, 1986–2004. *Source* Modified from PCARRD, 2008, p. 7

total reforestation efforts reached 15,088 ha with 61% of the area planted by the government and the rest by the non-government sector.[6] The gap in attaining the average rate of established plantations is a strong reason for further reforestation efforts and the establishment of new plantations.

Plantations established under the CBFM program are mainly dominated (80%) by local species such as *Swietenia Macrophylla (Mahogany), Acasia Mangium, Eucalyptus, A. auriculiformis, and Gmelina Arborea.* Agroforestry was also important, while assisted natural regeneration and enrichment planting played a smaller role. The main objectives were to re-green barren lands, produce timber, enhance watershed services and address upland poverty. The total area reported as planted from 1975 to 2002 was 1,597,472 ha, with the bulk (920,962 ha) planted by DERN; 100,485 ha by LGUs (*Local Government Units*) and OGAs (*Other Government Agencies*); 410,112 ha by timber license holders; 93,520 ha by other private enterprises and leaseholders; and 72,393 ha by private citizens and civic organizations. FSP I and II contributed to around one-third of the DERN reforestation in this period (Unna Chokkalingam et al. 2006: 21).

By the mid 1990s, advocates of CBFM from the government development agencies recognized the need to institutionalize the different people-oriented forestry programs and projects under the one umbrella to ensure their continuity and enhance effectiveness, reach and impact. To effect this institutionalization, President Fidel Ramos issued on 29 July 1995, Executive Order (E.O) No. 263 "Adopting Community-Based Forest Management as the National Strategy to ensure the Sustainable Development of the Country's Forestland Resources and Providing Mechanism for its Implementation". Section 3 of the order stipulates that local communities can obtain long-term tenurial rights to forestland "provided they employ environmentally-friendly, ecologically-sustainable and labor-intensive harvesting methods. Indigenous peoples also known as Indigenous Cultural Communities (ICCs), could also participate in the implementation of CBFM activities in recognition of their rights to their ancestral domains and land rights and claims (Sect. 4).

In 2004, President Gloria Arroyo issued Executive Order (E.O) No. 318 entitled: "Promoting Sustainable Forest Management in the Philippines", reiterating the government's confidence in CBFM as a means of achieving sustainable forest management. In the same year, DERN Secretary, Elisea Guzon, issued DERN administrative Order No. 29. The order replaced the 1996 rules and regulations for implementing the CBFM strategy and provides more flexibility to participating communities by reducing some bureaucratic requirements. The CBFM program in the Philippines is considered progressive because of its land tenure and resource use rights features (Utting 2000). In theory, the issuance of various tenure instruments under CBFM promotes a "win-win' strategy for both the government and local

[6]For further information about 'reforestation patterns' see The Philippines Recommends for *Reforestation, Tree Farming, and Plantation Development,* PCARRD Philippines Recommends Series No. 94. It was published by Department of Science and Technology (DOST), Los Banos, Laguna, 2008, p. 7.

communities. Granting of tenure to communities terminates the open access nature of forestlands. At the same time, it devolves the responsibilities of management and protection to the local communities at minimal costs (Pulhin 2007: 876).

As mentioned by an informant, the "community forestry" program highlights local farmers as the main actors in reforestation by planting local tree species, with the government, represented by FMB, acting as a facilitator of fund raising as well as a guarantor for seeds and fertilizer, monitoring and evaluation of the programs. The Philippines, India and Nepal provide excellent models of "community forestry" that actively engage local farmers and cooperate with other stakeholders (government, academics, NGOs, domestic and foreign funders).[7]

The program shows a great support for the planting of trees; however, farmers face many difficulties in realizing the cutting and transporting of trees, because of the long bureaucratic process from local to central areas and the high cost of transportation. It is often complained by small-scale farmers that if there is no 'timber certificate' issued by the FMB officer in Manila, the truck transporting the timber products will be stopped by the police.[8] The bureaucratic system for obtaining a 'timber certificate' must be simplified or included into the initial MOU of the reforestation program or every regional DERN office should be given the authority to issue the cutting permits.

In summary, from the late 1970s to 2000, a total of US$570 million for forest rehabilitation was invested through both large and small projects, and many different actors were involved. The major funding came from public investment, including foreign loans (93%). Foreign grants and private investment made up the remainder. DERN records a total of 5503 registered CBFM communities or POs (Peoples Organization) from 1975 to the present and around 2200 registered private initiatives (TLAs, TFs, ITPs, IFMA, and SIFMAs) (Unna Chokkalingam et al. 2006: 21).

Major Driving Forces for the Rehabilitation Program

The driving forces for rehabilitation of degraded forest lands were scientific, environmental, institutional and socio-economic:

Scientific Enquiry

By 1914, approximately 120 species had been planted in the Forestry School's nurseries and plantations, increasing to 600 species by 1916. The findings identified a number of species suitable for reforestation in the area. These include Molave,

---

[7]Interview with informant in UP, Los Banos, August 2, 2009.

[8]Interview with E. L. Tolentino, JR on July 31, 2009. He is Associate Professor of Silviculture & Resources Rehabilitation Division, Institute of Renewable Natural Resources, College of Forestry & Natural Resources, and University of Philippines (UP) Los Banos.

Narra (*Pterocarpus indicus*), Supa (*Sindora supa*), Para Rubber (*hevea brasiliensis*), Taluto (*Pterocymbium tinctorum*), Kalantas (*Toona calantas*), Malaruhat (*Cleistocalyx operculatus*), Teak (*Tectona grandis*), Mahogany, Ipil (*Intsia bijuga*), Lumbang (*Aleurites moluccana*), Banaba (*Lagerstroemia speciosa*), Agoho (*Casuarina equisetifolia*), Bitaog (*Calophyllum inophillum*), Baguilumbang (*Reutealis trisperma*), Akle (*Albizia acle*), Tindalo (*Afzelia rhomboidea),* Ipil-ipil and Kakaware (*Gliridia sepium*) (Orden 1960).

Promotion of Environmental Stability

Most reforestation efforts prior to the 1960s were located in established forest reserves, national parks, the watersheds of Agno, Pampanga and Cagayan in Luzon, and other places where problems of flooding, erosion, and soil and water conservation required attention (Orden 1960). Forest rehabilitation efforts in the 1970s and 1980s were carried out in state forest lands, especially in the upland regions. The government, private sector and local communities established plantations of mainly indigenous species. Hence, environmental considerations continued to be a major driving force of rehabilitation in succeeding decades. The Environmental Management Bureau (EMB) in 1990 estimated that 8.4 million hectares of land in the country are severely eroded. Soil erosion caused the siltation of rivers similar to that in the Cagayan Valley River basin (Conservation international, 2001). In this sense, loss of vegetative cover of watersheds resulted in extreme and unpredictable water flows, which eventually caused huge floods, such as in 1991 when the city of Ormoc, Leyte in the Visayas was inundated, claiming 4000 lives and leaving 2000 people missing (Vitug 1993). Another incident occurred in December 2004 in Quezon and Aurora provinces in Luzon where hundreds of people died and thousands were rendered homeless when heavy rains triggered landslides and flash floods (Unna Chokkalingam et al. 2006: 23). Many scholars, environmentalists and the government widely attribute the floods to forest destruction through logging (media reports for the period). These events spurred DERN to suspend logging concessions and incorporate greater rehabilitation into its 2005–2010 Plan of Action for the forestry sector.

Socio-economic Considerations

The emergence of people-oriented forestry programs in the 1980s and the 1990s shifted the emphasis from the traditional approach of "getting the trees on the ground" to "getting the livelihoods of the people off the ground" (Peluso 1992), through their involvement in reforestation and other forestry projects. Local farmers obtained additional income opportunities through participation in different reforestation activities—providing incentives and support to upland communities by providing tenure rights over reforested areas and livelihood support. With CBFM adopted as the national strategy for sustainably developing the country's forest lands, it would be safe to assume that most plantations were established by DERN from 1996 to 2002, totaling 185,407 ha.

Institutional Dimensions

At least eight major groups of actors drive the process and outcomes of forest rehabilitation/reforestation in the Philippines: the Congress, the President, DERN, LGUs, OGAs, upland farmers/local communities and the POs that represent them, NGOs and civil society, the private sector, academics and other research institutions and the donor community. Their varied roles are presented in Table 3.2. In this sense, rehabilitation is a complicated process given the presence of other stakeholders from various sectors and levels of society and the diversity of their personal and institutional interests and priorities. The dynamic interaction among them influences not only the form, but also the substance of forest rehabilitation.

The degree of engagement of various actors continues to evolve, shaping the process and outcome of forest rehabilitation along the way. For example, DERN's inability to promote successful reforestation by itself led to the involvement of the private sector and civil society in the different government-initiated reforestation programs starting in the late 1970s. On the other hand, the availability of funding support from the different financial institutions in the late 1980s to 2000 boosted the country's reforestation efforts, although funding availability did not necessarily result in project objectives being achieved (Korten 1994). Meanwhile, private sector involvement has declined in recent years due to an unstable policy environment and inadequate incentives.

The Role of the Government in Plantation Development

Introduction

The Philippine government established a good 'tree plantation' climate for investors. The government gives economic incentives, tax breaks, easy bureaucracy (permissions), and access to limited banking[9] to develop tree plantations and wood processing mills (sawn timber, plywood, veneer, pulp and paper, etc.). There are many schemes available for tree plantation investment, for instance, *IFMA/ITPLA*[10]

[9]Actually 'tree plantation' investment is costly, because of the long return on capital and high risk. As a consequence, limited banks provide 'credit' facilities to develop this sector. However, as the forestry sector is a significant contributor to national PDB (*Product Domestic Bruto*) and creates jobs, two national banks (*Land Bank of Philippines and Development Bank of Philippines*) provide access to credit for investors. This policy was launched in the early 2000s and is still relatively new, so the numbers of private sector companies that have received credit from both state banks are still limited. The interview was carried out with FMB Officers in Manila, on August 3, 2009.

[10]The government under DERN authority gives tree plantation concession for 25 years to the private sector under the scheme of IFMA or ITPLA (Industrial Tree Planting Lease Agreement) and this can be extended based on their performance. The area of IFMA reaches more than 5000 ha. Meanwhile, the area of SIFMA is less than 5000 ha. (Since the early 2000s, ITPLA was changed to become IFMA).

Table 3.2 Key players and their roles

Key players	Major roles in reforestation
The Philippine Congress	The Congress has the mandate to provide the legal framework for forest development and management including reforestation
The President	Of the country's five presidents since 1965, President Marco's Administration issued the most policies relevant to rehabilitation. The Forestry Sector Projects were initiated during the administrations of President Corazon Aquino and Fidel Ramos. The administration of President Joseph Estrada and Gloria Arroyo merely sustained their predecessors' initiatives, particularly FSP
Department of Environment and Natural Resources (DERN)	The DERN is the main government agency concerned with implementing forest rehabilitation initiatives. It promulgates rules and regulations that translate the generalities of law into concrete terms. The DERN Secretary is responsible for issuing Administrative Orders that guide the implementation of forest laws or decrees issued by the President
Private Sector	The private sector holds various leases of public forest lands such as the TLA, SIFMA, IFMA, Tree Farm Lease Agreement, Private Forest Development Agreement, etc. Different policies mandate these private individuals, corporations or legal entities to be involved in rehabilitating certain areas covered by their leases
Local Government Units (LGU)	The Local Government Code empowers LGUs to enforce forestry laws and implement reforestation and related forestry projects in partnership with the DERN and local communities. Some LGUs in Luzon and Mindanao have passed provincial/municipal resolutions appropriating funds to finance CBFM and reforestation projects
Other government and semi-government agencies	These include the National Irrigation Administration, national Power Corporation, and the Philippine National Oil Company, among others, which by virtue of legal arrangements with the Government are also tasked to engage in rehabilitation activities in watersheds under their jurisdiction. More recently, the Department of Finance has been involved in some initiatives in partnership with LGUs under its community-based Natural Resources Management Project supported by the World Bank

(continued)

Table 3.2 (continued)

Key players	Major roles in reforestation
Upland Farmers/Local Communities and POs (People's Organizations)	This group is composed of both indigenous people and migrants on the ground doing the hard labor of forest rehabilitation. Until the early 1970s they were simply hired as laborers in reforestation projects and did not have tenure security over the land they occupied. More recently, the government encouraged upland farmers and communities to organize themselves into People's Organizations and play a larger role in rehabilitation projects
NGOs and the rest of the Civil Society	NGOs and the rest of the civil society, such as religious groups, media, and others, operate nationally and locally. Their influence ranges from providing technical and financial support to Pos; policy advocacy; legal assistance especially to indigenous people; implementation, monitoring and evaluation of reforestation projects, etc.
Academic and other research institutions	Their main contribution lies in promoting science-based policies and programs; providing technical assistance and support; project monitoring and evaluation; critiquing government forestry policies, programs and projects, etc.
Funding Institutions	Multilateral and bilateral funding institutions act as global drivers of reforestation policies and programs in the Philippines. Their instrument of influence includes providing funds and budgetary and technical support. Chief among these are ADB, World Bank, JBIC, USAID, IFAD, ITTO, Ford foundation, and the governments of New Zealand and Germany

Source Modified based on Pulhin (2003). Cf Unna Chokkalingam et al. 2006, pp. 26–27

for the private sector and *SIFMA* for the cooperative sector, people's organizations (*PO*), and individuals. The response of the private sector and other stakeholders towards these government facilities are very positive, however the policy for obtaining banking credit must be approved in the field.[11] Therefore, to realize the efficient establishment of tree plantations, *DERN* cooperates with the private sector who have the technical capacity to provide and develop quality seeds. The local DERN, in Eastern Mindanao, covering Regions 10, 11, and 13 only has about 30–50 thousand hectares of forest land licensed for tree plantation development, while

[11]The policy of banking facilities to provide 'credit' for plantation forestry' is new and positive. Until 2006, the private sector in the forestry sector, whether in tree plantations or wood processing, found it very difficult to obtain 'credit' from government bank, but it was easier to access private banks, with higher interest rates. Therefore, CSDC Company always obtains 'credit' from private bank facilities for expanding plantation forestry and wood processing mill (Interview on August 6, 2009 with CSDC officer).

Fig. 3.3 *Eucalyptus* trees around 9–10 years old on a farmer's land in Butuan City. *Source* (individual photo document, August, 2009)

eastern Mindanao has been producing 60–70% of the national timber requirements and has developed a seedling research center.[12]

For example, DERN Caraga Region Officer developed quality seeds of 1200 plant species of *P. falcataria* with a local farmer (Mr. Ulip Sugano) who owned 1 ha of land in 2001 (Fig. 3.3). The *MOU (Memorandum of Understanding)* states that DERN is obliged to share technical assistance, advice, provide quality seedlings and consider mother trees (for obtaining new seeds to be developed) as the benefits. The local farmer received benefits from good maintenance, free seedlings, counseling and guidance, and harvesting the trees. Harvesting will occur in 2010 (9 years after planting), when the diameter of the trees has reached 35–45 cm. About 616 trees are still in good condition from a total of 1200 previously planted *Falcataria* trees. Under the MOU, the local farmer will leave 10% of 616 trees to become mother trees, with a diameter of more than 50–60 cm at cutting time after 15 years (2015) (interview on August 7, 2009).

[12]DERN Officer in Caraga Region 10 with a budget of 4 million Peso annually in 2008 developed 100,000 different seeds (such as *Falcate, Mahogany, E. deglupta, A. mangium, S. macrophylla, G. arborea*, etc.), and distributed almost 80% of quality seedlings for local communities, people's organizations, boy scout movements, etc., who would like to plant on their land. This interview with DERN Officer, Caraga Region was carried out in Butuan City, on August 7, 2009.

Table 3.3 Current and projected tree plantation establishment rates to achieve one million hectares of tree plantations for Eastern Mindanao by 2030

Tree planting target (in 000 ha)							
Species	09	010	011–015	016–020	021–025	026–030	Total
A. mangium	9	9	70	75	45	25	233
P. falcataria	27	27	70	75	45	25	269
E. deglupta	16	16	70	75	60	25	262
S. macrophylla	9	9	25	45	50	30	168
G. arborea	3	3	40	50	50	10	156
Other species	2	2	35	50	50	15	154
Total	66	66	310	370	300	130	

Source DERN Eastern Mindanao, September 2008

The response from the local farmer was positive. The MOU with DERN, in terms of free seeds, fertilizers and advice in initial planting and maintenance was significant to him.[13] Farmers with privately owned land and customary forest rights in rural areas also appreciate the MOUs. They say that the MOU with DERN and a company with a guaranteed market encourages local farmers to be actively involved in timber plantations, for the following three reasons: firstly, as a long-term investment, as the price of one cubic meter of wood (such as Acasia and Gmelina) is considered very stable at about 1000 pesos. Farmers can produce about 15–18 m^3 with 30–35 cm diameter timber on 1 ha of land; secondly, timber plantations, especially during planting and harvesting times, create jobs and improve 'income generation' for local farmers in rural areas; thirdly, the social and health impact is significant. For instance, local farmers can then afford to send their children to high school and college, creating better futures for their children.

The DERN regional units in Eastern Mindanao set a goal to move away from sourcing timber from remaining natural forests by working towards the development of one million hectares of more productive tree farms and plantations in Eastern Mindanao by 2030 (to date, the Philippines need to reforest 8.8 million hectares of classified forest land).[14] Table 3.3 presents the current and projected tree planting rates in order to achieve the goal of establishing 1 million hectares of tree plantations, while Table 3.4 presents the estimated amount of seeds required to meet the annual tree planting targets for the region. The estimated volume of required seeds only represents the requirements for commercial tree plantation establishment intended for timber production. Therefore, to achieve these trees planting rates will require a lot of inputs and investments which neither one public

[13]Interview with local farmer in Butuan city and San Luis town, August 6 & 7, 2009.

[14]For a better understanding of the policy on tree plantations of DERN in Eastern Mindanao, see Small-Medium Business Opportunities on Seed Production, Collection and Trade for the Tree Plantation Programs in Eastern Mindanao, published by DERN, AUSAID, and CSIRO, 2008.

Table 3.4 Estimated required amount of improved seeds to meet the target annual tree plantation establishment rates in Eastern Mindanao from 2009 to 2030

Seed requirement (kg)							
Species	09	010	011–015	016–020	021–025	026–030	Total
A. mangium	09	010	2800	3000	1800	1000	9320
P. falcataria	360	360	2800	3000	1800	1000	10.760
E. deglupta	5.6	5.6	25	26	21	9	91.70
S. macrophylla	13.5	13.5	38	68	75	45	252.00
G. arborea	2.55	2.55	34	43	43	9	132.6
Other species	3	3	53	43	75	23	
Total	1465	1465	5749	6211	3814	2085	231

Source DERN Eastern Mindanao, September 2008

sector organization nor a few timber companies can accomplish separately. This is a real opportunity for small-medium size businesses.

The planting rate for 2009–2010 is based on current DERN targets set for the three regional offices. To meet the tree planting targets of this period alone will require about 1.5 tons of seeds of the five identified priority species (*A. mangium, P. falcataria, E. deglupta, S. macrophylla, G. arborea*) being planted in the area every year.

One main reason for the establishment of tree plantations is to fulfill the requirements of the wood-based industry. Unless the establishment of industrial plantations and community tree farms is accelerated and given proper incentives from the government, wood supply from natural forests would fail to meet the needs of the country. In this sense, plantation trees are the main option for the government of the Philippines and other stakeholders of timber industries in order to ensure the sustainability of wood supply in the long run. There are at least five considerations highlighted for tree plantations: environmental, economic, availability of land, availability of funding and technical capacities.

Environmental Considerations

There is a current concern for global warming in the world. The role of tree plantations in the removal of atmospheric CO_2, for the build-up of their biomass during the process of photosynthesis is significant. The planting of trees to sequester atmospheric CO_2 has been considered to be the most effective long-lasting means and a significant approach to address the problem of increasing amounts of CO_2 in the atmosphere.

Also, the planting of trees significantly promotes environmental protection of watersheds in the production of water for domestic, irrigation, industrial, hydropower, and other important uses. Leaves and branches of planted trees intercept rainfall, thus reducing their erosive energy. The plants' roots and litter improve the soil structure and enhances infiltration of rainfall (PCARRD 2008: 3). Trees not only minimize the amount of eroded soil, but these also retard the rate of runoff. Erosion and subsequent sedimentation have a direct impact on a watershed's water yield and quality. Trees also give better flood protection than a natural covering of undesirable brush and grass species.

Economic Considerations

The development of timber industries such as sawn mill, plywood, veneer, lumber, furniture, construction for real estate and pulp and paper industries, have rapidly increased in the Philippines. Based on Forestry Statistics (2006), demand for timber annually reaches 13.5 million metric tons. However, the consumption of wood for timber industries is higher than the supply. Hence, the establishment of forest plantations provides an attractive and profitable undertaking for income 'opportunities' and 'employment' generation, while fulfilling the expected scarcity of wood from natural stands. Large volumes of round wood were being produced in the Philippines annually for various end products used by the wood-based industries (Tables 3.5).

A ready market for wood-based products, whether for domestic consumption or for export, is assured for those who want to invest in forestation. There are big markets for logs for lumber and construction, veneer and plywood, wood chips for pulp and paper production, poles and piles, furniture, fruit and vegetable crates, woodcrafts and wooden accessories, panel products (particle boards), wood wool cement board and firewood.

The 1950s to early 1970s were characterized by a logging boom in the Philippines. Japan, South Korea, Taiwan and the USA are recorded as major importing countries of the Philippine's logs. The area under *TLAs (Timber License Agreement)* more than doubled between 1958 and 1970, from 4.6 to 9.4 million hectares. Consequently, the annual allowable cut also more than doubled from 7.2 to 15.5 million m^3. However, timber started to run out in the 1970s, especially in some parts of Luzon. By the mid-1970s, logging areas in central and western Luzon were either abandoned or covered by logging bans. As an illustration, in 1974–1975 log exportation from the Philippines peaked at 6.84 million m^3 valued at US$283

Table 3.5 Round wood production: 1994–2004 (in '000' m^3)

Log					Fuel wood		
Year	G. Total	Total	Sawn log	Pulp-wood	Poles	Upland	Charcoal
2004	934	768	410	355	3	38	128
2003	699	506	349	151	6	39	144
2002	541	403	288	106	9	28	110
2001	713	571	319	241	11	58	84
2000	912	800	384	400	16	33	79
1999	860	730	568	160	2	49	81
1998	690	634	546	82	6	34	22
1996	804	771	400	365	6	33	–
1994	1063	957	805	149	3	104	2

Source Philippine Forestry Statistics (2004). G. Total: Grand Total

million. Conversely, log importation rose steadily until 1996 where it reached a maximum of 877,585 m^3 valued at US$127.4 million m^3 (Tolentino 2007). Therefore, the imminent timber shortage contributed to 'reforestation' efforts intensifying in the early 1970s. As already mentioned, the government initiated some policies and programs to encourage and support timber production:

(a) P.D 1153 (*Presidential Decree*) known as the "Tree Planting Decree".
(b) The Program for Forest Ecosystem Management that established one municipal nursery for each of the 1000 municipalities and increased the role of the Bureau of Forest Development in reforestation.
(c) The Energy Farm Program, which required each *barangay* (village) to plant at least two hectares as a community fuel service.
(d) P.D 705 and 1559 and Executive Order E.O 725 encouraged the establishment of ITPs, TFs and AFFs, and the reforestation of inadequately-stocked forest lands within forest concessions to help supply the raw materials needs of forest-based industries.
(e) The NFP (*National Forestation Program*) was to provide adequate 'industrial timber' and fuel wood supply in addition to its environmental and socioeconomic objectives. The NFP thus targeted reforesting 1.4 million hectares from 1987 to 2000 (Unna Chokkalingam et al. 2006: 24–25).

Availability of Public Lands

Currently, the country has enough potential lands for the establishment of quality forest tree plantations that can support the raw material needs of the wood-based industries. Based on the *Philippine Forestry Master Plan* (1990), the country needs to develop about 500,000 ha of quality forest tree plantations as the source of raw materials for the industry to be self-sufficient in plantation wood and surplus for export. However, both the government and private sectors must invest on a massive scale in the establishment of industrial tree plantations and the protection and improvement of existing stands.

Availability of Funding Mechanisms for Prospective Investors

The necessary financial institutional funding for investors in tree plantations is significant. There are banks (e.g., *Development Bank of the Philippines and Land Bank of the Philippines*) that provide loans to investors who would like to venture into reforestation, tree farming and plantation development. At present, there is a growing interest in tree farming and plantation development from domestic and overseas businessmen as well as, for example, SCG (*Siam Cement Group*) from Thailand, because of their bright economic prospects and growing income opportunities.

Availability of Tree Farming Technologies

Innovative technology that provides the best quality seeds is significant to bolster tree production in the future. There are technologies available, generated through research and development (R&D) by research institutions, academics and the private sector that can be used to enhance the production of forest plantations. Research institutions can be tapped to provide technical assistance in the production of high quality planting materials of forestation species and appropriate development and management techniques to improve survival and increase 'productivity' of forest plantations.

The 2003 Revised Philippine Master Plan for Forestry Development has targeted the establishment of 40,000 ha of commercial plantations per year, or a total of one million hectares over the next 25 years to meet the nation's timber needs. Therefore, the establishment of tree planting programs is necessary to be realized as below:

Tree Planting Trends in the Philippines

The government, through the authority of the *Forest Management Bureau* (FMB) Officer, gives tree planting concessions to the private and cooperative sectors. For instance, most planting has been conducted by government issued TLAs (*Timber License Agreements*) in 1970s–1980s. The private sector reforested large areas in three periods since 1975: from 1977 to 1984, 1988–1990 and 1994–1996. The total area planted was more than 93,520 ha. Generally, more than 20,000 ha were planted annually during these periods, while in the1990s TLAs changed to become IFMAs (*Industrial Forest Management Agreements*) and SIFMAs (*Socialized Industrial Forest Management Agreements*).[15] The aims of the establishment of IFMAs and SIFMAs were to revitalize the industrial forest plantation program and generate income for the private sector and smallholders in the local communities. A total of 8568 ha of trees were planted under IFMAs and SIFMAs and 3963 ha under TLAs from 2000 to 2002.

In general planting stocks, Tolentino addresses the significant benefits that could be obtained from seeds, vegetative produced stocks and wildlings, because seeds are the most common source of reproductive material when it comes to tree planting endeavors in the Philippines. The reasons are very obvious. First, seeds of the most popular and commonly planted tree species are abundant and widely distributed in many regions around the country, making them a more economical and practical source of planting stocks. Second, they are generally easier to handle and store for longer periods of time, unlike vegetative produced stocks (Tolentino 2007).

Historically, beginning in 1977, the participation of the private sector in tree planting through the years visibly manifested itself. From around 2000 ha, the aggregate accomplishments of the *Industrial Forest Management Agreement* (IFMA), tree farms and Agroforestry Farm Areas by 2004 had reached 824,000 ha (Fig. 3.4). Bulk planting initiatives contributed by IFMAs (86%), while agroforestry farms and tree farms accounted for 11 and 2%, respectively.

---

[15]Private sector who obtains IFMA concession from FMB officer could manage more than 5000 ha from state forest land. The concession is given by the government for about 25 years and could be extended for second terms based on his performance. Meanwhile, SIFMA for cooperative sector, the area concession below 5000 ha.

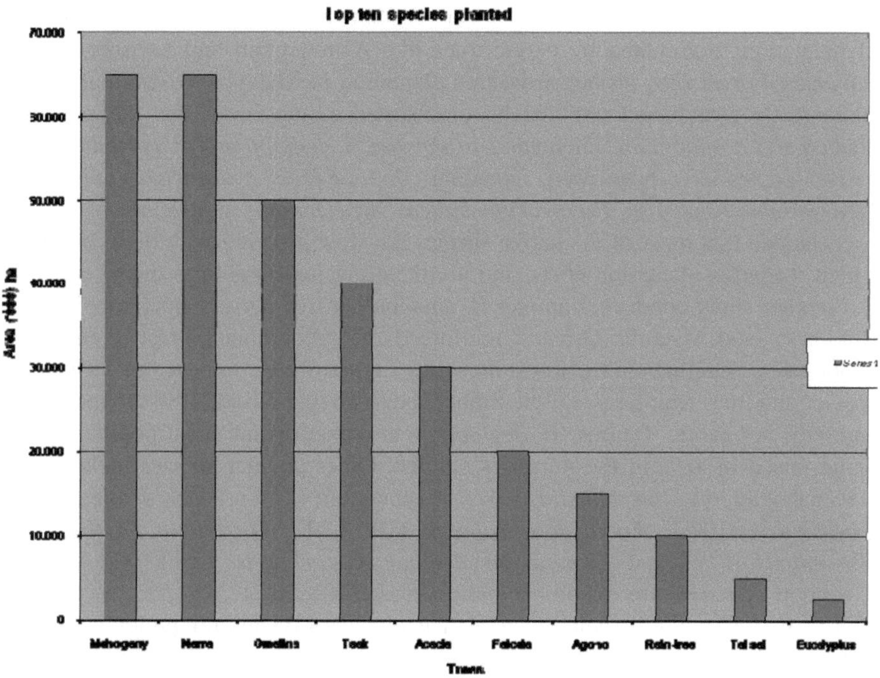

Fig. 3.4 Top ten species planted in reforestation projects in the Philippines (Forest Management Bureau, 2000). Cf: Tolentino (2008)

Species Selection in Plantations

To provide excellent tree species for tree plantations is very important. Species commonly used in reforestation are surprisingly few, considering the abundance of commercially valuable species used by the timber industry. The popularity of exotic trees in the Philippines as a reforestation species dates back to when reforestation started early in the 20th century. Apparently, of the top ten species planted in reforestation projects around the country, eight are exotic and only two are *ITS* (Indigenous Tree Species) (Fig. 3.4). Mahogany (*S. macrophylla*) and Gmelina (*G. arborea*) are among the dominant exotic trees planted. Narra (*Pterocarpus indicus*), a common ITS, comes a close second. Another ITS, Agoho (*Casuarina equisetifolia*) ranked seventh among the commonly planted species (Tolentino 2008: 321–322).

Reports on the plantations of private concessionaires showed a similar pattern: *Paper Industries Corporation of the Philippines* (PICOP) Resources Inc. (Surigao del Sur, Mindanao) have plantations of more than 40,000 ha, mainly *Paraserianthes Falcataria, Eucalyptus Deglupa and Acacia Mangium. Nasipit Lumber Company* (NALCO) (Agusan del Norte) has more than 4000 ha of exotic tree plantations. The main species planted are: P. f*alcate, G. arborea, Acacia auriculiformis, A. mangium, Pinus caribaea, S. macrophylla and Tectona grandis.*

Provident Tree Farm Inc (PTFI) (Agusan del Sur) has established another 6000 ha of plantations dominated by exotic trees like A. mangium and *G. arborea*. The *Bukidnon Forest Inc.*, an industrial tree plantation in Malaybalay (Mindanao), has successfully established 6367.32 ha of assorted exotic trees. The major species planted are: *A. mangium, Eucalyptus urophylla, E. deglupa and P. caribaea.* Some native species have been tried, including: *Pinus Kesiya, Casuarina equisetifolia, Lagerstroemia speciosa, Pterocarpus indicus var.echinatus* and *shorea contorta.* It was claimed that most of the native species are slow growing with high mortalities which increases plantation costs, and are therefore undesirable to management.

Another study conducted among 50 smallholder tree nursery operators in Cebu, Bukidnon, and Misamis Oriental reinforced the predominant practice of raising exotic trees. Seedlings in the forest nurseries studied were composed of 59% timber species and 36% fruit trees. Of the timber species being raised, 35% are indigenous and 65% are exotic. Bagras (*E. deglupa*) was ranked as the most popular species being raised in 48% of the nurseries studied. Other popular species include large leaf mahogany (*S. macrophylla*, 35%), A. mangium (21%), Black wattie (*albizzia lebekkoides*, 19%), *Eucalyptus robusta* (19%), E. *Torreliana* (17%), narra (*P. indicus*, 17%, and *Yemane or gmelina (G. arborea*, 15%). All, with the exception of *narra,* are exotic (Tolentino et al. 2008: 322).

The use of exotic species is not an exclusive silvicultural preference in the Philippines. In Southeast Asia, countries like Indonesia, Thailand, and Vietnam have developed extensive plantations of exotic trees like *S. macrophylla, P. falcataria, A. mangium. P. caribaea, Eucalyptus spp., and Casuarino spp.* (FAO Forestry Database). Even in Brazil, another country with active plantation activities, data as early as the 1900s revealed an inclination towards the exotic eucalypts over Brazilian timber species (Nararro de Andrade 1941). In fact, as of 2005, Brazil has an estimated 3.2 million hectares of eucalyptus plantations (Neto 2005), the exotic species most abundant in that country's plantation program (Mc Nabb 2005).

The following are some of the main reasons for why exotic trees dominate the country's tree planting program (Tolentino 2008: 323–324).

Wide Adaptability and Tolerance to Stress

It was recognized by most foresters and farmers that planting exotic trees provided certain advantages, such as the adaptability of exotics to degraded sites (e.g. acidic, low soil fertility, fire-prone areas) and their ability to colonize even marginal grasslands. As an example, the exotic legumes (e.g. *Acacias, Falcataria*) are fast growing trees and are nitrogen-fixing trees that permit optimum growth and development even in nitrogen-depleted soils.

Fast Growth and High Yield

Exotic trees are fast growing and provide a high yield. These characteristics make them very attractive for smallholder tree farmers desiring quick income and immediate returns on their investments. Some estimates revealed that the yield of exotic trees ranges from a low of 5 m^3 per hectare per year in poor sites to as much as 40 m^3 per hectare per year in good sites. However, most of these species exhibit impressive growth yields averaging from 30 to 35 m^3 per hectare per year. This is almost similar to the average growth performance of eucalyptus species in Brazil with is 20–40 m^3/ha/year.

Available Research and Technologies

Most research has focused only on a few economically important tree species, thus making available technologies for tree plantation development of these species easily accessible to tree growers. Apparently, many of these plantations species are exotics grown outside its native range. In the case of exotics planted world-wide, e.g. *Pinus caribaea, Eucalyptus grandis, and Tectona grandis,* available research, technology package and experiences allow many users to plant them with an acceptable degree of certainty.

Wood Processing Mill

Introduction

The development of the pulp and paper industry in the Philippines has not grown as fast as its ASEAN counterparts in Indonesia and Thailand (Table 3.6). There were at least three reasons why the pulp and paper industry in the Philippines is not as strong: (1) the government has not yet realized a good climate for investment in terms of facilities to obtain large credit in banks, lack of infrastructure, breakdown of tariff barriers and political instability; (2) lack of facilities from the government to encourage integrated pulp and paper milling with plantation forestry and (3) lack of new machinery equipment to establish new pulp and paper factories. The transfer of technology in this sector had been realized in Indonesia and Thailand, where at least 10 modern pulp and paper mills have been established, with production capacities ranging from 550 to 4000 tons per day (Table 3.7). Indonesia is ranked first for paper production in ASEAN. In 2002, the country produced 6.9 million m^3 and in

Table 3.6 Comparative pulp production for paper in ASEAN Countries (1998–2002) (000 Metric Ton/MT)

Country	1998	1999	2000	2001	2002
Cambodia	0	0	0	0	0
Indonesia	1895	1725	3726	5587	5587
Malaysia	112.7	119.7	123.7	123.7	123.7
Myanmar	17	39	42	47	47
Philippines	149	200	202	202	202
Thailand	772	836	844	999	999
Vietnam	133	294	314	314	314

Source FAO STAT, 2002

Table 3.7 Comparative paper and paperboard production across ASEAN countries (1998–2002) (000 MT)

Country	1998	1999	2000	2001	2002
Cambodia	0	0	0	0	0
Indonesia	5,487	6,978	6,977	6,995	6,995
Malaysia	761	859	791	851	851
Myanmar	41	37,4	39,2	41,6	41,6
Philippines	987	1,010	1,107	1,056	1,056
Singapore	87	87	87	87	87
Thailand	2,367	2,434	2,312	2,445	2,444
Vietnam	190	356	384	384	384

Source FAO STAT (Last accessed on 6 December 2004)

Table 3.8 Imports of waste paper from 1999 to 2004 (000MT)

Year	Volume	Value (US$)
2004	369,957	43,000,000
2003	18,942	1,997,691
2002	362,076	34,013,302
2001	354,103	37,197,208
2000	407,213	52,930,913
1999	368,547	40,824,307
1998	307,111	28,537,602

Source Philippines Forestry Statistics 1998–2004, PULPAPEL 2004

2005 to become 7.6 million m^3, and rapidly developed to become 9.2 million m^3 in 2009.[16] This is followed by Thailand with the production of 2.4 million m^3 in 2002 and in 2005 to become 4.3 and 4.9 million m^3 in 2009, followed by the Philippines at 1.056 million m^3 in 2002, and in 2005 1.6 million m^3 and to become 2.1 m^3 in 2009.

The pulp and paper industry consists of 42 operating mills in the country, mostly located in Metro Manila (14 units), categorized as non-integrated recycling paper mills; in region 3, Central Luzon area (4 units), while the rest are located in other regions, categorized as non-integrated pulp mills, and as integrated pulp and paper mills (1 unit—PICOP). The mill utilizes tropical hardwood for their pulp production. PICOP's production capacity is recorded at 171,000 tons per year for all paper grades (*newsprint, lines board and corrugating medium*). Unfortunately, PICOP filed for bankruptcy because of mismanagement and uncertain government regulations in 2005.

The paper industry in the Philippines is heavily dependent on 'recycled fiber' as a cheaper input than imported 'wood pulp'. The country's waste paper collection system is quite efficient with an actual rate of recovery of 65%. Table 3.8 indicates

[16]See other sources, *The Japanese Pulp and Paper Industry in Charts and Figures* (2005), published by Japan Pulp and Paper Co., Ltd. P. 45; FAO STAT 2002; APKI (Indonesian Pulp and Paper Association) 2009; *The Jakarta Post*, 16 August 2010.

the volume and value of waste paper imports from 1997 to 2004. According to Amando Rios, former President of PULPAPEL, the industry imported 369,957 metric tons (MT) of various grades of waste paper valued at US$43 million (Table 3.6). The largest sources of waste paper are USA, Europe, Japan, South Korea and Thailand.

The processing of pulp into paper could be derived from wood pulp, pulp from recycled fiber or non-wood pulp requires a lot of chemicals such as chlorine, calcium, sodium hydroxide and bleaching chemicals. Hence, the pulp and paper industry is highly dependent upon the chemical industry. Aside from the chemical industry, the pulp and paper industry also provides support to the following economic activities: publishing, printing, health/hospital and communication industries.

The emerging pulp and paper industries are significant in their impact on the country's economy. The industry's export of paper and paperboard products contributes an average of US$69.7 million per year to the country's economic growth and development. The export volume of Abaca pulp has reached amount to US $28.1 million per year over the last five years (1999–2003). Job creation absorbs about 5600 regular employees and around 1.6 million people derive their livelihood from wastepaper collection. The constraints pointed out by the private sector to invest even more in the pulp and paper industry include:

(a) tenure duration is too short to make long-term investments;
(b) obtaining credit is difficult;
(c) development and transport costs are too high to be financially viable, because mostly damage the street in rural areas;
(d) frequently changing policies affect plans and operations, particularly regarding timber harvesting rights and transporting;
(e) marketing support is low.[17]

DERN officers are aware of these constraints for plantation forestry investment, which is why, along with other regulations and infrastructure improvement, DERN engaged local farmers by giving SIFMAs and IFMAs for private sector concessions.

This paper aims to conduct a short review of paper industry performance and the role of the private sector, by considering the activities of a company actively engaged in plantation forestry in Caraga region 10. The company's name is *CSDC (Casilayan Softwood Development Corporation*), and is categorized as an integrated wood processing mill, producing veneer, clippings, lumber, etc., for distribution on the domestic market. A comparison will be made of the response of local farmers to the existence of the CSDC Company and academic discourse on the impact of ecological damage from the company's operations.

[17]This data is based on interviews with *CSDC* Company (Casilayan Softwood Development Corporation) Officer in Talacogon and Butuan City, Agusan del Norte on August 6 and 7, 2009.

Paper Industry Performance

Domestic Production

The Philippines' paper factories produce various paper grades, such as newsprint, printing and writing papers, craft/sack paper, corrugating medium and linerboards, tissues and specialty paper. In 2005, the total production capacity of the paper mills reached 1.4 million metric tons (MT), while pulp (abaca) production capacity was 24,000 MT. Production reached only 75% of its potential due to inefficient production, because most mills are old and need to be remodeled. Table 3.9 shows paper production by grades. From 1997 to 2003, total paper and paperboard production registered an increasing trend during 5 years, with an average annual growth of 10% except for the years 1996 and 2000 (which could be associated with economic crises, political instability, peso devaluation, the rising cost of fuel and other factors). Paper mill production eventually rose to 1051 million MT in 2001–2003. Packaging, paper and board constituted the largest proportion of production (45%), followed by printing and writing paper (28%), newsprint (24%), and sanitary paper, tissue and specialty grades (3%).

In general, per capita demand for paper consumption in the Philippines has been recorded at 14.4 kg, which is way below the world's average of 55 kg. However, there is more room for growth with the country's potential population of about 88 million. The population and economic growth of the country is at a rate of 2.3% per annum, an increase in per capita consumption by just 1 kg would require paper mills to produce an additional 85,000 tons of paper and paper board.[18] The industry is reported to be growing over the following years, due to increasing market size for both domestic demand and exports as well. For instance, sub sectors, including the food and packaging industries, are heavily dependent on paper and paper-based products. Hence, studies have shown that paper demand is expected to increase in the following years due to increased newspaper circulation, a high level of literacy and educational institution demand for material paper, packaging materials and other social and economic factors. Statistics have shown that from 1992 to 1999, in terms of share to total output, newsprint accounted for 20% of the industry's total production capacity, printing and writing grade 18%, and craft/corrugating medium/ sack grade has a total share of 57%, and the remaining 5% is devoted to tissue and specialty paper manufacturing (Table 3.10).

[18]There is optimism for the growth of the pulp and paper industries in Philippines in the next decade, while government decision makers are aware of the need for pulp and paper investment and have launched 'positive policies and facilities to the investors. Interview with *FMB* (Forest Management Bureau) officer on August 3, 2008.

Table 3.9 Paper production by grade (000 MT)

Paper grade	1997	1998	1999	2000	2001	2002	2003
Newsprint	154	156	174	275	258	258	258
Printing	126	296	296	135	296	296	296
Packaging	392	504	504	504	470	470	470
Tissues	21	23	27	38	27	27	27
Specialty grades	7	8	6	na	na	na	na
Total	700	987	1007	952	1051	1051	1051

Source FAO Statistic, 2003

Table 3.10 Paper and paperboard production and consumption 1998–2003 (000 MT)

Year	Production	Net import	Consumption
1998	987	418.1	1,405.1
1999	1,007	532.2	1,539.2
2000	952	507.0	1,459.0
2001	1,056	692.9	1,748.9
2002	1,056	609.7	1,665.7
2003	1,056	605.6	1,661.6

Source Philippines Forestry Statistics, 1998–2003 and FAO 2003

Trade

Export Orientation

The government has a policy to export paper production in order to be able to obtain foreign exchange earnings. Paper and paperboard exports have generally increased over the last 10 years. For instance, in 2003 the Philippines exported 150.000 MT of paper and paperboard valued at USD 71 million (Table 3.11). The predominant markets are Asian countries such as China, Hong Kong, Taiwan, Singapore, South Korea, India, and Malaysia.

The Philippines is well-known as the Abaca producer. Therefore, the country's pulp exports are mainly derived from Abaka pulp. It has been reported that around

Table 3.11 Export of paper and paperboard, 1997–2003 (000 MT)

Year	Quantity	Value (in US$)
1997	54,022	51,986,537
1998	113,500	78,587,056
1999	104,964	72,347,870
2000	134,097	91,398,350
2001	147,944	94,490,606
2002	175,376	72,522,174
2003	150,420	71,260,387

Source Philippines Forestry Statistics 1997–2003

70% of the world's abaca fiber requirements come from the Philippines. Table 3.12 shows the volume and value of Abaka pulp trading from 1998 to 2003. The target countries of Abaka exports are Japan, Netherland, United Kingdom, USA and France. Apparently, from 1998 to 2002, the country's export of Abaca pulp reached 78,334 metric tons valued at USD 128,597,666 (FMB 1998–2002).

Imports

As previously discussed, the Philippines still lacks sufficient timber production as the raw material for processing pulp. In this context, the Philippines imports pulp from other countries, mainly Indonesia, Thailand and China. Meanwhile, a small amount of paper and paperboard, such as printing and writing paper and other specialty paper grades, are still imported from other countries (Table 3.12). The reasons for this are: (a) the government tariff reduction program, especially for pulp and paper; (b) the declining prices abroad compared with the price of domestic paper; (c) the quality of local paper does not approximate the quality of prime paper produced in other countries (Aragon 1995).

Study of the CSDC's Profile: Wood Processing Mill

The company is familiarly called **CSDC**, an abbreviation of *Casilayan Softwood Development Corporation*. It was established in 1983 for plantation forestry, match manufacturing and wood processing. To date, the company manages four industrial forest plantations covering a total of 34,727 ha under contract through IFMA with the national government under DERN, in the provinces of Agusan del Sur and Agusan del Norte in Mindanao region (Fig. 3.5). Another new contract of IFMA 95,000 ha was received in 2009 located in the island of Samar in the Visayas.

CSDC provides raw material from plantation trees such as *Falcata, Gubas, Mahogani, A. Mangium and Gmelina* in many areas, such as 5000 ha in Mahayahai village, San Luis town and 12,000 ha in Lapaz (Fig. 3.5). These areas are mostly located in Agusan del Sur. The company used a strategy of making plantation forestry by encouraging the participation of local people as workers and providing a

Table 3.12 Export Abaca Pulp 1998–2003 ('000 MT)

Year	Quantity	Value (in US$)
1998	14,987	35,961.708
1999	13,881	31,627.331
2000	15,664	35,949.069
2001	15,164	29,601.293
2002	16,306	31,367.146
2003	2336	4293
Total	78,334	128,597.666

Source Philippines Forestry Statistics 1998–2003

Fig. 3.5 *Acasia Mangium* trees in CSDC's timber concession, in Mahayahai village (individual document, August, 2009). *Source* Wikipedia English 2008. CSDC owns 14,727 ha of timber plantations in this region in 2000s

guaranteed market at harvesting time of their tree products. The fees per hectare totaled 10,000 pesos and absorbed 10–15 persons for work. During harvesting (8–10 years after planting), the company gives contracts to groups which consist of 10 persons to cut down trees with chainsaw equipment. Hence, plantation forestry, especially in planting and harvesting time, creates jobs in rural areas and improves the incomes of rural individuals by synergic cooperation between company, local government and local people.

As an illustration, of a total of 5000 ha in Mahayahai village, 1000 ha belong to tribal leaders. The company has made an MOU with 89 tribal leaders. The aim of the MOUs is to improve the performance of the company in realizing input requirements of trees. The company provides a subsidy on good quality seeds, fertilizer and 2000 pesos per month per head for each tribal leader to develop their staple foods, such as rice, corn and cassava.[19]

[19]Interview with one of head of tribal leader in Talacogon, August 6, 2009.

Source: Wikipedia English, 2008.

CSDC owns 20,000 ha of timber plantations in this region.

In terms of transportation of products to the wood processing mill, there are two ways: (1) using land transport (trucks) in the dry season and (2) using the river in the rainy season.[20] Transporting wood from rural areas to wood processing mills causes very heavy traffic, destroying roads, especially during the rainy season (Fig. 3.6). Therefore, to overcome these difficulties, the company prefers to transport wood in harvesting time by the river, namely by rafting the timber down the river.

The CSDC holding company is JAKA Corporation, a business that produces safety matches, located in Agusan del Norte, provided with an ISO 9002

[20]Usually occurred in Philippines for dry season starts annually from the mid of November until April and for rainy season starts from May until the end of October.

Fig. 3.6 The infrastructure (road) to timber concession areas are bad (individual document, August, 2009)

certification and currently the country's leading manufacturer of high quality safety matches. The main shareholder is Juan Ponche Enrile.[21] The head office of the company is located in Makati, Manila. Juan obtains credit to expand his company (JAKA Corporation), but never from governmental facilities, in order to prevent any collusion with government officers or senators. Therefore, he prefers to gain financial access from private banks, namely RCBC (Result Commercial Banking Corporation) and BOC (Bank of Commerce).[22] JAKA Corporation also launched capitalization for new projects by selling shares in *Makati*. These shares were sold in order to expand the core business and other leading projects such as plantation forestry, wood processing mills (Table 3.13).

[21]P. Enrile was previously appointed as Defense Minister under the Ferdinand Marcos regime in the 1980s. In 2009, he was acting as Senator in the Upper House of the Philippines Parliament. JAKA Corporation is a synonym of the names of his two children, Jack and Catherina.

[22]The interview with MA Teresa FE C. Booc, Finance & Administration Manager was carried out on August 7, 2009 in Batuan City.

Table 3.13 Imports of pulp, paper and paperboard, 1996–2002 ('000 MT)

Year	Pulp		Paper and paperboard	
	Volume	Value (Mil. US$)	Volume	Value (Mil. US$)
2002	44	24	571	293
2001	56	32	533	309
2000	64	42	468	317
1999	73	41	477	257
1998	86	43	392	246
1997	107	57	458	297
1996	92	52	393	277

Source FAO STAT, 2004

Table 3.14 Plantation concession owned by CSDC

Concession	Hectares	Expires date
IFMA 03-2008	5000	January 1,2034
IFMA 019 (RMI)	5000	December 31, 2017
IFMA 07-2006	23,727	December 31, 2032
IFMA X-1994	1000	June 30, 2019

Source The Company's Profile, 2009

The 2009 financial report of CSDC indicates that the company is healthy and gaining good profits. Gross Revenues accounted for 108 million pesos in 2008, while costs accounted for 101 million pesos. Therefore, the company recorded a profit of 7 million pesos before income tax. The total income after tax was 5.25 million pesos, mostly from the wood processing industry. The company is profitable in its operations and is planned to be expanded with additional production by providing new machineries and raw materials.

The following statement from a CSDC officer was based on an interview with the writer concerning the company's responses to government regulations and the climate of investment for plantation forestry and its wood processing mill:

Do you feel that there is now a better climate for investors in 'plantation forestry'?

Yes, the company feels that the climate is good for plantation forestry. Although the regulation, infrastructure and other economic incentives must be largely given to the private sector in order to drive investment in the Philippines.

The company fulfils all the requirements of DERN, and is strong in terms of IFMA evaluations in *water, air, land, biodiversity and interaction with people* requirements, which are held twice annually by DERN.[23] Actually the company

[23]Forest Management Bureau (FMB) officers hold IFMA evaluations twice annually for all companies that have received plantation concessions. Interview with Ms. Nely Butic Head of Division on Environmental Affairs, FMB, on August 6, 2009.

Fig. 3.7 CSDC's wood processing mill for veneering products in Talacogon (individual photo document, August, 2009)

accepts that there is now a better climate for plantation forestry investment in terms of easier procedures for plantation concessions and tax breaks for the first five years of planting. However, the economic incentives through credit access from banks, reductions in machinery taxes, facilities for exports in the port, etc., are not assisted by the government. Therefore, the company established a timber processing mill in 2002, using second-hand machinery made in Germany for veneering (Fig. 3.7). This strategy was taken in order to reduce taxes by the use used machineries and efficiently spending money from the principal of *JAKA Corporation*. The most important one is the maintenance of machinery in order to keep them operating until now.[24]

How far does the timber processing mill create jobs in this municipality?

The existence of this company which allocates around 15 ha of land for timber processing creates jobs for 120 regular persons, 150 contract workers and largely

[24]The machineries for making veneering and lumber are still performing well in production timber, based on interview with Head of Wood Processing Mill on August 6, 2009 in Talacogon.

absorbs all work for rural villagers, especially in the planting, maintenance and harvesting of trees. The workers are derived 85% from local townspeople and 15% from other provinces. The allocation of 85% local and 15% from other provinces is based on an MOU (Memorandum of Understanding) between the company and the municipality of Talacogun. The company also pays above regional standard average salary. For example, for high school qualification and about 1–3 years working experience, the salary received is around 5000–6000 pesos/month and a college graduate with medium skills as a technical operator with 1–3 years working experience will receive about 8000–10,000 pesos/month, while a supervisor or head of division with 1–5 years working experience will receive 18,000–22,000 pesos/month and a manager with 1–5 years working experience, 35,000–40,000 pesos/month.[25] Also, all workers receive a one-month bonus at the end of year. Therefore, the flow of money is spread out in Talacogon municipality through property land taxes, business taxes from the gross income of the company (taken 0.5%) and spending money from the incomes of the workers and farmers surrounding the Talacogon area.[26]

What are the investment constraints you face and how much do you produce?

Actually, for foreign investors, plantation forestry and the timber processing mills are not attractive, because there are some constraints in this sector, such as limited land, banking facilities, unstable politics, expensive machinery taxes and lack of infrastructure. As an illustration, the timber processing mill produces 1950 m^3 (CM) per month of veneer. The company produced 1600 CM in 2007 and rapidly increased this to 1700 CM in 2008 and 1900 CM in 2009. Several veneer products such as Lowaan veneer costs 9000–10,000 peso/CM, Falcata veneer costs 6000–7000 peso/CM and Miselen veneer costs 9000–10,000 peso/CM. The marketing of these products could be conducted for the domestic market. For example, the company sends to the Mindanao plywood factories such as PSPI, Richmond, Emco, APC and APTCO and also veneer factories in Manila such as Winlex and Rewoodco. But the problem is to expand wood timber business and pulp and paper from viewpoint of investors need huge capital and could not be available in Philippines Stock Exchange Market and international exchange market for selling shares at moment. Besides conducive sphere for instance to be easy access for investment climate, to obtain credit, tax holidays, and infrastructure.

---

[25]Actually, the standard of regional fees (in Municipality/town) for high school graduate just receives 4000–4250 Peso/Per month. In case of income for managers around 40,000 Peso in wood processing mills is similar with the income of Professor in State University, such as University of Philippines. Currently US$1 is similar with 47.80 Peso. Then, 40,000 Peso is similar around with US$934.

[26]Based on interview with Manager of Wood Processing Mills on August 6, 2009.

The Response from Local People

The existence of the company (CSDC) is positive for both the local community and government, a tribal leader mentioned,[27] because the company encourages the "participation" of local people for the plantations and especially at harvesting time. At the same time, the company pays taxes to the local government, develops agricultural land (for staple food production) and repairs the road, especially during the rainy season in rural areas.

Three tribal leaders (Manubu, Man Waon and Ban Waon) consider that the population growth in their village (Mahayahai) is very high (300 households and 3945 people), with a strong inverse correlation between population growth and economic social welfare. If it cannot be managed properly, the community will suffer growing poverty. The reasons for the rapid population growth are: (1) a high birth rate per head of family (7–8 children) and (2) some tribal leaders are married to two women, and they usually bear many children. The customary forest lands are registered at about 21,720 ha, while 5000 ha is land concession belonging to CSDC, and 1000 ha is ancestral land. To manage and cultivate 21,720 ha, the land is divided between the 89 tribal leaders. One tribal leader represents 12–14 heads of households, cultivating 12–15 ha each. The system of working is "Pahina", which means one group cooperates with other groups without paying fees, just providing meals and water during work, from clearing land, cutting trees and planting seedlings. This Pahina system has been maintained throughout the generations until now and is still well practiced in rural areas. The main problem of the village is lack of educational infrastructure (*elementary school*) (Fig. 3.8), health and road facilities and sanitation.

Before CSDC, the PICOP Company carried out plantations and tree harvesting from local farmers as raw material for the pulp and paper industry. However, according to a tribal leader who witnessed the golden peak of production in the 1990s, there was only a guaranteed market for trees at harvest time, without any assistance with seedlings, free fertilizer and no cash payments for farmers. Farmers were paid two weeks after the trees were transported to the company. But, after PEACOP suffered bankruptcy in 2005, CSDC appeared as the best performance company and a true partnership with local farmers, especially in Sun Luis town and Mahayahai village.

Another positive of the program is that CSDC gives a bonus of around 2000–2500 pesos for every tribal leader in the New Year and also gives funds for the village committees celebrating the birth of the town and Independence Day of the Philippines.

What trees have local farmers planted?

Most plantations are occupied by *Falcata and Acasia magium* tree, because these trees are considered commercially viable and fast growing. These trees are used as

---

[27] The interview with Mr. Charlie, one of head of tribal leader was carried on August 6, 2009.

Fig. 3.8 Elementary School in Mahayahai village (individual photo document, 2009)

the raw material for pulp and paper and veneer products. One hectare consists of 400 seedlings, using 5 kg/ha of fertilizer and takes 8–10 years from planting until harvesting. At harvesting time, the trees reach around 30–40 cm in diameter. Production volume reaches 180 m^3/ha. The market price was about 1000 pesos/CM in 2008. Usually tribal leaders rent chainsaws during harvest, at a cost of 500 pesos/day. The operation of one hectare uses one chainsaw for three days. They collect the cut trees together into a log pond and the company's truck transports them to the timber processing mills. Generally, one hectare consists of 10–15 truckloads and one truckload consists of 18 m^3 of timber.

What is the contribution of the local government to the farmers?

The local government (San Luis town) contributes per head family seedlings and fertilizer for projects such as *rubber, falcate and Ruan* (*Local Philippino Mahogani*) trees to be planted in their customary forests. The purpose of planting trees is for 'forest conservation' rather than for commercial trees in order to prevent 'soil erosion and flooding' in the uplands. The government also gives non-organic fertilizer of 5 kg/tribal leaders. But, this gift in terms of seedlings and fertilizer is based on projects. It is not regularly distributed annually to local farmers.

Do you receive any benefits from synergic cooperation between the private sector, local government and local farmers?

Yes, we receive a lot of benefits, largely the opening of job opportunities and receiving income by planting trees and selling timber products at harvest time to the company. This has a positive impact on improving economic and social affairs for rural villagers. Actually, if the central government gives sufficient credit to local farmers from state bank in order to boost timber products in Philippines, it is the best way how to widely open job and improve income generating for rural communities.

How do they spend their income from harvesting trees?

At harvest time, the company pays money based on how many cubic meters of timber farmers produced, bolstering the incomes of heads of households by about 50,000–60,000 peso. This is a large amount by rural villager standards. About 20% of their income goes on daily necessities in town, 15% on replanting, 35% on educational and health fees for their children in high school and university and 10% for maintenance of their health, while 20% is saved in the bank for hard times in the future.[28]

Academics and NGOs Critiques of Environmental Disasters

During the mid-1980s and 1990s, stories on environmental disasters such as the following grabbed the headlines: floods—such as those that killed at least 5000 people in Ormoc, Leyte in November 1991 and drowned several hundred in Southern Luzon in November 1995, or Calauag, Quezon in late 1995, where hundreds perished because rising floodwaters could no longer be contained by treeless watersheds. These tragedies could be traced back to deforestation issues, strongly correlated with 'logging concessions and politics'. Companies' connections to former President Ferdinand Marcos who secretly owned nearly half of the lucrative firms, allowed them to flout environmental standards. In contrast, when Cory Aquino became President (1986–1990), the government took over and revised the policy of the company and appointed a new board to run it.[29] Actually the boom in Japanese demand for tropical *hardwood* (log) in the 1960s created enormous wealth for those who found themselves in control of the timber concessions,

[28]Educationally, their children are currently performing better, because they mostly continue through high school graduation to colleague/university. So, parents spend more money to send them to the municipality and city. This is the case in Charlie's family, as one of the tribal leaders. Three of his daughters are in high school and one in university, while Charlie himself only graduated from high school.

[29]Caronel (1996)

especially in Agusan, north Mindanao. Some logging holders brought the amount of money for political elite as paying 'tribute' for their sustainable concession and dealt with the-powers-that-be in Manila and with their buyers in Japan. This is a common portrait of logging holders' behavior in connection with politics in the Ferdinand Marcos regime.

To overcome the impact of *A. mangium* and *Eucalyptus* trees plantations due to 'shortage of water', the government, represented by the Forest Management Bureau (FMB), included an *environmental performance monitoring (EPM)* assessment on several elements, such as the quality of land, air, biodiversity, water and interaction with people in the SIFMA and IFMA assessments held twice annually. So far, based on the evaluations, both the private and cooperative sectors are standing up to scrutiny on 'environmental assessment'.[30] Local farmers in Caraga region 10, which make up almost 40–50% of national tree production gave a positive response on 'plantation forestry', consisting of several trees such as *Gubes, Paraserianthes falcata, A. mangium, E. deglupa, S. macrophylla and G. arborea.* They do not talk any more about the negative impact of *eucalyptus* trees on the ecology. They prefer to consider these trees as commercial trees that generate an income for people in the short term.

The following argument is used to support the idea that *eucalyptus* trees could be cultivated and interplant with native species *(fungi)* in order to make fertile land.[31]

Why do you criticize eucalyptus trees as monoculture plants?

E. deglupa is disadvantageous as a monoculture plant as it absorbs much water. Therefore, to maintain biodiversity, reduce water depletion and improve fertility of the soil, farmers should interplant eucalyptus with native species (*ectomycorrhizal fungi*). The aim of interplating is to restore back water, reduce soil erosion and increase biodiversity, eventually restoring the entire wildlife and ecosystem.

Do you agree that eucalyptus and other fast growing trees (P. falcate, Acasia mangiun, G. arborea, etc.) can bring about a positive impact through income generation for farmers and the private sector?

Yes, I agree that these trees could be beneficial to local farmers and the company (private sector), as raw material for pulp and paper, plywood, sawn timber and house construction.

[30]An interview with Head of Environmental Affairs Ms. Nely Butic of Forest Management Bureau (FMB) on August 5, 2009 in Butuan City.

[31]An interview with Dr. Nelson M. Pampolina, he is a lecturer in Department of Forest Biological Science, College of Forest, UP Los Banos, on August 2, 2009. Mr. Nelson wrote Ph.D. thesis in Murdoch University, Australia entitled:" Ecology of Ectomycorrhizal fungi in eucalypt plantation in Western Australia and the Philippines," in 2000. Currently, *E. deglupa* is mostly planted by local farmers in Caraga region 10, because of its fast growing and high yield.

Are there indications of a connection between deforestation and poverty?

Yes, there are activities that lead to deforestation, such as illegal logging, 'shifting cultivation' (*Kaingin*) in the uplands and encroachment on state forest land for agricultural conversion. Therefore, to prevent these activities, the government and other donor countries need to cooperate with local farmers to realize *forest rehabilitation* and *reforestation* in many regions, ensuring that local people could get income by planting trees and harvesting timber as a counter to poverty in rural areas.

How is your view of the ADB (Asian Development Bank) as a funder of planting eucalyptus trees as raw material of the pulp and paper companies?

I do not agree with the policy launched for the private sector by the ADB. ADB must actively engage other *stakeholders* (*local farmers, cooperative sectors*, etc.) to plant several types of trees. The Philippines is occupied by a very rich diversity of local species. The ADB policy only selected one species of trees (*E. deglupa*) without encouraging other local species trees to be planted as commercial trees for the domestic market.

Actually, the ADB has supported two plantation projects in the Philippines. Both projects created problems. The first was approved in 1983 and suffered from deficiencies in project design and implementation according to the Bank's Project Performance Audit Report. The project was redesigned in 1988, after a typhoon hit the project area. Instead of planting a different tree species, as initially planned, "the project adopted a strategy of near monoculture plantations of eucalyptus camaldulensis." The plantations were poorly maintained and "were characterized by a highly uneven and low tree growth rate." As a result, the bank failed to monitor the project adequately.[32] The second project by the ADB in the Philippines was the Industrial Forest Plantation and also faced problems. The project started in 1991, and aimed to establish 30,000 ha of industrial tree plantations. In fact, 6100 ha were planted. The cost of establishing the plantations was higher than expected, because of "the cost of settling disputes over the land for the plantations" according to Alastair Fraser, an ADB consultant.[33]

Concluding Remarks

The Philippines' forests have been facing deforestation over the past 40 years and more. From the view point of 'political ecology', which emphasizes *stakeholders,* the role of government in the period of Ferdinand Marcos (1965-mid 1980s) was the peak of 'deforestation' (Table 3.1). There were, and are, two driving factors causing 'deforestation' in the Philippines' forests. First is the political factor that

[32]See Chris Lang, "Plantations, poverty and power: Section 3", p. 11 (http://chrislang.org/2009/02/06/plantations-poverty-and).

[33]Ibid.

forests are treated as commodities and political interest. Logging concessions were given by Marcos to his cronies and political elites supporting his power. This led to large scale deforestation (165,000 ha annually), because most logging holders did not practice sustainable forest management. Second is the conversion to agricultural land (2.8 million hectares) due to population growth and growth in agriculture in general. In many developing countries—such as the Philippines—there is great pressure imposed on forest lands by increasing populations, relentlessly damaging the ecosystem, especially shifting cultivation (*kaingin*) in the uplands, conservation and protected forests. The impact of the deforestation criticized by NGOs and academics was the creation of greater environmental disasters occurring since the mid-1980s–1990s.

Observing the above environmental disasters, the Philippine government, especially under Cory Aquino and Fidel Ramos, invited other **stakeholders** (*academics, congressmen, NGOs, local governments, local people,* etc.) and even donor countries to actively design various programs to protect and conserve the remaining forest by launching 'forest rehabilitation and reforestation'. These programs included the Integrated Social Forestry Program (ISFP), Upland Development Program (UDP), National Forestation Program (NFP), Forest Land Management Program (FLMP), Low Income Upland Communities Project (LIUCP), Community Forestry Program (CFP), Regional Resources Management Project (RRMP), Forestry Sector Project (FSP), and finally revised to become the Community-Based Forest Management Program (CBFMP).[34]

CBFMP, introduced in 1995, in particular, recognized the indispensable role of local people in managing forest resources in the country. Many forest analysts say that theCBFMP program, which has been practiced for at least two decades in the Philippines, is the best model of sustainable forest management and has inherently improved the socio-economic situation of many local farmers. The success of the CBFMP is in the active *participation* of other stakeholders, such as local people, local government, academics, NGOs and donor countries in the program.[35] The focus and aim of forest management has thus shifted from technical commercial forestry to a more people- oriented social forestry. However, improving the socio-economic condition of the rural populace (*particularly small holders*) remains a great challenge for the Philippine government, because 30–40% (Statistics, 2008) of Filipinos remain in *rural areas*, and are mostly categorized as 'poor' compared with people who live in urban areas. However, the main problem in reforestation issues within the CBFMP is that at harvest time, the government is reluctant to provide permissions to farmers to cut and transport their timber, even if it is in the MOU that farmers may cut their trees at harvest time with a cycling cutting system from one block area to another. The argument of the FMB officers is that large cutting of trees by farmers in certain areas (such as the uplands) subsequently affects soil erosion and flooding in the downstream areas.

[34]See Harrison et al. (2005)

[35]Interview with informant in UP, Los Banos, July 31, 2009.

The demand for timber has reached 13 million m^3 annually. Therefore, to fulfill this high demand for timber, such as pulp and paper, sawn timber, plywood and house construction, the Philippine government issued 'timber plantation' concessions for 25 years that could be extended based on their performance. There are many schemes available for tree plantation investment, for instance IFMA/ITPLA for the private sector and SIFMA for the cooperative sector, people's organizations (PO) and individuals. The main constraints to investors, both foreign and domestic, is that the government of the Philippines does not provide economic stimulus and incentives and good infrastructure services (roads and ports) for integrated investment from timber plantations (downstream) to timber processing mills (upstream). For instance, the lack of economic stimulus and incentives means that state bank officers are reluctant to give big credit to investors, because bank officers still consider the forestry sector as high risk and a very long-term return. Even fast growing trees (*A. Mangium, Eucalyptus*, etc.) need 8–9 years and the price of timber products is unstable. Besides, infrastructure (roads and ports) in rural areas are in very bad condition. The central government still does not pay enough attention to infrastructure investment in rural areas. Even better performance of infrastructure could easily mobilize investors on transporting valuable commodities such as agricultural products, cash crops, timber from rural to urban areas and exporting to overseas from the domestic port. The limited infrastructure facilities are a constraint, especially for transporting timber to processing mills. Empirically, the Philippines is considered 'backwards' within the forestry industry compared with other fast growing forestry industry ASEAN countries such as Indonesia and Thailand. As a consequence, foreign investors are reluctant to enter and invest in the forestry sector. As an illustration, domestic investors find it very difficult to obtain credit from state banks or the stock exchange to obtain fresh capital, as told by CSDC Company officer in Butuan city.

Therefore, the Philippines government, especially *FMB* under *DERN* (*the Department of Environmental and Nature Resources*) should create a political commitment to cooperate with other institutions, such as banking institutions, to create a strong foundation for the forestry industry, originally based on the *reforestation* programs under the **CBFMP scheme**, where timber plantations to timber processing mills are categorized as a *strategic industry* in the Philippines for the following reasons: firstly, the forestry industry could largely contribute to obtaining foreign exchange earnings and national GDP of the Philippines. Secondly, the forestry sector could create many more jobs in rural and urban areas, creating long-term income generation for local people. Thirdly, it is highly recommended that the government and other stakeholders practice sustainable forest management in the Philippines as this will help to prevent soil erosion, flooding, and excessive negative impacts from typhoons. Therefore, to realize and facilitate the above conditions, the Philippines government should revise the regulations on timber plantations, tax breaks, law enforcement and access to credit from institutional banks. These changes could actively engage stakeholders of the forestry sector to invest and expand their businesses in the Philippines in the long run.

Chapter 4
Indonesia

Review of Industrial Timber Plantation (HTI)

Introduction

Industrial Timber Plantations (**HTI**) can be utilized as a means of equal land distribution between local people and company holders, utilized as a source of production and improving the socio-economic condition of local communities. Walhi NGO officers highlight that there are two main conflicts concerning **HPH** (logging concessions) and **HTI** and local communities (Hidayat 2004: 152). First is the conflict of space, as the government[1]claims, through Forestry Act number 5/1967 and number 41/1999 that all forests belong to the state and therefore the government ignores the customary forest ownership of local people. Most of their lands are generally not certificated by the government, leading to land conflicts between logging and HTI holders and locals in many districts of Indonesia. Under the Soeharto regime, there were 1700 land conflict cases (*Tempo*, 24 September 2000). Second is discrimination in the provision of concessions. Most logging and HTI concession holders occupy 51.6% of production forests. For instance, Sinarmas Forestry Company owns more than 900,000 ha of HTI concessions, while the number of landless farmers is rapidly increasing. A recent study reveals that the percentage of landless farmers on Java increased from 3.2 to 14.9% between 1973 and the 1980s, with the percentage of farmers owning less than 0.5 ha rising from 45.7 to 63%. [2] Hence, potential causes of further conflict in forestry development, both logging and HTI concessions are: (1) how the government manages equal land distribution to stakeholders (local communities and private company holders) and

---

[1]Reformation era (1999–2009) is the period of government administration after Soeharto stepped down (in May 1998), governed by B. J. Habibie, Abdurrahman Wahid, Megawati, and Susilo Bambang Yudoyono (SBY), still defending that all forests belong to the state (interview with Department Forestry Officer (April 30, 2010).
[2]For further information, see Colchester (1986, pp. 61–67).

H. Hidayat, *Sustainable Plantation Forestry*,
https://doi.org/10.1007/978-981-10-7653-4_4

(2) how the government regulates provision of concessions for logging and HTI's holders and local communities. In order to improve conditions in Indonesia, the government should first revise the regulation of HTI concession holders and actively engage individual farmers and households as actors in plantation forestry and second, the government should commit to drastically reducing poverty in Indonesia, under the 13–14% (30.1 million) registered in 2009 (*Kompas*, 22 September, 2010). Therefore, HTIs must be utilized as a strategic means of shifting people into better socio-economic conditions. The government should provide economic incentives to stakeholders (local farmers and private companies), through access to credit, encourage contract farming between farmers and the private sector and ensure guarantees in the market for timber by companies).

The government played a significant role in facilitating investment in **HTI** (*Industrial Timber Plantation*), issuing simplified procedures, economic incentives, friendly trade regulations, credit facilities, land for HTI plantations and reforestation policies. However, many companies are using these reforestation funds, while others, such as APP, Toba Pulp Lestari, Indah Kiat, Wira Karya Sakti and RAPP, are not using government reforestation funds, preferring to use their own funds for their timber plantations.

During Pelita I[3] (1969–1975) there were just six factories in operation, with total production reaching 9000 tons annually. However, by 1987, there were 36 pulp and 41 paper factories, increasing again in the 1990s to 82 approved pulp and paper companies. In line with the development of pulp and paper factories, the demand for sustainable timber supply also grew. Therefore, the Ministry of Forestry began an **HTI** program in the 1980s with a regulation to plant 200,000 ha of trees for every pulp and paper factory in order be able to provide their raw material needs. As a result, these HTI produce short fibers (*serat pendek*) as raw material for the pulp industry. However, long fibers (*serat panjang*) are still imported from overseas.

This paper highlights some issues: the government's policy of HTI; private company responses to this policy; the adoption of contract farming, sustainable forest management and the preservation of primary forest biodiversity and academic and NGO critiques concerning the ecological damage created by the APP mill in Riau province.

Government Policy on HTI, Ministry of Forestry

All forests in Indonesia belong to the state under the management of the Ministry of Forestry. According to the Ministry of Forestry, in 1999, Indonesia forest cover was

---

[3]Pelita (Pembangunan Lima Tahun/Five Years Development). The Soeharto regime launched 'Program for Economic Development' continuously consistent for every five years during his presidency period from 1969 to 1998.

133.8 million hectares, consisting of: (1) production forest of 60.8 million hectares; (2) estate plantations of 22.7 million hectares; (3) protected forest of 30.5 million hectares and (4) conservation forest of 19.8 million hectares.[4]

The Ministry has the authority to classify the allocation of forest use and issues concessions for *industrial timber plantations* (**HTI**) on critical lands of production forests. According to Schlager and Ostrom (1992), **HPH** (*forest rights concessions*), called "access and withdrawal", manage timber concessions in production forests and are provided by the Department of Forestry, and only timber is allowed to be harvested. The government has the authority to withdraw HPH concessions, if HPH or HTI holders break regulations (Kartodihardjo 2006: 65).

Based on regulation P.19/Menhut-II/2007, article 4 (requirements of applicants), four (4) categories of entities can apply for IUPHHK-HTI concessions[5] (1) a cooperative; (2) Indonesian private sector; (3) state enterprise business (BUMN) and (4) district enterprise business (BUMD). The size of HTI concession areas is between 100 and 100,000 ha. In practice, the majority of HTI concession holders are large companies, such as Sinarmas, Musi Hutan Persada, RAPP forestry, Inhutani,[6] etc., who manage more than a hundred thousand hectares of HTI areas each. According to the regulation (article 9), HTI holders of more than 10,000 ha should perform an **AMDAL** analysis (*Analisis Mengenai Dampak Lingkungan*) (*environmental impact analysis*) and those of less than 10,000 ha should perform a **UKL** analysis (*Upaya Kelola Lingkungan*) (*environmental management analysis*).

The Department of Forestry, in an effort to improve investment in the sector, provided easy access to timber concessions to the private sector. As a result, some companies manage large HTI concessions. For instance, Sinarmas manages more than 900,000 ha in many districts in Indonesia and, in the initial establishment of some pulp and paper factories in Indonesia, the government also gave permissions to import the necessary equipment and tax exemption to the companies. The government and the private sector in Indonesia considered pulp and paper industries as strategic for future foreign exchange earnings and the Indonesian government considered the growth of these industries as investment in the economic development of rural areas (interview, 30 April 2010).

In the 1980s, when HTIs were initially launched, the government instigated a fund which was paid for by 40% of shares in the HTI. The fund was originally taken and collected from logging concession holders for reforestation. It aimed to

---

[4]For further information see (http://www.dephut.go.id).

[5]The abbreviation of IUPHHK-HTI is (*Izin Usaha Pemanfaatan Hasil Hutan Kayu-Hutan* Tanaman Industri) or (Timber Utilization Products Concession for Industrial Timber Plantation). The concession issued by Ministry of Forestry based on their requests and must fulfill several requirements such as administration and technical requirements. This regulation has been revised by regulation (P.11/Menhut-II/2008), but no fundamental changes are included).

[6]Inhutani is represented of State Forestry Enterprise. It is categorized as **BUMN** (*Badan Usaha Milik Negara*) or (*State Enterprise Company*). Inhutani owns HPH and HTI's concession areas mostly located in outer island (Sumatra, Kalimantan, Sulawesi, Maluku and Papua). The management of Inhutani is the responsibility of the Ministry of Forestry.

Table 4.1 Development of HTI areas and units from 1989 to 2009

Year	Area (ha)	Units
1989	30,000	1
1989–1995	600,000	14
1996	1,742,509	24
1997	2,752,392	58
1998	3,877,897	89
1999	4,028,754	93
2000	4,133,391	95
2001	4,210,713	97
2002	4,236,225	100
2003	4,314,565	102
2004	5,491,170	111
2005	5,548,285	114
2006	6,616,400	133
2007	6,790,287	163
2008	6,835,687	165
2009	7,830,289	195
2010	8,763,555	207

Source Directorate general of bina produksi hutan (*Timber Production Resources*) (2009), the ministry of forestry

actively motivate private companies to replant trees for their industry. However, since the reformation era, starting in 1998, the reforestation fund was returned to the government budget, but it was difficult for the Ministry of Forestry to request the funds from the Ministry of Finance. As a result, private companies were forced to borrow from banks or issue stocks on the stock exchange in order to fundraise to develop their industrial timber plantation (HTI).

As an illustration, the government had issued 600,000 ha of HTI from 1989 to 1995, 4.2 million hectares by 2001, 7.8 million hectares in 195 units of HTI for all of Indonesia's districts by 2009 and finally 8.7 million hectares of forest in 207 units by 2010,[7] from a total of 60.8 million hectares of production forest (Table 4.1). The growth of HTI units was in accordance with the development of pulp and paper factories. For instance, in 1987 a total of 36 factories were registered, increasing to 41 in 1990, 57 in 1998 and 69 in 2008 (Tables 4.2 and 4.3).

[7]Based on interview with HTI officer held on April 30, 2010 in the Ministry of Forestry Office in Jakarta.

Table 4.2 Annual pulp production and wood consumption of Indonesia's pulp industry 1988–2000, with projection to 2005, 2010

Year	Pulp production (000 tpa)	Wood consumption (000 m^3)
1988	368	1805
1989	461	2261
1990	697	3415
1991	850	4165
1992	870	4263
1993	900	4410
1994	1314	6439
1995	2022	9908
1996	2561	12,549
1997	3048	14,984
1998	3430	16,807
1999	3400	16,660
2000	4140	20,286
2005	5790	28,945
2010	6715	33,605

Source Indonesian pulp and paper association (APKI) for 1988–2000
Jaakko Poyry (1998) for 2005–2010 projections

Table 4.3 Private companies planting in HTI in 2000 (ha)

Company	Location	Land reserved	Realization
Using reforestation fund			
Alas Helau	Aceh	96,899	24,630
Musi Hutan Persada	South Sumatra	296,000	205,084
Surya Hutan Jaya	East Kalimantan	183,300	125,642
Tanjung Redep	East Kalimantan	180,330	77,342
Finantara Intiga	West Kalimantan	299,700	33,268
Acindo Foresta	East Kalimantan	201,821	29,016
ITCI Hutani Manuggal	East Kalimantan	161,127	88,181
Indonusa Indrapuri	Aceh	111,000	30,600
Menara Hutan Buana	South Kalimantan	268,585	113,952
Non-use of reforestation fund			
Toba Pulp Lestari	North Sumatra	269,060	49,117
Indah Kiat	Riau	299,975	181,313
Wira Karya Sakti	Jambi	172,978	84,703
Riau Andalan Pulp and Paper	Riau	159,500	101,327
Total		2,700,675	1,114,175

Source Ministry of forestry, Directorate General Bina Produksi Hutan (*forest production management*), 2000. Data consult research paper, 2000

Regulation for Providing Raw Material

The government initially issued a policy obliging every pulp and paper company to plant around 200,000–300,000 ha of HTI. These investments for HTI usually reach ten percent (10%) of capital for every factory, worth US$1.5 billion, to establish a new factory. For example, the harvesting times of *Acacia mangium* and *Eucalyptus* trees are an estimated 6–7 years after planting and will produce 150–200 m^3 of wood per hectare. Therefore, 25 m^3 of wood is produced annually per hectare. If the area of HTI is 200,000 ha $\times$ 25 m^3, then five million m^3 (5,000,000 million m^3) of wood can be produced. This is roughly equivalent to one million tons of pulp. There are many ways to obtain raw material: (1) the pulp and paper companies plant trees on their plantations (*A. mangium* and E*ucalyptus pelita*/EP05); (2) the companies conduct contract farming where the company guarantees a market at harvest, provides fertilizer and seedlings for farmers and provides technical extension to farmers; (3) the companies buy logging waste and wood cuttings from land clearing, such as from oil palm and transmigration areas and (4) companies collect recycled paper.

The HTI plantations and other materials provide *short fibers* for the pulp industry. However, long fibers are still imported from overseas, because the material is not available in the domestic market. To illustrate, in 1993, pulp companies imported 705,700 tons of long fibers, producing 900,000 tons domestically. 123,600 tons of pulp was exported and domestic consumption was 1.4 million tons, rapidly increasing in 1998 to 4.3 million tons of pulp production, with 1.6 million tons exported, 839,510 tons imported and domestic consumption at 2.6 million tons (*Directory Indonesia Pulp and Paper,* 1999).

APP (Asia Pulp and Paper) Mill

Profile of Sinar Mas Group

Asia Pulp and Paper (APP) is one of the holding companies of **Sinar Mas Group** (SMG). SMG was founded in 1962 by **Eka Tjipta Wijaya** (Chinese name is **Oey Ek Tjhong**), a prominent Indonesian-Chinese businessman. According to Globe Asia Magazine on May 31, 2010, Eka Tjipta Wijaya was the second richest man in Indonesia with personal assets of US$4 billion or IDR 40 trillion (*Kompas*, June 6, 2010). Annual sales of his companies total IDR 20.2 trillion (US$8.5 billion) with more than 200 affiliated companies (Sato 1998). Currently, SMG holds assets worth a total of over US$20 billion and 150,000 employees, recognized by *Forbes Magazine* as one of the largest consortiums in Indonesia. The core business of SMG consists of Sinarmas forestry, pulp and paper, agriculture and food processing, especially the palm oil estate called **Golden Agri Resources** (GAR), finance and real estate (http://www.appbrasil.com). This study focuses on two emerging industries under SMG, namely pulp and paper (APP) and Sinarmas forestry, mainly located in Riau province, Siak district.

Study Sites

Interviews were conducted with Department of Forestry officers, academics and NGOs, the Sinarmas plantation and APP mill executive directors, with observations of the plantation forestry program, research and development department and Arara Abadi Company (affiliation company with Sinarmas Forestry) cooperation with local farmers in Siak district, Riau (Fig. 4.1).

Siak district is where the APP mill and most of the forest plantations of Sinarmas Forestry is located.

APP Mill

The Company's Response to Government Policies

According to company officers, the climate for plantation forestry is conducive, because the company can obtain easy access to timber concessions in many districts. Although land conflict with local farmers still occurs in rural areas, to overcome these conflicts, the company conducts negotiations and shares production fees with local farmers for planting trees. (Interview, 29 April 2010).

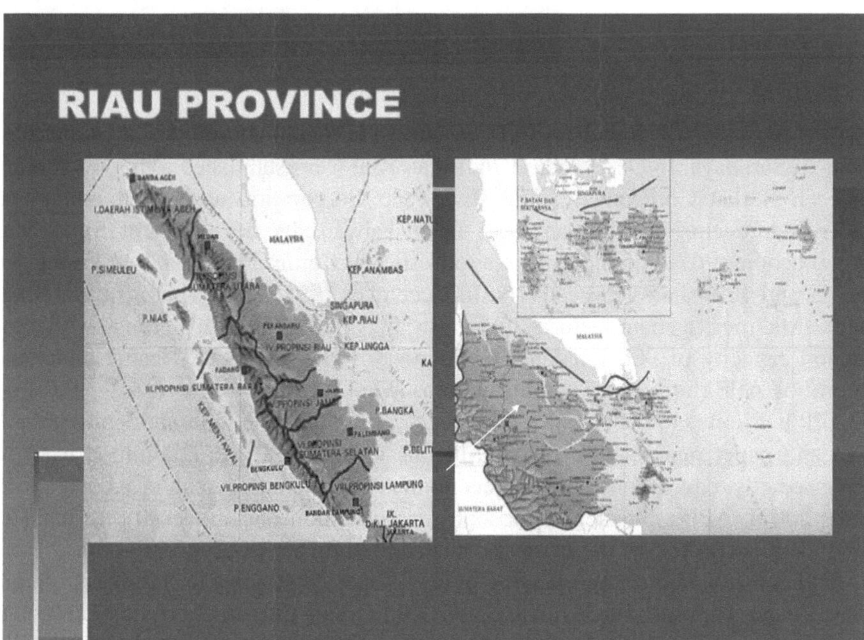

Fig. 4.1 Riau province is located in Central Sumatera Island. Siak district is where APP mill and most of the forest plantations of Sinarmas Forestry is located

Fig. 4.2 Integrated company offices of Sinarmas Group within APP company located in Jakarta downtown. The picture was taken in May, 2010

APP began in 1972 with a small caustic soda manufacturing plant on the outskirts of Surabaya. In December 1976, Indah Kiat was established as a joint venture between Berkat Company, Chung Hwa Pulp Corporation and Yuen Foong Yu Paper Manufacturing Company Ltd from Taiwan. Its development Sinar Mas Group acquired 67% of Indah Kiat's total shares. Chung Hwa and Yen Foong Yu had 23 and 10% shares respectively. In December 1991, Indah Kiat acquired Sinar Dunia Makmur, a manufacturer of industrial paper located in Serang, with production capacity of 900 tons/day. In February 1994, Pindo Deli came under the control of APP (http://www.appbrasil.com) (Fig. 4.2).

APP's vision is to become the 21st century's number one, international standard, pulp and paper manufacturer. To fulfill this vision, APP has committed itself to being socially, environmentally and economically sustainable in all of its operations (APP 2007). APP's business philosophy is: "tradition and modernity go hand in hand" and is the key to the company's success (http://www.appbrasil.com).

APP consists of the top 5 leading pulp & paper companies in Indonesia.: Indah Kiat, Lontar Papyrus, Pindo Deli, Tjiwi Kimia, and Ekamas Fortuna (Table 4.4) (Interview with APP officer, May 6, 2010). APP has its principal operations in Indonesia and China, with total assets of over US$10 billion and annual production capacity of over 13 million tones and over 930,000 plantations. Since 1992, APP

Table 4.4 APP mill sales and production in 2007

Mill	Sales (US$ thousands)	Production
Indah Kiat	1,856,307	3,806,366
Pindo Deli	1,262,999	1,693,243
Tjiwi Kimia	1,153,500	1,335,000
Ekamas Fortuna	69,453	178,923
TOTAL	4,342,259	7,013,532

Source APP 2007. **Pindo Deli** pulp and paper mills owns 80% of Lontar Papyrus pulp and paper industry. Sales for these entities are consolidated for reporting

China has been investing a huge amount of capital into the establishment of pulp and paper industries in the Yangtze and Pearl Delta and expanding, with the establishment of joint ventures, such as Ningbo-Zhonghua Paper, Gold Huasheng Paper, Gold Hong Ye Paper, Gold East Paper and Hainan Jinhai Pulp and Paper (Fig. 4.3). The headquarters of APP China is in Shanghai. Gold East Paper and Ningbo Zhonghua Paper were ranked first and second respectively in the "Top 100 Most Prominent Chinese Enterprises" for 2000–2002 (http://www.appbrasil.com). Currently APP China has shares in over 20 pulp and paper companies, as well as more than 20 forestry plantations in China with approximately 28,000 employees and a total asset value of US$8.9 billion (RMB 66 billion). The annual production reaches over 6.4 million tons. In 2006, APP China revenue was an estimated US$2. 28 billion (RMB 21.4 billion). APP in Indonesia and China combined has an

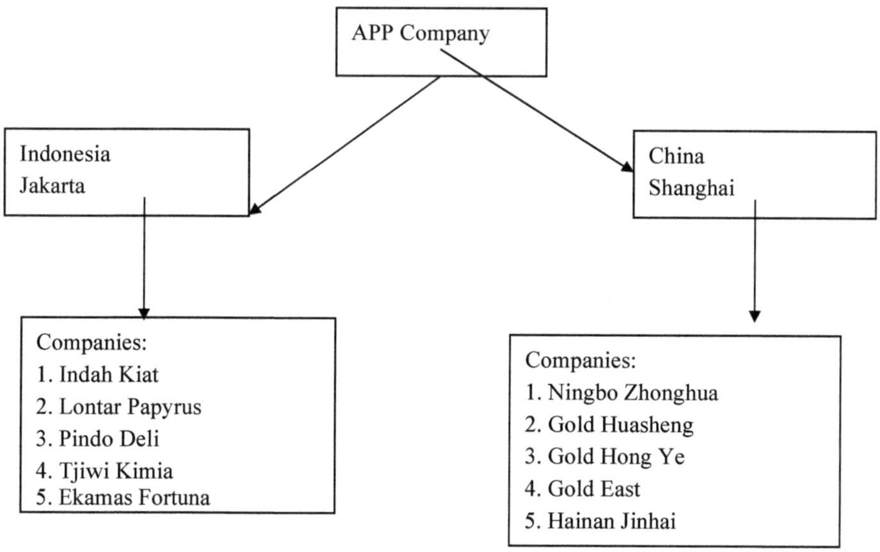

Fig. 4.3 APP company's in Indonesia and China

estimated total asset worth of over US$10 billion and an annual production capacity of over 13 million tons of pulp and over 700,000 ha of man-made plantations.

Raw Materials

How Does the Company Ensure Constant Timber Supply?

The Sinarmas Forestry Company (SMF) partners with other stakeholders to provide timber for the mills. SMF affiliates with other companies owning private estate plantations, invites farmers to enter into contract farming and also contracts land from farmers. SMF provided nearly 60% of the total timber used in the production of APP paper products during 2007. Also, recycled paper from domestic and overseas sources reached 23% recycled at the mills and 10% from non-fiber fillers, with 10% purchased paper pulp—primarily softwood (APP 2007).

APP Provides Capital and Technology

The initial establishment of APP was supported by the government through tax relief for imported machinery, while the company obtained credit from banks and the stock exchanges in Jakarta, Singapore, Tokyo and New York by selling shares to the public. As APP is a public company, responsible to shareholders, it is assessed by public accountants, with annual assessments and reports to shareholders. In terms of technology, APP's manufacturing facilities are from Scandinavia (Finland and Sweden), USA, Canada and Japan. Wastewater management and air emissions and effluents at the mills meet all Indonesian and most international standards. The Company has competitive advantages with other foreign companies in terms of providing the best quality seedlings that grow faster in tropical zones and labor costs are lower (*interview with APP Officer, May* 6, 2010). According to the 2008 annual report, there was an "increase of 21.2% in consolidated net sales, from US$1879.4 million in 2007 to US$2277.0 million in 2008, [an] increase of 31.3% in consolidated operating income, from US$224.6 million in 2007 to US$295.0 million in 2008 and [therefore an] increase of 120.5% in consolidated net income, from US$91.8 million in 2007 to US$202.4 million in 2008" (Indah Kiat 2008). This profit was mainly caused by increasing sales prices of the company's products and its total sales volume of pulp, enjoyed from the beginning of the year until the third quarter of 2008 (Indah Kiat 2008).

The Marketing of Products

The marketing division of APP actively engages in domestic as well as overseas marketing. Jakarta is appointed as the central office for APP marketing management over all of Indonesia's districts and provinces. APP has also established marketing in Singapore, the Middle East and Japan. The Company exports the majority of its products to different countries in Asia (Southeast Asia, Japan, China and Taiwan), the Middle East, Europe, USA, Australia and Africa. The greatest sales occurred in 2007 and 2008 consecutively, when the company received the **Primaniyarta Award** as one of the best exporters, provided by the National Export Development Bureau (*Badan Pengembangan Ekspor Nasional*) (Indah Kiat 2008). This success, according to an APP officer, was due to the professionalism and performance in improving products and quality services for customers, providing maximum benefits to the shareholders, while providing strong attention to environmental preservations.

APP and Sustainability

APP has a high commitment to sustainability. The company operates in compliance with the national laws on sustainability, fiber procurement, environmental protection and health and safety. In 1996, Pindo Deli, which is located in Banten province, became the first mill in Indonesia to receive ISO 14001 *Environmental Management System* (**EMS**) certification, an environmental compliance and wood supply traceability audit carried out by **AMEC Simons** Forest Industry Consulting in November 2001. The findings highlighted the group policy adopted by APP and Sinar Mas Group (SMG) that had resulted in the following: (1) implementation of comprehensive ISO 14001 environmental management systems for all mill and forestry companies, certified by internationally leading certification companies; (2) working within the Indonesian government forest law and Department of Forestry regulations concerning operational licenses for conversion of land for industrial pulpwood plantations and legal documentation required for wood delivery (3) and commencement of implementation of forest management practices in accordance with *Forest Stewardship Council* (**FSC**) forest certification principles and criteria.[8]

The other two company's mills, located in Perawang—Riau and Serang—Banten respectively obtained ISO 14001 certification in 1997 and 2004. The company also has been **SMK3** (*Occupational Health and Safety*) certified (Indah Kiat 2008: 22). Another APP mill, Tjiwi Kimia in East Java, received Green Seal Environmental Certification in December 2009 for its recycled photocopy paper products under the category of Printing and Writing Paper.[9] A Green Seal is provided by the *American National Standard Institute* (**ANSI**), which is a US and Canada-based independent, non-profit organization with the aim of safeguarding

[8]For further information, see report of *APP Pulp Mills & Sinar Mas Group Forest Companies: Environmental Compliance & Wood Supply Audit*: Executive Summary. Reviewed by AMEC Simons Forest Industry Consulting, November 21, 2001.

[9]For further information please see Stakeholder update of APP (environment@app.co.id).

the environment and transforming the marketplace by promoting the manufacture, purchase and use of environmentally responsible goods and services. Green Seal utilizes a life-cycle approach, which means it evaluates the product from raw material composition (forest plantation) processing through to the manufacture and end use of the product and then on to recycling and disposal. The final products can only become Green Seal certified after rigorous testing and scientific evaluations based on internationally accepted methodologies.

Production and Employees

APP production during 2007 was more than 7 million tons of pulp, paperboard, stationery, tissue products and packaging (Table 4.4). In total, this represented a 4–5% increase over production levels in 2005 and 2006. All of APP's mills are compliant with ISO 9001 Quality Management System (QMS) standards and with ISO 14001 Environmental Management System (EMS) standards.

In 2007, sales for the five APP companies were estimated at US$4.3 billion and employed more than 37,000 workers. The workers mostly come from Indonesia and other countries such as Singapore, Malaysia, Philippines, Finland, China and Taiwan. Meanwhile, in APP mills, most expertise comes from Taiwan, because Taiwan's company owns 23% of shares. Mill operations created in excess of 25,000 indirect (part time and temporary) jobs, and fiber suppliers under the flag of **Sinarmas Forestry** (*SMF*) utilized an estimated 6900 employees and accounted for nearly 1600 indirect jobs. Total employment in APP mills and Sinarmas Forestry (SMF) reached more than 71,000 persons (Table 4.5) (APP 2007).

Sinarmas Forestry Response to Government Policy

Sinarmas Forestry (SMF) conducts sustainable plantation forestry based on economic, social and ecologically feasibility in the field. SMF is one of the leading plantation managers in Indonesia. SMF manages timber plantation affiliations with

Table 4.5 Distribution of APP mill employees, mill indirect jobs and fiber supplies jobs in 2007

Location	M. employee	Indirect job	Sinarmas forestry	Temporary employee	Total
Java	26,877	6041	0	0	32,918
Sumatra	10,712	19,501	5909	735	36,857
Borneo	0	0	1054	855	1909
Total	37,589	25,542	6963	1590	71,684

Source APP 2007. **Borneo** refers to the Indonesian provinces of East and West Kalimantan. M: Mill

Table 4.6 Sinarmas forestry concessions area of December 31, 2007 (in hectares)

Island	Total area	Conservation	Planted	Degraded	Fire damaged
Sumatra	1,865,142	655,142	639,550	148,936	412,220
Kalimantan	522,620	296,679	121,034	2,503	102,404
Total	2,388,468	961,821	760,584	151,439	514,624

Source APP 2007

Table 4.7 Accumulative size of SMF plantations until September 2009

Province	Width (ha)
Riau	311,781
Jambi	197,176
South Sumatra	255,408
West Kalimantan	55,329
East Kalimantan	110,607
Total	930,301

Source Sinarmas company, 2009
Note Total amount of area 930,301 ha, includes all partnerships (contract farming with farmers) with the company

many affiliated companies, such as Arara Abadi (Riau), Satria Perkasa (Riau), Bina Duta Laksana (Riau), Mutiara Sabuk Khatulistiwa (Riau), Wira Karya Abadi (Jambi) and Finantrata (in West Kalimantan). The aim of SMF is as follows: (1) the company is committed to implement sustainable forest management, feasible from a social, economic and environmental viewpoint; (2) the company provides sustainable raw material by plantations and contract farming and (3) the company provides the best quality seedlings, such as *E. pellita* (EP05) and *Acacia crassicarpa*, based on intensive research in the R&D center in Perawang.

The Sinarmas Forestry plantations consist of 961,000 ha (40%) of conservation area, reserves for community use, indigenous species and infrastructure; 760,584 ha of plantations (32%) and the rest either degraded forest (151,439 ha) or bare land, scrub lands, waste lands or fire damaged areas awaiting reforestation (514,624 ha) (Table 4.6).

A Sinarmas forestry officer mentioned that the plantations grew to 930,301 ha by 2009 in order be able to provide timber for APP mills (Table 4.7). The forestry industries, which were initially categorized as unstrategic, became strategic after the signing of an EPA (Economic Partnership Agreement) between Indonesia and Japan in 2007, as pulp and paper, plywood and furniture industries could obtain much more foreign exchange earnings and produce more jobs than before.[10] As an

[10]An interview with Sinarmas forestry Executive Director, on April 29, 2010 in Jakarta, whereby the argument made was that they needed to extend the area of timber concession in order be able to export wood-chip to overseas. However, when cross-checking with a Ministry of Forestry officer, this was refuted, because concessions had been issued for 8.7 million hectares, but planting of trees had occurred in less than half by 2010.

illustration, Sinarmas Forestry in Riau Province absorbed 254,289 workers, produced a net income of IDR 11.5 trillion (6.97% from total Riau's income), IDR 92.6 billion in government taxes and export earnings of US$837.4 million.

Sustainable Forest Management

SMF is committed to three goals in its vision of sustainable forestry: (1) to practice sustainable fiber productivity (growth rate and fiber yield); (2) to provide competitive fiber cost and (3) to use methods that are environmentally acceptable and socially compatible. To achieve these goals, Sinarmas Forestry is working with LEI (*Lembaga Ecolabel Indonesia*) or (*Ecolabel Institute of Indonesia*), which has recently adopted standards for both a regular and phased approach to fully sustainable forest management certification. LEI first produced its principles, criteria and indicators in 1998, based on the International Tropical Timber Organization (ITTO) Guidelines for Sustainable Forest Management and has developed certification standards for industrial plantations in Indonesia, which were implemented in 2004. Currently LEI Standard 5000-2 is the only credible national forest certification scheme. Because SMF and APP both agreed to ensure that all timber brought into its two pulp mills in Perawang and Jambi are verified legal and of non-controversial origin, system performance and improvements are verified regularly through third-party auditors, such as LEI.

SMF and Conservation Areas

Biodiversity should be maintained and developed in conservation forests. The commitment of Sinarmas to conservation is realized through practicing sustainable forest management, which highlights ecological issues. According to Sinarmas, buyers from Europe, Japan and the USA have misconceptions about the pulp and paper industry in Asia, especially Indonesia.[11] Sinarmas has a high commitment to protecting biodiversity and endangered species in Indonesia. All initiatives, according to a Sinarmas officer, are co-managed with other stakeholders, such as academics and government institutions, local and international NGOs and other private sector organizations.

Arara Abadi, one of the affiliated companies of Sinarmas Forestry (SMF), made significant investments in conserving species and maintaining forest diversity through actively engaging: Arboretum in Rasau Kuning sub district; the Giam Siak

[11]For further information see Stakeholder Update on "APP Surprises the Heart of Europe with Nature Conservation Projects", in (environment@app.co.id).

Kecil—Bukit Batu Biosphere Reserve in Siak district; the Taman Raja Conservation Area in Jambi province; the Senepis Buluhala Tiger Sanctuary in Riau and the Kutai Orangutan Conservation Group in East Kalimantan.

Sustainable Forest Management (SFM) could be implemented in forest conservation for instance in Arboretum area in lowland forest in 1990s and manages about 173 ha in Rasau Kuning Minas sub-district, Siak district. The aims of the arboretum are: (1) conservation of flora and fauna species; (2) to maintain elephant habitat and breeding; (3) to maintain the watershed system; (4) to keep exotic local species and (5) to develop medicinal plants (interview with SMF officer, May 2, 2010). In addition, the arboretum implements joint research studies with universities, research institutes and the forest department as an educational, research and recreational center. According to an arboretum officer, "we have built various nature trails with most of the trees labeled with identification tags both in local and botanical names. In terms of flora, there are 135 species of flowering plants and 16 species of palm found here." The endemic flora species growing in this area are: *Shorea sumatrana* (Meranti balau), *Shorea leprosula* (Meranti pirang), *Shorea parvifolia* (Meranti bunga) (Fig. 4.4), *Dipterocarpus crintus* (Keruing Bulu), *Dryobalanops oblongfolia* (Kapur), *Stryax benzoin* (Kemenyan), *Aquilaria* species (Gaharu), *Dyera costulata* (Jelutung), *Macaranga* spp. (Mahang), *Endospermum malaccense* [Sondok-Sendok (Fig. 4.5)], *Durio carinatus* (Durian hutan),

Fig. 4.4 *Shorea leprosula* (Meranti Pirang) in Arboretum's collection

Fig. 4.5 *Endospermum malacensis* (Sendok-Sendok tree) in Arboretum

Kompasia spp. (Kempas), *Ganosty macophyllus* (Ramin) and *Tetramerista glabra* (Punak). The arboretum also holds 26 species of animals comprising of mammals, reptiles, birds and fish. The large endangered mammals found are *Elephas maximus sumatranus* (Gajah Sumatera), *Helarctos malayanus* (Beruang Madu), *Maccaca* sp (Kera ekor panjang), *Hylobates agilis* (Siamang), and *Sus scrova* (Babi hutan) and *Tupaia splendidula* (Tupai). Birds categorized as endangered are: *Bucheros rhinoceros* (Rangkong), *Ichthyophaga ichthyaetus* (Elang), *Alcedo* (Raja Udang), *Gracula* R (Burung Beo), *Dicrurus* A (Srigunting) and *Nectarinia* (Burung Madu). Endangered reptilian species such as *Python reticulatus* (Ular Sawah/Phyton), *Chendopyton leichardii* (Ular Hijau), *Naja* sp (Ular Kobra), *Salvator veranus* (Biawak) and *Chitra indica* (Labi-Labi/Bulus) can also be found in this area.

The **Giam Siak Kecil-Bukit Batu (GSK-BB) Biosphere reserve was established** in 2009. The core reserve zone of 179,000 ha consists of natural forest and a peat lake, designated as a Biosphere Reserve by UNESCO in 2009. Some core stakeholders are the *Forestry Agencies* (Province and District Level), the *Lembaga Ilmu Pengetahuan Indonesia* (LIPI/Indonesian Institute of Sciences), *Balai Besar Konservasi Sumber Daya Alam* (BKSDA/the Natural Resources Conservation Agency (Department of Forestry), the Center for Biodiversity Conservation (University of Riau) and APP and SMF (APP 2007). The GSK-BB Biosphere reserve encompasses two wild life reserves of nearly 100,000 ha and SMF, together with its partners, has contributed more than 72,000 ha from its production forest concessions in Riau. These locations will serve as an ecological corridor between two wildlife reserves. In 2007, Arara Abadi Company, in cooperation with LIPI, funded a US$24,500 biodiversity study to create baseline data for an initial management plan. Arara Abadi also funded a US$34,000 socio-economic study, also conducted by LIPI during 2007. According to research findings there are at least 195 species of plants (of which 173 are species of woody plants), 162 species of moths, 150 species of birds, 30 species of fish, 10 species of mammals and 8 species of reptiles. SMF, in 2010, issued a budget of IDR500 billion for the management of this reserve and is working in partnership with other stakeholders, especially the Riau provincial governor, to develop landscape level management and conserve biodiversity (Interview with SMF Officer, February 11, 2010).

Taman Raja Conservation Area in Jambi Province is estimated to be more than 16,400 ha, but excludes about 6,800 ha inhabited by several local communities and 3600 ha already deforested by illegal logging prior to the granting of the concession. The remaining 6000 ha of natural forest is a preserve for biodiversity.

SMF invited other stakeholders with collaborative agreements, including APP, the Sumatran Tiger Foundation, the Community Alliance for Pulp and Paper Advocacy (CAPPA), the Forum Komunikasi daerah (FKD) and The Natural Resources Conservation Agency (BKSD). SMF provided US$17,059 during 2007 to develop management planning frameworks and for a Social and Biological Survey of the proposed reserve. The list of IUCN endangered species found in this area include: the Malayan Sun Bear (*H. malayanus*), Clouded Leopard (*Neofelis nebulosa*), Malayan Pangolin (*Manis javanica*), Pig tailed Macaque (*Macaca nemestrina*), Hornbill Rhinoceros (*Buceros rhinoceros*), Great Hornbill (*Buceros bicornis*) and the Great Agus (*Argusianus argus*) (APP 2007).

Finally, **Senepis Buluhan Tiger Sanctuary,** with a total area of about 106,000 ha, is aimed at protecting wildlife in a production forest that still supports a viable population of the Sumatran Tiger. APP and SMF provided US$26,800 in 2007 for the development of an initial management plan, aimed at enhancing the remaining tiger habitat. An additional US$5100 for operational costs was provided by SMF for the working group, comprising of Yayasan Penyelamatan dan Konservasi Harimau Sumatra (Foundation of Sumatran Tiger Conservation and Protection), the Riau Province Forestry Service, BKSDA, the Forestry Service of Rokan Hilir and Dumai Regencies, the Wildlife Conservation Society-Indonesia, Arara Abadi Company, Diamond Raya Company and Suntara Gajapati (APP 2007). Jusuf Anwar, Ambassador of the Republic of Indonesia for Japan, on Independence Day celebrations on August 17 in Tokyo, said "Indonesia is firmly committed to their preservation through a mix of strategic policy, regulation and action such as relocating and rehabilitating tigers back into their natural habitat" (*The Japan Times*, August 17, 2010). The plan targets to preserve habitat requirements of the Sumatran Tiger, put forth practical actions that will be required to ensure the long-term viability of the current tiger population and identify risks and opportunities at the landscape level.

Research and Development (R&D)

The R&D division of Arara Abadi was established in 1988 in Perawang, Riau Province. The mission of the R&D is to improve plantation productivity through development and application of leading technologies. The budget allocated to develop R&D was IDR 40 billion in the 1990s, declining to IDR 20 billion in the 2000s. The R&D division is comprised of 8 senior researchers, 44 researchers, 118 assistant researchers, 26 support staffs and 400 field workers (Interview, May 4, 2010). R&D encourages producing best quality seedlings in the following species: *A. mangium*, *A. crassicarpa*, A. hybrids, *E. pellita*, E. hybrids, *Acacia auriculiformis*, *A. aulacocarpa*, *Eucalyptus urophylla* and *Gmelina arborea* (Table 4.8). R&D also conducts research on the following local species: *Shorea* spp. (Meranti), *Palaquium* sp. (Nyatoh), *Calophyllum* sp. (Bintangur) and *Laphopetalum* sp. (Perupuk).

Table 4.8 Tree improvement program

No	Species	No of trials	No of clones
1	*A. mangium*	34	1113
2	*A. crassicarpa*	7	327
3	A. hybrid	4	120
4	*E. pellita*	12	169
5	*E. Urophylla*	3	28
6	E. hybrid	2	139
	Total	62	1896

Source http://www.pdf4free.com (*Google* Sinarmas forestry)

In 1993, the company carried out research in cloning to develop *E. pellita*, succeeding in creating the EP05. In 2000, most plantations have been substituted with EP05 (Fig. 4.6). The company attained property rights (*Certificate for Plant variety Protection*) for EP05 from the Department of Agriculture, Center for Plant Varieties Protection, on January 5, 2007 and subsequently developed 7 million seedlings from 2007 to 2009. 1666 EP05 are planted per hectare, reaching a size of 270 m^3 after 6 years. Meanwhile, *A. crassicarpa* grows to 150 m^3 with a density of 2000 seedlings per hectare. To produce 1 ton of pulp from EP05 we need 4.15 m^3 of timber, but *A. mangium* needs to supply 4.3 m^3.

Fig. 4.6 *Eucalyptus pellita* (EP05) in R&D location in Perawang

Fig. 4.7 EP 15147 tree species mostly develop by SMF in 2010

The R&D division is constantly trying to improve their seedlings, and have since produced EP 499, EP 15147 and EP 5193 (Fig. 4.7). These new seedlings have been trialed and performed well (Interview, May 4, 2010).

The R&D division collaborates with domestic and foreign research institutes in producing better seedlings, such as the Center for Research for Forest Biotechnology and Tree Improvement Research and Development, Jogyakarta, Bapedal (Environmental Assessment Board) in Riau and Jakarta and the Research and Development Center of Forest and Nature Conservation, Department of Forestry or Jaako Proyri (1991–1997), Beijing University (1994), Michigan Technology University (1995–1998), Taiwan Chung Hsing University (1997–2000) and the Chinese Academy of Forestry (1998). (http://www.pdf4free.com).

Local Farmers Response to Plantations

Partnership Model: Village Cooperatives

The existence of a company in a district should benefit the communities in employment opportunities and connection to markets. The company has a cooperative sector program (*Koperasi Desa*), such as the one called 'Bunut Abadi', which was established in 2002 by villagers of Bunut village, Pinang Sebatang, Tualang sub-district, Siak district. The Bunut Abadi Cooperative conducts a

Table 4.9 Realization of community development SMF 2009 in Siak district

Field activities	Siak	Pelawan	Bengkalis
Education	257,550,000	267,210,000	177,140,000
Economics	863,778,588	34,900,000	175,843,988
Socio-cultural	401,825,590	126,085,000	181,927,970
Religious	4,500,000	23,300,000	43,000,000
Infrastructure	15,640,000	687,337,069	142,973,583
Total	1,543,294,178	1,138,832,069	720,885,541

Source Sinarmas forestry, CD officer, 2009

Eucalyptus Nursery Program in partnership with Arara Abadi to provide *Eucalyptus* seedlings to the company (APP 2007: 112). The cooperative was established in this village because it directly borders a plantation area. SMF allocated about IDR 4.3 billion for its CD program in 2008 and IDR 3.9 billion in 2009. The CD program's budget of IDR 3.9 billion in 2009 covered 9 districts in Riau Province: Siak, Pelalawan, Bengkalis, Kampar, Dumai, Rokan Hilir, Rokan Hulu, Indragiri Hulu and Indragiri Hilir. The CD program works in five main areas: education, economy, social, religious activities and infrastructure. Siak District receives more of the budget allocation than any other district, receiving IDR 1.2 billion in 2008 and IDR 1.5 billion in 2009 (Table 4.9), because this district is where the mill, factory and staff accommodations are located. The cooperative of Bunut Abadi was categorized as an economic activity (Interview with Arara Abadi Officer, May 5, 2010).

The development of Bunut Abadi cooperative sector in cooperation with Arara Abadi was progress in 2004, to guarantee market and credit fund Rp. 30 million rupiah of Arara Abadi (Fig. 4.8) The Company eventually led to cooperate with Bunut Abadi cooperative to establish nursery project to produce quality seedlings such as *A. crassicarpa* reaches 80% and *A. mangium* reaches 20%.

According to an Arara Abadi nursery unit officer, acacia crasicarpa, produced by the Bunut Abadi cooperative, fully meets the quality standards set by the company and they hold a contract with the cooperative to supply seedlings. In 2005, they started with a monthly request of 100,000 plants, rising to 650,000 plants from 2006 to 2007 and again to 1,000,000 plants monthly from 2008 to 2009 (Interview with Bunut Abadi officer, May 4, 2010). The estimated costs per plant are IDR 175, if the company provides fertilizer and *polibags*,[12] but through a different scheme, the estimated cost per plant is IDR 225, where all fertilizer and polibags are provided by Bunut Abadi cooperative itself. According to the 2008 financial report, the cooperative gained IDR 39 million in profit, with total assets of IDR 600 million and sales (*income*) of plants of IDR 882 million (Fig. 4.9).

Cooperative membership in 2008 reached 183 persons, consisting of 123 males and 60 females. Members have to pay IDR 75,000 per person for membership and

[12]Polibags are used in developing nursery projects for holding individual seedlings.

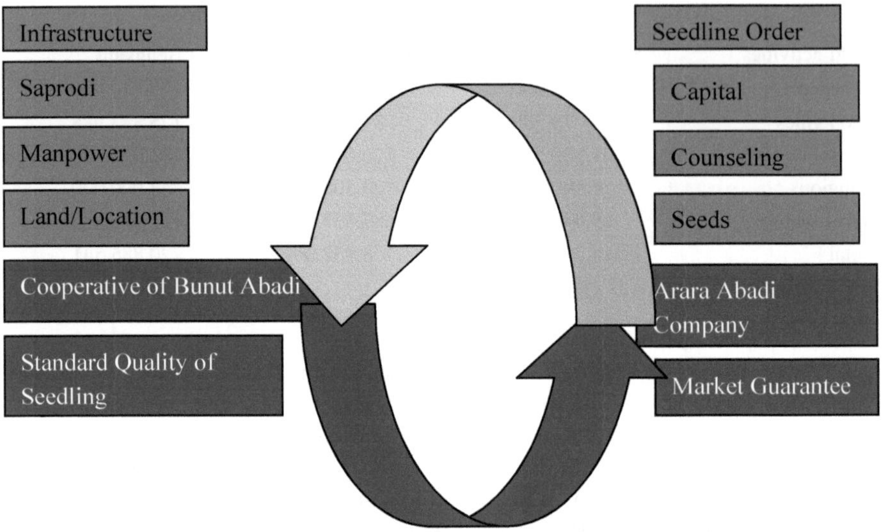

Fig. 4.8 Mutual cooperation and partnership between the cooperative and the company. *Source* Bunut Abadi cooperative booklet (2009)

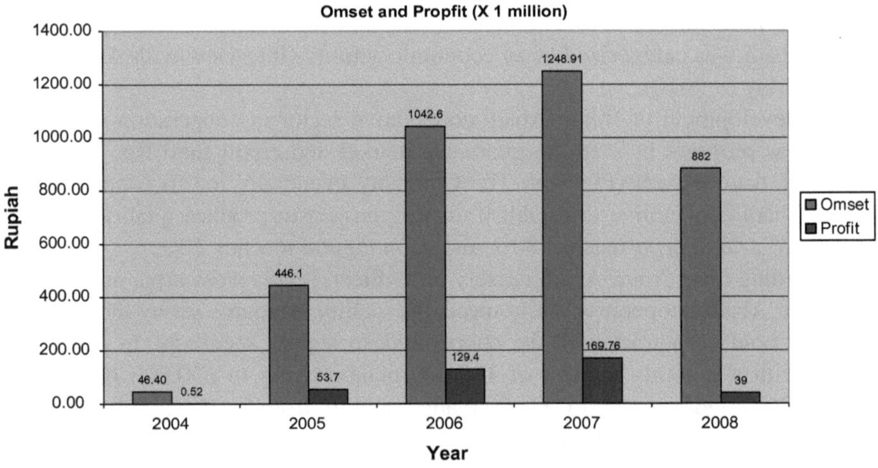

Fig. 4.9 Omset (Income) and profit of bunut abadi cooperative (X 1,000,000). *Source* Bunut Abadi cooperative (2009)

IDR 5000 a month. The cooperative has extended business to include a shop, transportation (2 small trucks) and drill machines for making fertilizer. Based on this performance, Arara Abadi expanded the capital to IDR 75 million, to be paid back within 2 years.

Job Creation

The cooperative provided jobs in rural communities, mostly for the unemployed in Pinang Sebatang village (10 persons in 2004, 51 in 2005, 82 in 2007, falling to 62 persons in 2008). Salaries are based on total achievement of work per oneunit plant —IDR 42–45, but usually they work in groups consisting of 10–15 persons. One group can collect 1000 plants a day, getting around IDR 42,000—IDR 45,000. The total monthly income averages IDR 1 million/per person (Fig. 4.10).

Achievements

Bunut Abadi cooperative officers and its members work hard in managing the cooperative with intensive guidance and counseling from Arara Abadi officers. The cooperative was deemed second best management performance in 2005 by the local Siak District and in 2006 and 2007, it was deemed best performer in Siak District. The Ministry of Cooperative and Small Business Holders in October 2009 selected the cooperative as winner for best performance, based on their accounting reports, in Samarinda, East Kalimantan.

Community Development on Economic Sector

In 2007, Arara Abadi established a Community Training Center in Pelalawan district. The Company provided US$61,207 for the construction of the training center and other facilities. The Center provides local people with training in cultivation of thirteen types of fruiting plants and trees, harvesting and marketing of food crops, as well as in aquaculture techniques. The Center's orchard largely produces quality seeds that are distributed to the 520 farmers living near Arara Abadi area (APP 2007: 112).

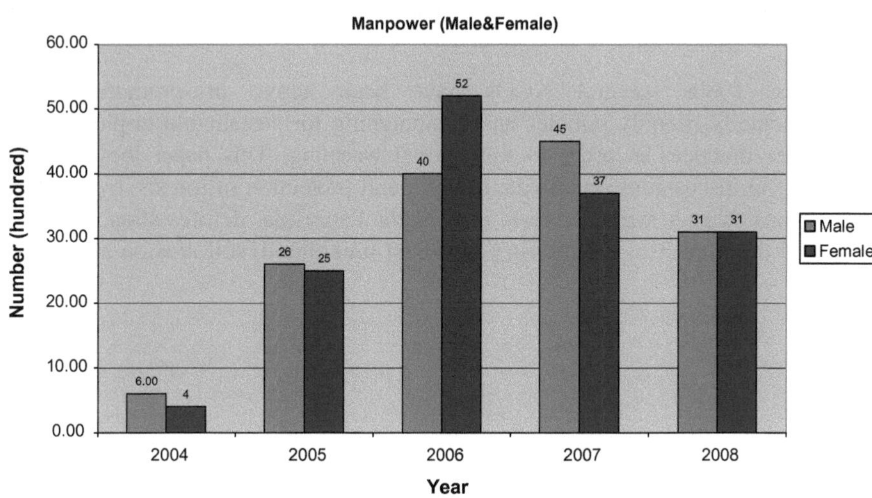

Fig. 4.10 Manpower absorbed by Bunut Abadi cooperative. *Source* Bunut Abadi cooperative (2009)

From Conflict to Partnership: Farmers and SMF

The issue began in 2000 with land conflict concerning about 854 ha in Kuala Gasib village, Tualang Sub-district, Siak District. Arara Abadi had received an HTI concession from the Ministry of Forestry No.743/Kpts-II/1996 on November 25, 1996, but the land was occupied by local farmers.[13] To overcome the land dispute, the company and farmers agreed to utilize the land as an industrial timber plantation (HTI), planted with *A. crassicarpa*, whereby farmers would receive compensation and production fees of about IDR 10,000/ton when harvested by the company. Arara Abadi initially provided IDR 40 million as requested by the head of Kuala Gasib village on December 3, 2002. The money would be repaid through the compensation and production fees paid each year to the village. After 7 years, in August 2009, Arara Abadi made the first payment of IDR 381,007,000 to the villagers of Kuala Gasib in a ceremony held in the Aston Hotel, Pekanbaru on August 25, 2009, attended by the head of Kuala Gasib village, the head of the informal village leaders, other villagers and SMF and Arara Abadi officers. Nazaruddin, as public affairs officer of Arara Abadi, said "a social partnership through close relations between the company and farmers is the best method of cooperation in the future". Meanwhile, Basri Hasan, head of Kuala Gasib village, commented that "a good relationship between the company and the people must be defended and improved, because a poor village such as Kuala Gasib needs the cooperation of the company to share in construction". The amount of money based on the calculation of tree production reached 38,100 tons from 515 ha at harvest. The remaining 178 ha was still not harvested yet, planned to be harvested in 2011 (Interview with Arara Abadi officer, May 5, 2010).

NGO Critiques Concerning Ecological Damages

Since the 1980s, several NGOs have been active in promoting more environmentally-friendly policies and campaigning for sustainable tropical forests and timber practices in order to stop global warming. This paper looks at how NGOs act in discussions on climate change, and protection of forests, by adopting case studies of peat swamp forests in Kampar Peninsula, deforestation, land disputes and the impact of the industry on forest fires, flood, soil erosion and more.

[13] For further information, see "Agreement letter between Arara Abadi Company with Kuala Gasib villagers, Tualang sub-district, Siak district about "Agreement to Overcome Land Occupation", Number 001/NKB/AA-KG/XII/2002.

Climate Change

Climate change is now widely accepted to be the greatest threat humanity has ever faced. Southeast Asia is one of the regions most vulnerable to the impacts of climate change, according to the *Asian Development Bank* (**ADB**). It warns that the poor—and especially women—are the most threatened. Approximately 2.2 billion Asians are subsistence farmers; they are already experiencing falling crop yields caused by floods, droughts, soil erosion and other factors. (http://www.greenpeace.org/international/press/reports/forests_for_climate_factsheet).

The ADB has highlighted the driving factors as: the burning of fossil fuels, such as oil, coal, and gas and forest fires are causing global warming, while *deforestation* is considered to contribute to about a fifth of global greenhouse gas emissions. At the United Nations Climate Summit in Copenhagen in December 2009, there was agreement among advanced and developing countries on how to significantly reduce greenhouse gas emissions from both fossil fuels and deforestation. The **REDD** (*Reduce Emissions from forest Destruction and Degradation*) concept was one of the options chosen to reduce gas emission and they agreed on master plan to end global *deforestation* before 2020. Advanced countries have committed to provide 30 billion Euros annually to a forest fund to boost reforestation and afforestation in developing countries with tropical forests, such as Brazil, the Democratic of Congo and Indonesia for them to guarantee to conduct forest conservation and protect biodiversity and indigenous peoples' rights.

From his speech on last the Copenhagen meeting, in December 2009, President Susilo Bambang Yudoyono (SBY) said that Indonesia was targeting a greenhouse gas emission reduction of 26% by 2020. Even with the fund from international donor institutions, Indonesia could improve this target to 41%.[14] The commitment from Indonesia exceeds the developed countries' commitment, raising the questions: how will it implement the 2020 target in annual activities and where does it start so that the commitment can be realized? A 'roadmap' must be designed, which contains strategic policy and procedures to guide this process, with measurable progress of **REDD** implementation in Indonesia.

Protecting Forests Stops Climate Change

According to FAO (2005), 'more than one million hectares of forest, mostly located in tropical rainforest, is destroyed every month—that is an area of forest size of a football pitch every two seconds' (www.fao.org/forestry/fra/fra2005/en/). Forests and their soils are huge carbon stores; they contain nearly 300 billion tons of carbon. That is 40 times more carbon than we currently emit to the atmosphere

[14]For further information, see *Warta: Forum Komunikasi Kehutanan Masyarakat* (Social Forestry Communication Forum), February edition, 2010, p. 14.

every year. Hence, deforestation and forests degradation drive 'climate change' in two ways: Firstly, land clearing and burning of forests release gas (carbon dioxide) into the atmosphere. Second, forests, which absorb carbon dioxide, decline. Indonesia was categorized as having 'one of the fastest rates of deforestation', emitting so much CO2 that Indonesia is the third largest climate polluter, after China and the USA.[15]

The Case of Land Clearing for HTI

Walhi criticizes natural forest conversion for industrial timber plantations (HTI) carried out by big companies, such as Sinarmas, RAPP and Musi Hutan Persada. According to Deddy Ratih, a Walhi officer, when land clearing for plantations occurs, forest soil becomes dry and there is a loss of biodiversity (interview, April 29, 2010). HTI and natural forests have totally different functions. The natural forest has a complex ecosystem, which monoculture plantations do not. For example, natural forests are self-regenerating: fertilizing the soil, conserving water and producing a microclimate, creating energy and sheltering fauna and flora. Therefore, sustainable forest management is an ecological balance within a natural forest, while HTI areas are one type of tree only and it is more appropriate to call them **"Kebun Kayu Komersial"** (*Commercial Tree Gardens*) than forests. HTI areas can have a negative impact on ecosystems, because they reduce soil fertility, consume much water and encourage loss of biodiversity in flora as well as fauna. However, the greatest difficulty, according to Walhi investigations, is that almost all HTI areas are located in previously production forests, which were intended for reforestation programs.

The Case of Peat Swamp Land

The largest contribution of CO_2 is from illegal logging in conservation and protected forests, land clearing in natural forests in *peat swamps* and conversion of forests to palm oil and HTI. For example, land clearing of around 13,000 ha of natural forest in a peat swamp in Kampar conducted by APP affiliated companies such as Bina Duta Laksana (PT BDL) and Mutiara Sabuk Khatulistiwa (PT MSK) in 2004. According to **Eyes on the Forest** (EoF),[16] natural forest clearance by both

---

[15]See WRI (2008).

[16]Actually, Eyes of the Forest (EoF) is a coalition on Environmental NGOs in Riau, Sumatra. It consists of: Friends of the Earth Riau Office, Jikalahari "Riau Forest Rescue Network", WWF-Indonesia and Riau Program. See Investigated Report EoF December 2009, published in April 2010.

APP affiliated companies in Kerumutan forest in Kampar District are legally questionable based upon the following laws and regulations:

(1) Forest was cleared on peat with a depth of more than 3 m deep which is not allowed to be converted into plantations;
(2) Natural forest with a dense canopy cover was cleared, which is not allowed to be converted into plantations;
(3) The majority of the concessions overlap with national Protected Areas and
(4) Some of the concessions overlap with provincial Protected Areas.[17]

BDL Company is one of 14 companies that the police carried out an investigation into in 2007–2008 for alleged involvement in the spread of illegal logging by the pulp and paper industry in Riau. The government set up an inter-department team and recommended that 14 companies should be processed thoroughly by the law. Strong statements by President Susilo Bambang Yudhoyono have been made, ordering the presidential *Judicial Mafia Eradication Task Force* to probe illegal logging practices, including that of BDL. Kampar peninsula registered only around 400,000 forests covered in 2007, which was 700,000 ha in 2002. NGOs and the government of Indonesia consider that Kampar peninsula is a conservation site and must be protected. A local NGO, **Jikalahari,** signed an MoU with the Siak and Pelawan District administrations at a side event of the Bali COP last year (http://jikalahari.org/index). Meanwhile, WWF included these forests in Sundaland Rivers and Swamps ecosystem of its Global 200 priority eco-regions and proposed to the Ministry of Forestry to protect Kampar (http://www.panda.org) (Interview with WWF officer, April 30, 2010). **WWF** findings indicate that tigers have rapidly declined to only number 60 in 2009, from an estimated 400 in 2002. Other NGOs such as Wetlands International and Bird Life International have designated it as an important bird area and identified Kampar as one of the highest priority areas for inclusion in the protected area network. According to **Greenpeace,** the government of Indonesia has strict laws to protect these carbon-rich peat areas in Kampar, but it fails to enforce the law and even continues to give permission to companies to destroy peat land.[18] Under Indonesian law, it is prohibited to develop or clear the forest and to drain any peat if it is deeper than 3 m.[19] While over 80% of Kampar peat is deeper than that, companies are still granted licenses for this area. Therefore, there is obvious collusion and corruption occurring on the issuing of concessions for peat lands between the Department of Forestry, head of the district and private companies. Only one person has been brought down for this: Azmun Jaafar, head of Pelalawan District, issued licenses for peat land to Putra Riau Perkasa Company

---

[17]Ibid.

[18]For further information see Greenpeace comments on “Indonesian’s Rainforests and the Climate Crisis”. Human Impact Report, Climate Change: Global Humanitarian Forum, Geneva. The Anatomy of a Silent Crisis, May 2009 (www.ghfgeneva.org/Portals/O/pdfs/human_impact_report.pdf).

[19]See Ministerial Decree 14/Permentan/PL No. 110/2/2009.

and was subsequently jailed by the Corruption Eradication Commission (KPK) due to allegations of issuing logging licenses against existing laws.[20]

Impact on Kampar Peat Ecosystem and Global Climate

Construction of new roads, canals on both sides, and all the natural forest clearance involved, has drained water from the peat soil and caused serious levels of CO2 emissions. According to EoF, efforts have not been made by APP's affiliates to minimize the destructive hydrological impacts of their canals. In GPS location 2, the investigators calculated the canal width to be around 6 m and estimated the depth to be 7 m. The water surface was around 2 m from the top of peat banks. This condition could cause enormous CO2 emissions from decomposition of dried peat.

The report by WWF-Indonesia and peat scientists concluded that average annual CO2 emissions from deforestation, forest degradation and associated peat decomposition and fires in Riau between 1990 and 2007 was equal to 122% of the Netherlands total CO2 and annual emissions, 58% that Australia, 39% that of the UK and 26% that of Germany. EoF has called on APP to stop further clearance of natural forest in the Kampar peninsula due to its negative impacts on *climate change*, threat to critically endangered Sumatran tigers and questionable legality of their activities. EoF has called on APP to immediately: (1) dismantle the road to prevent illegal logging, encroachment and poaching from entering the heart of Kampar; (2) dismantle the drainage canals to stop further draining of Kampar peat domes and (3) set aside the two concessions at the centre of Kampar fully for conservation.

Impact on Kampar Peat Land of Forest Fires

There is a strong correlation between land clearing on peat swamps in Kampar carried out by Sinarmas and forest fires. 51% of Riau Province consists of peat swamp land. Converting this peat swamp into usable land for palm oil and HTI, means creating canals that can affect the water content, drying the land out. This causes the land to be more prone to forest fires, which is, according to Sudirno, Vice Head of the Forestry Agency in Riau, "it is very difficult to extinguish, because fires could develop underground, even though on the surface, it looks like there is no fire'. The only tool to extinguish fires in peat swamp, according to Sudirno, is to flood all the peat swamp areas (*FKKM Riau*, No. 3, December 2006–March 2007: 6). During the massive fires of 1997/1998 over 3.3 million hectares of forest were destroyed in Indonesia, including parts of 17 protected forest areas. Riau was one of largest contributors of forest fires in Sumatra. Based on police investigations in

[20]*Tribun Pekanbaru*, published in Saturday 15 December 2007, p. 1.

1997, Arara Abadi, one of affiliation of wood supply of Indah Kiat pulp and paper factory, was named by the Forestry Department as one of 176 companies whose concessions affected by fire, based on monitoring from August 1 to September 15, 1997. As a result, the Ministry of Forestry temporarily suspended their timber concession on October 3 for failing to submit reports proving that they had not started fires in their area.[21]

According to the report of the Department of Forestry in 1999, Riau province suffered forest fires on 852 ha, declining to 422 ha in 2001 and growing again to 2211 ha in 2002, falling again to 750 ha in 2003[22], rising again to 7189 in 2004 and again to 15,476 fires in 2005 and 1419 in 2006 (*FKKM Riau, Op Cit*: 8).

Impact of Forest Fires on Physical Environment

The impact on the physical environment from forest fires is land, air and water damage. The discussion will highlight: (1) land damage because of forest fires. It is widely known that fires damage the characteristics of the land, causing loss of plants and biodiversity. If followed by rain, this causes soil erosion, creating infertile land; (2) air pollution, causing smog, which affects people's health (*asthma, bronchitis, pneumonia, eye irritation and skin,* etc.) and stopping airplane flights, business and school activities; (3) water preservation, through the elimination of plants that function as water holders, impacting on water flows, leading to soil erosion and floods.

Land Disputes

There are some land conflicts occurring between Arara Abadi and local communities. The most significant of these are in Bengkalis and Kampar and Pangkalankuas sub districts.

Arara Abadi initially attained a HTI concession of about 299,975 ha in 1991 in Bengkalis and Kampar districts, which includes Tarik Serai, Tasir Serai Timur, Melibur, Minas, Mandiangin, Pinang Sebatang Barat and Koto Garo villages. The company invested in these areas, created job absorption (15,000) and obtained foreign exchange earnings. However, the local people claim that part of the area belongs to customary forest rights and people withdrew from cutting a part of the acacia plantation in Tasik Serai and Tasik Serai Timur in protest, causing the company a loss of more than IDR 10 billion. Unfortunately, the government

---

[21]For further information see Friends of the Earth for the planet for people on Briefing "Asia Pulp & Paper" (www.foe.co.uk).

[22]The hotspot means where original forest is derived from. Forests in Kalimantan the hotspot originally the fires come from the coal mining under soil or the leaves of trees in the surface because of sun heating. These factors are natural and happen in summer time. But, they are also manmade; illegal logging caused fires, because the previous wood cutting burnt by local farmers.

responded very slowly to land conflict between the company and the farmers. According to Yuwilis, an Arara Abadi attorney, the company manages the conflicts in accordance with the law.[23]

Also, however, Arara Abadi, backed by state security forces, has seized land for the plantations from indigenous communities, such as 3000 ha of the Sakai people's forest gardens,[24] causing serious clashes with local villagers in Mandiangin, Betun, Pangkalankuas sub-district, in Riau Province on February 3, 2001.[25] Hundredsof club-wielding company militia attacked residents, seriously injuring nine and detaining sixty-three. These villagers were eventually released following representations by the Pelelawan community leaders and NGOs. The conflict was a result of the villagers blockading the road through the village leading to the pulp plant's feeder plantation. The blockade occurred as a result of community frustration at the damage being caused to the forest. Human Rights Watch requested donors, at their upcoming Bali meeting, to call for a complete and transparent audit of all military and police activities and the legality of steps taken by the Indonesian government to address tenure disputes on state forest land. The Indonesian government should appoint an independent land claims board or ombudsmen to deal with compensation disputes over seized forestry. Human Rights Watch also recommended that pulp and paper companies, such as APP, should establish and effectively enforce performance standards for both private and state security personnel, using the Voluntary Guidelines on Security and Human Rights developed by the U.S State Department and the British government as a foundation.

Concluding Remarks

The rapid development of pulp and paper companies in Indonesia has had a positive impact on economy of the country, where Indonesia is now the largest pulp and paper producer in Southeast Asia. Paper production reached 7.6 million tons in 2005,[26] reaching 9.2 million tons by 2009,[27] while pulp production went from 5.2 million tons in 2005 to 6.4 million tons in 2009. According to the Ministry of Industry and **APKI** (*The Indonesian Pulp and Paper Association*), the pulp and paper sector is definitely included among the country's top 20 commodities,

---

[23]For information, see 'Conflict between Arara Abadi and farmers, while Customary Rights forest Threaten investment" (http://srikat-tani-nasional.blogspot.com/2008/03.riau-konflik-ptAraraAbdi-masyarakat).

[24]Ibid.

[25]See "Indonesia: Paper Industry Threatens Human Rights" (http://docs.hrw.org/embargo/indon0103/index.htm).

[26]For further information, see *The Japanese Pulp and Paper Industry* (2005). It was published by Japan Pulp and paper Co., Ltd., p. 45.

[27]See the Statement of APKI (*The Indonesian Pulp and Paper Association*) officer, in the *Jakarta Post*, August 16, 2010.

generating from US$4.8 billion in exports in 2007 to US$5.3 billion in 2008. Indonesia's timber demand was estimated to have reached 33–40 million m^3 by the end of the 1990s, putting a severe strain on the natural forests, due to deforestation caused by over logging and illegal logging practices. The Ministry of Forestry initially provided easy access to **HTI** (*Industrial Timber Plantation*) concessions and economic incentives through access to banking in the 1980s. Unfortunately, the government lacked the capacity to actively engage individual farmers and households, and even the forestry regulations on timber concessions does not include individual farmers as actors, which would have allowed individual farmers to improve their socio-economic performance. Instead, the government and private sector developers have progressively planned and developed forest plantations as the backbone of agro-industries since the early 1990s. 1.7 million hectares of HTI in 1996 became 4.2 million hectares in 2001 and 8.7 million hectares in 2010, producing employment, but not own-employment.

The response of private companies and state forest enterprises (Inhutani) towards the incentive to replant through the 40% of shares payment for HTI holders was positive, but APP did not use government funds for replanting, preferring to obtain credit on the stock exchange or from private banks.

Through ensuring a continuous, guaranteed supply of timber (60% of mill demand), conducting research and development to improve the quality of seeds and cooperating with other companies to open up domestic and overseas markets, APP ensured a solid and growing business. In order to ensure these overseas markets, APP attained internationally recognized levels of certification, such as ISO 9001, ISO 14001, LEI, etc., in, concentrating on sustainability and environmental issues.

NGOs have criticized the ecological damage created by APP in Riau, such as the report by Walhi that there has been an ecological functional transformation affecting forest soil, becoming dry and losing biodiversity, due to the transformation from *multiculture* natural forests to *monocultures* (*A. mangium* or *Eucalyptus*). **Eyes on the Forest** (EoF) has also criticized the cutting of peat swamp land in Kampar district. This report is legally strong, as: (1) the company cleared natural forest on peat with a depth of more than 3 m deep, which is not allowed to be converted into plantation and (2) the majority of the concessions overlap with national protected areas.

Therefore, APP must be consistent in its stand on ecological issues, such as the conservation of species and forest biodiversity, consistently practicing sustainable forest management. APP should also ensure that it follows all the laws, especially concerning the cutting of *peat swamp forest* with a depth of more than 3 m.

Chapter 5
Vietnam

Review on Forest Planation Policy

Introduction

Forestry resources have traditionally been a significant mainstay of the livelihoods of many rural communities in Vietnam. Hence, a number of government programs have been introduced, encouraging communities to participate in a plantation forestry program. The government believes that 'forestry and agricultural products' can improve the livelihoods of rural communities, thereby increasing national food security in the future. However, in reality, the forest areas of Vietnam have declined sharply in the past century due to three interrelated reasons, namely: over exploitation for timber production; shifting cultivation practices among local communities in upland regions; and wood extraction for fuel. As a result, forest degradation has seriously affected water quality and supply and exacerbated flooding and soil erosion. The decrease in forest area has led to a reduction in the biodiversity of the country. A variety of animals and plants found naturally in forest ecosystems of Vietnam are now facing extinction.[1]

To overcome this deforestation, the government has taken two different approaches. Firstly, the Vietnam government and donor communities have established a partnership to promote 5MHRP (five million hectares program), carrying out reforestation and forest rehabilitation. The main objectives of the program are protecting the environment and biodiversity, producing raw material and creating employment, particularly in rural communities, to reduce poverty. Secondly, the Vietnam government, in cooperation with other stakeholders, is carrying out 'sustainable forestry' in production forests, to ensure a continued, sustainable supply of raw materials for forestry industries, such as pulp and paper, plywood, sawn timber, furniture and more. An FAO (2001a) report indicates that this

[1]Hieu (2004), p. 85.

H. Hidayat, *Sustainable Plantation Forestry*,
https://doi.org/10.1007/978-981-10-7653-4_5

program has established 600,000 ha of sustainable plantation forestry in 1990, increasing to 1.7 million ha in 2000.[2] The government shifted the land management policy from collectives through centralized management in the early 1980s to one of contracted farming through a free market system by providing 'timber concessions' to stakeholders. This has reduced the role of the state as the main actor and supported the emergence of other stakeholders, such as cooperatives, farm households, individual farmers and private companies. Recently the timber demand in Vietnam has increased, because of rapid development of pulp and paper, sawn timber, furniture and plywood industries.

Forestry contributed to 3% of GDP in 1990, fell to 1.4% in 1995, and the value of exports from forestry fell from 5% of total export value in 1990 to less than 3% in 1995. Currently, forestry contributes only about 4% of total agriculture value and 1.8% of GDP and provides jobs for 3.9% of the labor force (FAO 1999: 142). Most of the poor in rural communities derive their living from plantation forestry. The development of pulp and paper industries in the 1980s and 90s reinvigorated the industry, providing job creation and producing paper to fulfill the rising demand for paper in Vietnam, expanding to the export of paper, earning the country much needed foreign exchange earnings.

This paper aims to examine the role of stakeholders in the government's forestry policies, including the reforestation and rehabilitation program and timber concessions for stakeholders through a case study on land distribution and plantation forestry carried out by the local government in Phu Tho province. This paper looks at how private companies, household farms and the cooperative sector have responded to government policies on timber concessions and the pulp and paper industry.

The Significance of Forests and Other Resources

Generally, the *forestry sector* was categorized in the agricultural sector in Vietnam (Hieu 2004: 87). In the reform period from 1989 to 2001, Vietnam had high GDP growth of about 7.1% annually and the role of agriculture in creating employment in rural communities was significant. As shown in Table 5.1, in this period the growth rate of the agricultural sector remained stable at an average of 3–5%, while the growth rate in the industry sector was higher, at about 8–14%. From this stable and high growth in both sectors, the economic structure of Vietnam gradually shifted into 'industrialization', which is addressed via the reduction of the agricultural sector contribution to GDP from 33.1 to 23.6% in 2001.[3]

The main social implication from the reform period was a dramatic reduction in the poverty rate. The expenditure per capita below the poverty line fell from 70% in

---

[2]FAO (2001a).

[3]See GSO (General Statistic Organization) (2003), and also in Hieu (2004: 86).

Table 5.1 Annual growth rates of industrial and agricultural sector GDP and rural population share in Vietnam 1990–2001

Year	Industrial growth (%)	Agricultural growth (%)	Population (%)
1990	2.27	1.00	80.5
1991	7.71	2.18	80.5
1992	12.79	6.88	80.1
1993	12.62	3.28	80.0
1994	13.39	3.37	78.9
1995	13.60	4.80	80.0
1996	14.46	4.40	78.9
1997	12.62	4.33	77.3
1998	8.33	3.53	76.9
1999	7.68	5.23	76.4
2000	10.07	4.64	75.8
2001	10.32	2.79	75.2

Source Hieu (2004: 86)

the mid 1980s to 58% in 1993 and to 37% in 1998 (World Bank 2000). According to a World Bank report, Vietnam is one of the few countries that have succeeded in the twin objectives of increasing the growth rate and reducing poverty. However, over 75% of the population lives in rural areas (Table 5.1), and about three-quarters of the rural labor force work in the agriculture and forestry sector, of which about 25 million people rely on non-timber forest products for their livelihoods.[4] On the other hand, development of the industrial sector has mainly focused on low job-creatable fields; hence this sector is not strong enough to attract surplus workers from rural areas (CIEM 1999: 27). As a consequence, the slow movement of workers from rural to urban areas and declining contribution of agriculture to GDP have led to a widening of the income gap between urban and rural workers, and 90% of the poor now live in rural areas (FAO 1999: 160).

Administration's perspective

The *Ministry of Agriculture and Rural Development* (MARD) plays a significant role in managing forest resources, the agricultural sector and rural development in Vietnam. At the central level, MARD is responsible for forest resources sector administration, including managing land functions as production, protection and *special use* forests.[5] Plantation forestry, launched as a national policy to provide timber supply for the forestry industry, is located in production forests. The Forestry Department (DoF) and Forest Protection Department are MARD's agencies tasked with focusing on forest administration.

---

[4]See CIEM (Central Institute of Economic Management) (1999), and see in Hieu (2004: 87).

[5]The decree divides special-use forests into: (1) national parks; (2) natural reserves, sub-divided into natural reserves and fauna and flora habitat reserves, and (3) historical, cultural and environmental relics or landscape-protected areas, see Wil de Jong et al. (2006, p. 12).

The organizational structure of Vietnam's forest sector administration has four administrative levels: the central/national level, provincial level, district level and commune level. At present, Vietnam has 64 provinces, about 600 districts and 10,000 communes. All administrative levels are under the control of the state.

At the provincial level, two forest administration agencies are under the control of the Provincial People's Committee (PPC). The first of these is the Department of Agriculture and Rural Development (DARD), in which the Forestry Sub-Department operates as a specialized agency to assist the Director of DARD in forestry activities. At present, Vietnam has 34 Forestry Sub-Departments with a total 530 employees.

At the district level, the Economic Division for Agriculture and Rural Development is under the control of the District People's Committee (DPC) and employs one or two forestry staffs responsible for monitoring forestry activities.

At the commune level, as regulated by the Forest Protection and Development Act, communes with forest cover are obliged to recruit forest employees, while Forest Protection Units assign one forest ranger to work in one commune/commune group (Interview, March 5, 2010).

Trends in Forest cover

In 1943, Vietnam had 14.3 million hectares of natural forests, accounting for 43% of the country's area. Since that time, forest cover has decreased dramatically, especially during the 1976–1990 period. During that period, about 98,000 hectares were annually contracted for logging (Table 5.2). By 1973, the proportion had declined to 29, 27.2% in 1990, but increased again to 28% in 1995. The main causes of forest cover decline between 1943 and 1990s were: (1) land conversion for farming; Vietnam's accelerated population growth during much of the second half of the twentieth century and its persistent poverty levels and (2) devastation by war, including two anti-invasion wars, from 1945 to 1954 and 1961 to 1975. During these wars Vietnam lost nearly 2 million hectares of forest (Wil de Jong et al. 2006: 13). The establishment of plantation forestry, protected forests and special-use forests increased, causing the total forest area to first stabilize and then increase. As of 2004, Vietnam's forest cover had reached 12.3 million hectares, or 36.7% of the country's total area (Table 5.2). This phenomenon was appropriate with the data gathered by FAO (1990) that plantation forestry has rapidly developed in Vietnam from 661,000 ha in 1990 to 1.7 million hectares in 2000 (Evans and Turnbull 2004: 34).

By 2008, Vietnam had a forest area of 13,118,773 million hectares. It consisted of 10,348,591 ha of natural forests and 2,770,182 ha of plantation forestry. The main role of special-use forest is nature conservation, protection of historical and cultural relics, tourism and to some extent, environmental protection. It is important to maintain protected forests to protect water streams to prevent soil erosion and militate against natural disasters. On the other hand, *production forests* have the main aim of providing timber and non-timber forest products (NTFP), but also to provide environmental protection (Table 5.3).

Table 5.2 Vietnam's forest cover throughout different periods (1000 ha)

	1943	1976	1980	1985	1990	1995	2000	2004
Total area	14,300	11,169.3	10,608.3	9891.9	9175.6	9302.2	10,915.5	12,306.7
Natural forest		11,076.7	10,016.0	9308.3	8430.7	8252.5	9444.1	10,088.2
Plantation		92.6	422.3	583.3	744.9	1,047.7	1,471.3	2,218.5

Source Wil de Jong et al. (2006), pp. 13

Table 5.3 Forest land of Vietnam by its function (2008)

Forest classification	Area (ha)
Production forest	6,199,294
Protection forest	4,739,236
Special-use	2,061,675
Other outside three type uses	118,568
Total	13,118,773

Source Dien Tich Rung va Dat Lam Nghiep (Vietnam Forest Statistic), 2008, pp. 13

According to the draft National Forest Strategy 2020, the total area of land with forest cover is to be increased to 16.2 million hectares, consisting of 5.7 million hectares of protection forest, 2.3 million hectares of special-use forest and 8.2 million hectares of production forest.

Study Sites

Interviews among stakeholders, such as government officers, **RIPPI** (*Research Institute of Pulp and Paper Industry*) officers, Vietnam Pulp and Paper Association members, academics and NGOs were conducted in Hanoi. Meanwhile, the field study sites were located in Phu Tho province, Phu Ninh district (Fig. 5.1). In-depth interviews with informants, such as the Department of Forestry officer were conducted in Phu-Tho, Vinapaco, research institute of Vinapaco officers and with local farmers.

Timber Consumption Demand

The most rapid development of forestry industries occurred in the 1990s. Its impact on economic development was assisted by Vietnam joining the **WTO** (*World Trade Organization*). According to the Ministry of MARD on the Anniversary of 50 years of Forestry in Vietnam, export of wood products such as furniture, veneers, plywood and paper reached US$2.5 billion in 2009. A total of US$1.1 billion was produced from furniture and veneer industries, which accounts for over 42% of the country's export value of the wood processing industry (Nong Nghiep and Nong

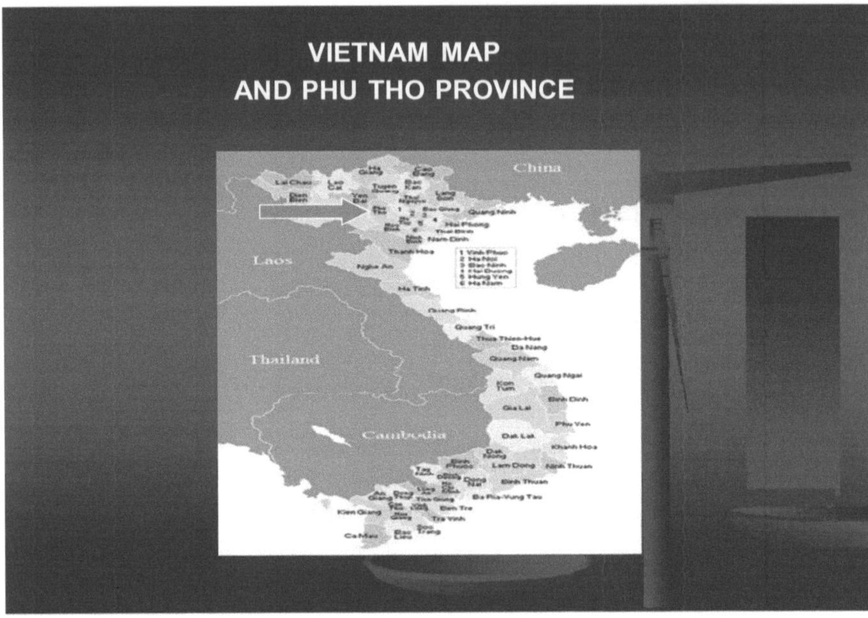

Fig. 5.1 Vietnam, where field work is located—Phu Tho province and Phu Ninh district

Thon 2009: 2).[6] For this reason, the forestry sector should be maintained and developed for future sustainable economic benefits. The impact of sustainable forest management was positive in shifting the livelihoods of local farmers, creating new jobs and attaining foreign exchange earnings.[7]

There is a strong correlation between the rise of timber demand with the growth of sustainable timber supply through rapid development in forestry industries, such as pulp and paper, plywood and sawn timber in Vietnam. From the 1960s to 1980s, Vietnam harvested about 2 million m^3 of timber annually for civil and industrial purposes, excluding firewood, rattan and bamboo. Paper consumption reached 1.7 million tons in 2010 and is expected to rise again to 2.6 million tons in 2015 and 4 million tons in 2020 (Table 5.4). As a result, timber consumption for paper products, pulpwood, plywood, sawmill, furniture, etc., will grow to about 3.8 million m^3 in 2010 and 5.5 million m^3 in 2015 (Table 5.5). As an illustration, Bai Bang Paper Company upgraded and expanded production to annual output of 100,000 tons of paper and 61,000 tons of pulp in 2006 and again to 250,000 tons of paper in 2007 (http://www.baibang/evn_td.htm).

[6]For further information "The Wood Processing Industry in Vietnam" (http://www.ambhanoi.um.dk/en/menu/CommercialServices/.

[7]The speech was delivered from Ministry of MARD in celebrating 50 years of Forestry in Vietnam.

Table 5.4 Forecasted paper consumption (1000 tons)

Products	2003	2010	2015	2020	Yearly growth (%)
Newspaper	54.8	92.8	133.4	192.0	8–9
Writing paper	159.9	295.2	451.0	690.6	9–11
Cardboard	680.1	1240.9	1880.9	2856.4	9–11
Others	75.8	138.3	209.6	318.4	9–11
Total	970.6	1,767.2	2,674.9	4,057.4	9–11

Source Wil de Jong et al. (2006: 20)

Table 5.5 Forecasted timber and forest product demands (1000 m^3)

Products	2003	2010	2015	2020
Timber: domestic & export	7420	14,004	18,620	22,160
Large timber for industry	4561	8030	10,266	11,993
Small wood for wood based	1649	2464	2992	1682
Panels				
Pulpwood	1150	3388	5271	8283
Pitwood	60	120	160	200
Export value of timber products	721	2400	3200	4000
And NTFPs (million USD)				
Wood products	567	2100	2600	3200
NTFPs	154	300	600	800

Source Wil de Jong et al. (2006: 20)

The timber for this industry originally came from both natural and plantation **forests**, but the government encouraged a shift towards plantations. The government provided easier procedures for obtaining timber concessions, easier access to credit from banks, and other financial institutions and tax reliefs for the import of machinery, upgrading *infrastructure* (port and high ways), the planting of trees, etc. (Interview, March 5, 2010).

Reform of Forestry Policy

The Vietnam government consolidated as one nation (North and South) at the end of the war with the USA in 1975. The communist single party system was adopted as state ideology. Prior to the 1980s, the characteristics of Vietnam's economy were essentially those of a centrally planned economy (CPE). The economy was heavily distorted in resources allocation with poor incentives and restricted information flows. As a consequence, the economy suffered from persistent shortages with low levels of per capita consumption and inefficiency of investments (Taylor 2004: 64). The Sixth Party Congress in December 1986 was a starting point in a shift in Vietnam's economic policies. The government recognized the existence and the essential role of a multi- ownership structure in Vietnam's economy and subsequently declared the approval of the **doi moi** (renovation) program by the Congress. In March 1989, Vietnam adopted a radical and comprehensive reform package aimed at stabilizing and opening the economy (Ibid: 65). The issue of collectivization and centralized management underwent modification at the end of the 1980s, with the introduction of a contract system, through Resolution No. 10, which officially ended the *collectivization period* and was followed by official acknowledgement of *a free-market system*. Forestry was viewed as a "source to be used for construction of the country, with its generation not being given adequate attention".[8] Changes made since 1990 have caused a transition in forestry, from a centrally planned sector with the state as the major actor, to a multi-sectored participatory sector. As a result, the state is no longer the main actor in plantation forestry, but other stakeholders have emerged: cooperatives, state forest enterprises, farm households and private companies. The government issued the 5MHRP (million hectares reforestation program) during 1998–2010. The program aims to protect the environment, provide jobs and reduce poverty for rural people, while increasing the wood supply for industry and households (Ohison et al. 2005: 253). The program has three main purposes (ICARD 2003):

(a) Increasing the proportion of forest area nationally to 43%, assuring environmental security for soil and water, as well as preserving biodiversity;
(b) To supply domestic demand for firewood, providing materials for paper production and artificial board production, timber quantities for export, promoting forestry to be an important sector in the national economy; and facilitating socio-economic progress in mountainous areas;
(c) Creating 2 million more permanent jobs, increasing the incomes of people living in forests, helping them to eliminate hunger and reduce poverty and stabilizing politics, society and national defense, especially in the mountainous and border areas.

The government also issued a forestry policy to maintain forest resources and a tree- planting program to increase timber resources. Being aware of the necessities of natural forests, the government of Vietnam implemented a range of policies to highlight the role of forests, reduce exploitation, promote tree planting in production forests and restore/rehabilitate natural forests, especially protected forests. Of the policies created since 1990, about 150 important policies related to forestry have been created in the following ways:

1. The Forest Law, issued by the National Assembly and adopted by the government
2. Decisions by the prime minister
3. Regulations and circulars
4. Ministerial directives

[8]See Ohison et.al. (2005), p. 248.

As a result, the following are the main forestry policies:

(1) The 1991 Law on Forest Protection and Development of Forests (reviewed and amended),
(2) The Land Law (1993), reviewed and amended in 1998–2000,
(3) Government Resolution 01/CP (1995), on the allocation and contracting of land for agriculture and forestry to state enterprises,
(4) Prime Ministerial Decision 661/QD-Ttg (1998), on the objectives, tasks, policies and organizations for the establishment of five million hectares of new forest,
(5) Government Decree No. 163/1999/ND-CP, concerning the allocation and lease of forestland to organizations, households, and individuals for long-term forestry purposes,
(6) Prime Ministerial Decision No. 08/2001/QD-TTg (2001), which issued regulations on management, rules for special-use forests, protection forests, and production forests,
(7) Prime Ministerial Decision No.178/2001/QD-TTg (2001), on the rights and obligations of households and individuals that are allocated and contracted forest and forest land for benefit-sharing,
(8) Other policies on development investment credit, taxes and extensions related to forestry activities.[9]

The major trends in forestry policies have been focused on three main objectives:

1. To invite stakeholders to develop social forestry as an integral part of establishing socio-economic entities and conservation of protected and special forest use;
2. To emphasize a transition in the forestry sector from harvesting natural resources, especially timber, to sustainably managing and utilizing forests;
3. To address the need for a market-oriented economy and foreign exchange earnings, create employment and improve the socio-economy of rural communities.

Land Tenure

In Vietnam all forests are owned by the state under the management of the *Ministry of Agriculture and Rural Development* (**MARD**). The *Department of Forestry* (DoF), which is under the MARD, has the authority to manage and issue regulations regarding forests. It emphasizes in articles 22, 24 and 26, Decree number 108/2006 ND-CP on 22/9/2006, detailed regulations and guidelines for the implementation of the law on investment. Forest rent is stipulated in Decree number 23/

[9]For further information see Sam and Trung (2003).

2006/NS-CP, dated 03/3/2006, concerning implementation of the law on protection and forest development in 2004 (interview with DoF officer, March 5, 2010). The DoF is to allocate production forests through timber concessions for three main actors: namely private companies, state enterprises and individuals or households (under cooperatives). The government provides incentives to the private sector in the form of easier procedures for obtaining timber concessions, easier access to financial institutions and tax holidays for the import of machineries for establishing the pulp industry. The government also provides subsidies to households and individual farmers for planting trees by proving seedlings and fertilizer, with the aim of actively engaging farmers in planting trees (Interview March 5, 2010). As a result, the programs eventually lead to a boost in economic development and job creation in rural communities.

Generally, land tenure for the private sector is provided for 50 years, with a possibility of extension based on the stakeholder's performance. The size of land that private companies could obtain ranges from 5000 to 10,000 ha; for households from 10 to 30 ha and for individual farmers 2–5 ha. The regulation of timber concession sizes, monitoring and penalties are managed by the DoF at the provincial level, because management was *decentralized* (Interview March, 5, 2010).

According to a MARD report, nearly 8.8 million hectares of forest (50%) have been allocated to various targets, of which 2.61 million hectares (29.7% of the area allocated) have been allocated to 450,000 households. An additional 50,000 households have received a total of over 1.86 million hectares of forest on contract for *State Forest Enterprises, protection and special-use forest management*, as well as foreign companies or joint venture companies (Sam and Trung 2003: 161).

Benefit-sharing Policy

The model used for encouraging rural communities to manage and develop protected and special-use forests is called "the benefit-sharing policy". In 2001, the prime minister issued two significant decisions on the management of natural forests and benefit-sharing covering: (1) the rights of households who sign contracts for and invest in protection forests. This regulation highlights that households have an obligation to plant, protect and regenerate forests, in accordance with the contracts signed with the forest management boards. As a reward, households have the right to collect fuel wood and non-wood forest products under the forest canopy (20% in timber forests; 30% in bamboo forests) and 85 to 90% of harvested products, after tax) and (2) special use-forests. The forest owners (households or individuals) are allowed to conduct harvesting, research and ecological tourism activities, in accordance with the laws and regulations.

In brief, the benefit sharing policy deals with protected, special-use and watershed forests. It is significant in supporting forest conservation and improving the socio-economic conditions of rural communities in Vietnam. Hence, the response from people, especially households, has been positive and they have actively engaged with the program.

Social Forestry Development

After the allocation of forest land, cooperative relationships in forest production have arisen in many districts. The implication for the social participation of communities has been great. Many households took the initiative and invested capital and labor to organize production activities, such as tree planting, protecting and practicing agroforestry. Below are some examples of what activities households have participated in:

(a) In Yen Bai province, 9500 farms combining forestry and agriculture have been established, representing 11.9% of the total farming households in the province;
(b) In Lao Cai province, 1500 farms producing annual revenues of 15 to 20 million VND each (US$1200 to $1600).[10] Every farm has an average of three hectares of fruit trees or industrial crops;
(c) In Thanh Hoa province, over 13,000 households are engaged in forest gardens and forest farm production. The area of each farm is about five to ten hectares, with annual incomes of 5 to 20 million VND per farm;
(d) Many households that were allocated land that included denuded hillsides, depleted natural forests, have planted trees and are able to produce enough firewood and timber for their own use. This effort has made significant contributions to forest conservation, protecting biodiversity and the environment, soil improvement and much more (Ibid).

Case Study of DoF in Phuto-Tho Province

As mentioned previously, the technical operation in terms of implementation of regulations and monitoring of land tenure depends on the DoF at the provincial level. Therefore, Phu Tho province was chosen to carry out field research. Phu Tho province has an integrated allocation of households' model for the establishment of *cooperatives* for planting trees and State Forest Enterprises using contracts with farmers and households, such as Vinapaco and Bapaco. The annual budget of the government office covers 1 billion VND to manage forests, plant trees and conduct monitoring (Interview, March 5, 2010). The amount of forest registered is 353,261 ha in Phu Tho province.[11] The DoF in this province has given 195,000 ha for timber concessions to households and 144,000 ha to private companies. In the case of the State Forest Enterprises, represented by BAPACO, it has received 60,000 ha for plantations. The area is divided into 50,000 ha where BAPACO cooperates with

[10]One U.S dollar is equivalent to 12,000–13,000 VND in 2003. But in March, 2010 rapidly developed that one U.S dollar is reached to 19,000 VND.

[11]See *Dien Tich Rung Va Dat Lam Nghiep*-Nam 2008 (Forest Area and Forestry in Vietnam in 2008), that forest area in Phu Tho Province is categorized very limited compared with other East North provinces such as Cao Bang 672,462 ha and Lang Son 830,348 ha.

households and has established cooperatives, while the remaining 10,000 ha is owned by VINAPACO[12] for planting trees.

Under the 661 projects policy, the government is obliged to implement *sustainable forest management*. The national budget to cover 661 projects is 20 billion VND, with a target of planting 10 million hectares of trees located in production, protected and special use forests. In the case of Phu Tho, sustainable forest management is being implemented with the planting of 17,000 ha of trees in special-use forests and 33,000 in protected forests. The purpose of planting trees in protected forests is aimed at forest conservation, reducing soil erosion and protecting biodiversity. The districts within Phu Tho province have the responsibility to monitor plantations, while the province levels functions as a general manager.

What about economic incentive?
The government provides credit for stakeholders to plant trees as forest resources are considered *strategic* commodities to be exported as paper products. In the case of 195,000 ha of planted trees, Vinapaco Company provides credit, seedlings, fertilizer, technical assistance, training and a guaranteed market for timber products to households and farmers. The Company issued 10–15 million VND per hectare in the form of seedlings, fertilizer and maintenance until harvesting after 8 years. Tree planting started in 2008 in Phu Tho and will be harvested after 8 years (in 2016).

Eucalyptus Plantation Development

The Vietnamese forestry sector has never in its history witnessed such an active participation of the people at large (farmers, the cooperative sector, private companies, etc.), as during these last 10 years (1990s) in *Eucalyptus* plantations. The driving factors of this are: (1) Vietnam launched a strategy to recover denuded hill areas for planting trees and forest conservation. The effort in tree planting in the Northern Provinces started in 1975; (2) the government has enabled foreign investors to actively engage in plantation forestry to provide timber for forestry industries and conservation as well. This policy brought about a positive response from foreign companies and NGOs to conduct tree planting and (3) economic reform in Vietnam shifted from centrally planned systems to mechanisms oriented to a market economy (liberal trade). The main target in the forestry sector has been the application of a policy of land and forest allocation for farmers, households (the cooperative sector) and private companies. This led to encouragement of stakeholders to aim for a better socio-economic life in the future.[13]

Since their development in 1985, Eucalyptus plantations have been developing nation- wide from north to south under two methods of establishment: First is

[12]VINAPACO is also State Forest Enterprise major task is providing for planting trees for Bai Bang Paper Company.
[13]See Tran (2005).

Fig. 5.2 Profile of Vinapaco Company in Phu Ninh district, Phu Tho province. The photo was taken in March, 2010 as personnel achieve

concentrated industrial plantations, which have been carried out by private companies and State Forest Enterprises. Second is scattered tree planting practices, which have been conducted by households, farmers, people's leagues, etc.

The areas planted with Eucalyptus have become larger than any planted with other tree species, such as *Manglietia glauca, Pinus spp, Styrax tonkinensis, Tectora grandis*, etc. According to a DoF officer, most local tree species are prioritized for planting in protected forests because they absorb much water and it is better for conservation and protection from soil erosion. Meanwhile, Eucalyptus trees, as fast growing trees, are planted in production forests to provide timber for the pulp and paper industry (interview March 9, 2010).

It seems that Eucalyptus plantations have been playing a significant role in providing timber for the pulp and paper industry, with about 77,291 ha planted over the period 1986–1990 (Fig. 5.2). As an illustration, for the period 1986–1990, 629,000 ha of plantations of all species were established; 50% by State Forest Enterprises and 50% by other entities such as private companies, farmers and households and 165,000 scattered trees (equivalent to 1,020,000 ha, counted on the basis of 2000 trees for 1 ha) planted. According to MARD, it is estimated that over 50–60% of the full plantation area was planted with Eucalyptus (300,000–400,000 ha) and 70–80% of the scattered tree areas were planted with Eucalyptus (700,000–800,000 ha) (Tran 2005: 6). Data on Eucalyptus plantation areas established by state-owned enterprises during the period 1986–1990 are outlined in Table 5.6.

Table 5.6 Eucalyptus plantation area 1986–1990 (Area-ha)

Species	1986	1987	1988	1989	1990	Total
All tree species	99,280	64,477	63,262	35,637	40,830	303,486
Eucalyptus	18,084	7,058	15,045	17,104	20,000	77,291

Source 30 years (1961–1990) of Forestry Development; Department of Forestry (DoF), 1991

The table above shows that State Forest Enterprises are categorized as the lowest in the planting of eucalyptus, as they are located in mountainous areas in which Eucalyptus is not possible. However, the trend is towards more eucalyptus plantings in State Forest Enterprises, such as in the central northern zone, coastal midlands of the north east, coastal hillsides of northern central Vietnam, coastal hillsides of central Vietnam, south- eastern Vietnam and south-western Vietnam.

Main species used

Some 30–40 eucalypt species have been introduced into Vietnam, but in almost all areas only one species, namely *Eucalyptus camaldulensis*, from Petford district (Australia) has been used for practical plantations. In its development, *E. tereticornis* has also been used for some small plantations in North Vietnam and more extensively in central and southern Vietnam. *E. urophilla* has shown good performance on some degraded hillsides of North Vietnam, as it was recommended by experts from the Forest Research Center (FRC) in Phu Tho to be most appropriate for infertile land (Interview, March 10, 2010). On the other hand, *E. exserta* is popular and used for fuelwood production by the people in the northern midlands.

According to Stahl (1991), a SIDA consultant: "The main weakness of the eucalypt plantation program in Vietnam is that only *E. camaldulensis* has been used and the seeds only come from the district of Petford, Australia". Meanwhile, Gray (1991), an FAO energy consultant, also said in his report on biomass energy for the "Forestry Sector Review" project: "The plantation program in Vietnam relies only on a few tree species and provenance, in particular on Eucalyptus" (Tran 2005: 7–8).

Which lands can be planted with *Eucalyptus urophylla* and *Acacia mangium*? Based on an investigation carried out by DAO Dinh Sam, former Director of Forest Science Institute of Vietnam, farmers should plant acacia mangium in good soil conditions and *E. urophylla* in degraded soil,[14] because acacia mangium has bigger leaves, with an open crown, therefore the sunlight can directly penetrate the soil, eventually affecting soil fertility in the future. Therefore, acacia mangium and acacia hybrids have only been planted in southern Vietnam. In contrast, *E. urophylla*, which has smaller leaves and the sunlight, does not affected soil negatively, then eventually the *hurmus* content in the soil will grow. This natural forest process will improve previously degraded soil, helping it to become fertile again.

[14]See Sam (2006).

The Response from Stakeholders

State Forest Enterprises

Vinapaco Company focuses on plantation forestry. The company is the second largest State Forest Enterprise (Fig. 5.1), after Bapaco Company. The main task of the company is to provide chip wood as raw material for Bai Bang Company. According to Vu Ngoc Pha, Manager of Silviculture for Vinapaco, the company created cooperation for the plantations with household farmers under the contract farming scheme in many districts. This scheme engages 30 households and the profit agreed by both parties is 50% for the household and 50% for Company at harvest time. The company also contracts independent farmers to purchase their timber, while the company's own plantations total 80,490 ha. The timber concessions are enough to last 50 years and could be extended again, based on the company's performance. The company pays tax of 4% for land use and the government provides special treatment for Vinapaco of only 6.9% interest per annum to the government bank in order be able to accumulate capital for planting trees in huge areas.

According to an officer of Vinapaco, there are some constraints to the development of the company:

(1) The limitation of production forest land in Vietnam. The Company cannot expand within Vietnam, but must expand in neighboring countries such as Laos and Cambodia;
(2) land dispute with local farmers. Farmers who had been occupying the land previously, although they had no certificate from the provincial DoF;
(3) contracting with households at the district level is sometimes difficult, because households sometimes use the land for agricultural purposes rather than plantations and
(4) lack of sufficient capital to expand plantations. Therefore, the relationship with Banka and other institutions such as the stock exchange is necessary to obtain fresh capital.

Vinapaco functioned as a *Forest Research Center* (**FRC**) in Phu Tho in order be able to produce better quality seedlings. The FRC cooperates with ACIAR (Australian Centre for Agricultural Research) and SIDA (Swedish International Development Agency) to develop quality seedlings, with funding and research. The research findings highlighted *E. urophylla* (PN2) and acacia mangium as the plants most suitable for the soil and climate in northern Vietnam (Fig. 5.3). Also, FRC developed and produced 3 million seeds of local trees, such as Sua, Cho Chi, Tram trang, Moraceae, Carabian, Mercury and Masrina. These local trees were ordered by the provincial DoF and distributed in protected forest areas for conservation aims. The production of commercial trees reached 4 million VND annually in 2009,

Fig. 5.3 *Eucalyptus urophylla* seeds in FRC. The photo was taken as personal archive, in March, 2010

growing again to 5.5 million seeds annually in 2011–2012. Seeds for the domestic market reached 75 and 25% were exported. The domestic market is mainly companies, households and farmers planting for commercial use and the 661 projects. The export market of seeds is largely for Taiwanese investors of forest plantations in Vietnam and Cambodia (50,000–60,000 acacia mangium seeds annually). The cost of seedlings is 700 VND/seed for *E. urophylla*, while the price for acacia mangium is 600/seed. Individual farmers order between 2000 and 3000 seeds, but households for the cooperative sector purchase between 15,000 and 25,000 seeds, while companies purchase on average between 200,000 and 500,000 seeds each year. (Interview, March 10, 2010).

The constraints faced by the Forest Research Center (FRC) (Fig. 5.4) are as follows:

(1) a lack of budget for new technology development research. So far, the annual budget of about 15 billion VND is insufficient;
(2) the attainment of intellectual property rights for new tree species findings, which creates difficulties in own-funding of activities through the sale of products in the market and
(3) an inability to expand international cooperation to improve seed quality (Ibid).

Fig. 5.4 FRC Office in Phu Ninh district, Phu Tho province The photo was taken in March 2010, as personal archive

Models of Partnership Company with Farmers

Company with Cooperative Sector

Vinapaco has been working with 32 cooperatives under a contract farming scheme in many districts. In one example of this, fourteen (14) households have established a cooperative with a total area of 34.3 ha, called Commute Tu Da, Phu Ninh District, Phu Tho Province. The initial planting of *E. urophylla* started in 2005. Ten Lem, the female head of the cooperative, owns approximately 2–4 ha of this land. The land was originally owned by a community group in the village and was infertile and very difficult for growing vegetables and cassava. The community group then shifted to growing trees by planting *E. urophylla* (Fig. 5.5). In the eight years it took for these trees to grow to harvestable size, (2013), farmers plant vegetables and rice on their, other, more fertile land, maintain livestock, such as cows and pigs, and fish in the river.

The Head of community group Tran Thi Thanh Lam approached Vinapaco Company and the reply from Vinapaco was positive to establish contract farming, providing seedlings, fertilizer and credit for households and guaranteeing the market for their products (Interview, March 8, 2010). According to Ten Lem, an informant, the planting fee for 34.3 ha was 96,562,801 VND and 64,859,105 VND for maintenance, coming to a total capital expense of about 161,421,905 VND.

What about production?

According to a Vinapaco monitoring officer, production reached 70 m^3/ha, coming to a total production of (70 m^3 $\times$ 34 ha) = 2380 m^3, which will be harvested in

Fig. 5.5 *Eucalyptus urophylla* planted by households in the formation of a *cooperative* in Tu Da village, Phu Ninh district, Pu Tho Province. The portrait is personal archive, taken in March, 2010

2013. The Vinapaco Company has guaranteed the market for harvested trees, which will be determined in terms of price per m^3 of trees based on the spot market; based on the estimation of tree quality and climate, the tree production will be fruitful and farmers will obtain a profit from this business.
The profit share, according to a cooperative member, will be reinvested into the next planting of trees and other socio-economic needs of members, such as education and health for their children and savings in the bank (Interview, March 8, 2010).

Independent Farmers
The Company has also cooperated with 70 independent farmers who own about 10–20 ha of land for tree plantations. Buo Vaw Thu is one of these independent farmers who owns 17 ha of land in Phu Ninh District and has been working for several years within the planning division of Vinapaco Company.

The land was previously in a critical condition, originally owned by local farmers, but appeared unproductive and infertile, even with cassava crops. When Bu Vaw bought the land five years ago (2005), the price was 20 million VND/hectare. He used about 30% of his savings, while the remaining 70% of the price was obtained from a commercial bank with interest of 1 percent/month. He says that the total capital he used was 400 million VND to buy the land and manage the planting of trees in 2005 and the credit was to be paid back within 10 years, but he is optimistic that the plantation will provide enough of a profit. He planted mostly acacia hybrid on his land in 2005 (Fig. 5.6), a little acacia mangium and bamboo as fencing. He was provided with the seedlings by the **Forest Research Center** (*FRC*) at a price of 600 VND/seed and bought fertilizer at 2800 VND/kg from the market. Buo Vaw hopes that his plants will be harvested after 8 years (2013) and he estimates that at harvesting time, he will produce 70 m^3/ha, totaling 17 ha × 70 m^3 = 1190 m^3 (Interview, March 8, 2010). He estimates the price of timber will be 620,000 VND/m^3, if the product is directly transported to the Company; however, based on information from a company officer, if the company has to buy the product on the spot (in the plantation site), the price will be reduced to 590,000 VND/m^3.

Buo Vaow has decided to sell his timber through directly transporting to the Company, so the price will be about 620,000 VND/m^3. The total amount he will make will be 1190 × 620,000 VND = 737,800,000 VND, earning a profit in 2013 of 737,800,000 − 400,000,000 = 337,800,000.

DoF in Phu Tho Working with Cooperatives

The role of the provincial DoF in protected, special-use and community forests is significant, because the DoF has an annual budget for conservation forests from the central government. The DoF has a program to actively engage households to plant trees through 48 cooperatives that manage households to actively participate in tree plantation in Phi Tho province. One of these, "cooperative X", in Thu Nich village, Phu Ninh district, manages 18 households that own around 0.75–1 ha each.

The 661 projects scheme obliges all households to plant trees to fulfill a target of 5 million hectares of new forests. The regulation requires the following of households involved in the scheme: (1) the management of land under scheme must be conducted by a cooperative, consisting of several households; (2) the size of land must be at least 1–2 hectares per household; (3) the DoF officer will check the status of the land in order to obtain the government subsidies, such as 1600/hectare of free seedlings and fertilizer.

The land registered for the project is 16 hectares and was initially planted in 2009 in Phu Nich village, Phu Ninh District. According to Nguyen Ngoc Than,[15] head of the cooperative, the program has had 3 positive impacts: (1) by planting fast growing trees in 2009, such as *E. urophylla*, they are preventing soil erosion and protecting biodiversity; (2) the farmers get the profit from timber harvests after

[15]Nguyen Ngoc Than was born in 1951 and is a retired Vietnam soldier from the 1960–1970s war with America. He was elected as head of the cooperative in Phu Nich because of his leadership qualities and wisdom.

Fig. 5.6 Acacia hybrid planted in critical land belong to independent farmer. The Photo is taken as individual archive, in March 2010

7 years, estimated to occur in 2016 and (3) it will improve the socio-economic condition of the local community in the future (Interview, March 9, 2010).

However, there are some constraints faced by households in developing tree plantations:

(1) There is no guaranteed market for the timber. Until now, no company has provided a guaranteed market for their timber. In this case, the role of the DoF as a mediator between households and companies such as Vinapaco and Bapaco is necessary;
(2) More seedlings and fertilizer for households are needed, requiring assistance from financial institutions, such as banks, aside from the DoF;
(3) They also require expert guidance for better performance of the plantation;
(4) The commitment of member households within the cooperative must be improved, following the regulations on attending regular meetings to solve their problems and innovate planting in the future.

Development of Pulp and Paper Industry

Problems: Challenges and Opportunities for Development (1970s–1990s)

Vietnam's first paper machine was established in 1912, with a capacity of 2500 tons annually (Le 1995: 57). In the 1970s, there were three large paper factories in North Vietnam: (1) Viet Tri, with a capacity of 10,000 tons per annum, constructed with the assistance of the Chinese government; (2) a 5000 tons per annum capacity mill and (3) a mill south of Hanoi producing wrapping paper (Jerve et al. 1999: 48).[16] There are now also three state owned pulp and paper mills in Vietnam: (1) Bai Bang (55,000 tons per annum) in Phu Tho province in the north Vietnam; (2) Dong Nai (20,000 tons per annum) and (3) Tan Mai (48,000 tons per annum), both in Dong Nai province in the south. In addition there are 100 small-scale pulp and paper mills around the country (Pesonen 1995:17).

Bai Bang Company

Bai Bang company was established in 1974 through SIDA[17] funding of US$170 million (World Wood 1974: 3). The project was the subject of huge debates in Sweden, especially during the 1970s and 1980s. In order to overcome the negative perception of this project, SIDA published an evaluation concluded that "Bai Bang has proved to be an example of a sustainable development cooperation project". In total, SIDA contributed US$1 billion to improve Bai Bang mills to producing 55,000 tons per annum. Swedish experts are still obliged to develop Bai Bang mills, because Vietnam does not have enough qualified technicians to conduct the mill operations and provide the necessary spare parts and chemicals (Sayer 1991: 239). The first paper machine was completed in December 1980, the second in March 1982 and the pulp mill in September 1982 (Hamilton 1989: 12). In April 2000, the mill employed 3500 workers. In December 2001, it was announced that a further expansion of the mill would be conducted, expanding the plant from a capacity of 55,000 tons of paper per annum to 100,000 tons. Annual pulp capacity would be increased from 48,000 to 61,000 tons. This was the first stage of a plan to increase the mill's annual paper capacity to 200,000 tons and pulp capacity to 150,000 tons (Vietnam Panorama www 1).

The products of Bai Bang Mill are mainly distributed to the domestic market and regional markets in Malaysia, Hongkong, Taiwan, Sri Lanka, etc. Vietnam is expected to meet the demands for foreign currency by exporting paper, and SIDA was actively engaged in promoting this export trade to Korea, Taiwan and Japan.[18]

---

[16]See also, Lang (2000).

[17]SIDA (Sweden' government Agency for giving Aid) to developing countries, include Vietnam.

[18]Cf Chris Lang, "Vietnam: Deforestation, reforestation and industrial plantations". *Loc Cit.*

Tan Mai Paper Company
In the early 1990s, Tai Mai Mill and Dong Nai Mill were established in Dong Nai Province in the South. Tan Mai company is a state-owned business belonging to Vinapimex with a mill paper capacity of 10,000 tons per annum in 1990 that was later expanded to 48,000 tons per annum in 2000. Newsprint production accounts for 50–60% of the company sales. The mill originally sourced pine from Lam Dong province, however, as there was not enough pine to supply the mill, machinery in the mill was adjusted to process eucalyptus. From 1999 to 2001, the Tan Mai company posted more than US$1.1 million in losses, as a result of its dependence on imported raw materials and the fall in the price of paper (Saigon Today 19 June 2001).

Vinapimex Paper Company
Vinapimex is a state-owned paper corporation, established at the end of the 1990s. It is categorized as the largest single producer of paper products, because it owns 11 factories, producing a total of about 170,000 tons a year. In September 2001, Vinapimex announced a plan to expand the pulp and paper industry in Vietnam, investing US$1.9 billion and involving 16 new pulp and paper production projects and an additional 693,000 ha of plantations, thereby increasing Vinapimex's annual paper production capacity from the current 170,000 to 419,000 tons. According to industry forecasts, demand for paper in Vietnam is predicted to increase by more than 10% each year. In 2010, demand is estimated at 1.25 million tons. Vinapimex estimates that US$3 billion needs to be invested in buying new machinery and plantations over the next 10 years, "to bring the industry up to scratch" (*Saigon Times* 30 July 2001); (Chris Lang 2000: 100). According to the *Saigon Times* (August 2001), each year the industry is short by 189,000 tons of pulp and recycled paper, and relies on imports to fill the gap. Domestic pulp and recycled paper prices are about three times world prices, a fact that the Saigon Times blames on Vietnam's small production capacity and non-automated mills with obsolete and decaying machinery (*Saigon Times* 29 August 2001).

Paper Industry (2000s)

Since 2004, Bai Bang Paper has upgraded and expanded production to annually 100,000 tons of paper and 61,000 tons of pulp at internationally competitive quality and environmentally improved to national standards. The Vietnam Paper Corporation, in November 2006, announced plans to invest almost US$300 million in expansion of the Bai Bang Company, while Vinapaco plans to establish a 250,000 tons per annum pulp production line at the Bai Bang site. A further US $100 million will be spent on "material forest zones" to expand industrial tree plantations, covering a total of 160,000 ha in five provinces.

In May 2006, An Hao Pulp Factory in Tuyen Quang Province was established, planning to produce 130,000 tons a year. Japan's Marubeni Corp won a US$130 million contract to build the mill, which is now in the second phase and the government has approved an area of 380,000 ha of forest to supply the mill. Another project, sponsored by the Saigon Export-Import Company, is an investment of US$150 million to build a 115,000 tons per annum pulp mill in Nui Thanh district, in the central province of Quang Nam. To supply the raw material for this mill, Quang Nam authorities have allocated 30,000 ha of land for acacia and eucalyptus plantations. Also in September 2008, another company, the Tan Mai Paper Company, received permission to build four new pulp and paper operations in Quang Ngai province: a paper mill in Dong Nai province, a pulp mill in Lam Dong Nai province, and pulp and paper mills in the central highlands of Vietnam. The projects will produce 550,000 tons of paper and 460,000 tons of pulp per year. Hence, the Tan Mai Paper company plans to establish 10,000 ha of plantations in Lam Dong province to feed its pulp and paper operations. The company is also carrying out a US$30 million plantation project in Di Linh district in Lam Dong Province.[19]

In January 2007, Hong Kong based Lee & Man Hau Giang Paper Manufacturing, plans to build a 330,000 tons per annum containerboard paper mill and a 150,000 tons per annum pulp mill in Vietnam. In line with the construction of paper mills, Lee & Man Hau Giang is also reported to be investing in plantation projects in Vietnam. Below are the investors in the pulp and paper industry (Table 5.7).

In brief, the government policies on providing easy access to banking, infrastructure, investment mechanisms and tax breaks is encouraging investors to invest in Vietnam. Therefore, there is a correlation between economic growth and the consumption of paper in Vietnam. Consumption in 2008 reached 2 million tons and increased to 2.2 million in 2009. Meanwhile paper production increased from 1114 million tons in 2008 to 1133 million tons in 2009 (Table 5.8). In general, paper production in 2009 increased only 1.74% compared to 2008 due to a reduction in newspaper production by 66% compared to 2008. In 2009, only 40% of the total capacity was utilized, although printing/writing paper production and consumption increased by 3%. Most producers sold their products at production price, breaking even, or at a loss, for many months from the fourth quarter to the early third quarter. On the other hand, packaging production moved up by 15%, as most packaging enterprises are small-scale, therefore, it is easy to restructure operations to match market conditions. Tissue paper production dropped 6.44% as the two tissue paper machines of New Toyo, with a combined capacity of 30,000 tons/year, were stopped for repairs, pulling the total output down by 15,000 tons. However, since May 2009, the newly installed tissue machines began operating again and have covered this downturn (Table 5.9).

[19]Cf Chris Lang "Vietnam: Paper Shortages, price increases, new mills and more plantations", (http://chrislang.org/2008). Also see Thanh Nien News, 28 June 2008, "Southern firm to build US $36 million pulp factory in Central Highlands".

Table 5.7 Private companies investing in pulp and paper projects and planned capacity (1000 tons/year)

Company	Pulp	Paper	Types of product	Year
1. Vietnam Paper Corp				
Bai Bang Project-the II Phase	250		Bleached chemical hardwood pulp	2011
Than Hoa Paper JS Co	100	100	BCTMP, DIP, newsprint, PW	2011
Paper machine upgrading		30	Printing/writing paper	2010
Bai Bang Paper JS Co		50	PW, newsprint, LWC	10/2009
2. Tan Mai Group Co				
Long Thanh Paper Mill		150	Newsprint	2010
Kon Tum Paper Mill	150	200	BCTMP, coated paper	2011, 2012
Quang Ngai Paper Mill	40	70	CTMP, PW, newsprint, packaging	2010
3. Lee & Man Hau Giang Co	330	420	BJKP; packaging	2011
4. An Hoa Paper Co	130, 60	130	BHKP, coated paper, BCTMP	2009, 2010
5. Quang Nam Pulp& Paper Co	100		BCTMP	2012
6. My Huong Paper JS Co		45	Packaging	
7. Sai Gon Paper JS Co		230	Packaging, tissue, coated paper	
8. Phuong Nam Co	100		BCTMP	2011
9. Viet Thang Paper Co		50	Coated paper	2009

Source Vietnam Pulp and Paper Association (2009)

Table 5.8 The comparison of paper consumption and production (tons)

Unit	2008	2009	2009/2008 (+/−) (%)
Production	1,114,416	1,133,831	1.74
Import	1,006,394	1,141,190	11.81
Export	117,000	60,000	−48.72
Consumption	2,003,810	2,215,021	9.54

Source Vietnam Pulp and Paper Association, 2009

Economists quote a government report saying that it will boost the paper industry in an effort to make it a key economic sector in the next decade in order to fulfill paper domestic consumption. The strategy will apply two approaches: inviting foreign investors to build pulp and paper mills in Vietnam, and direct investment by the government itself. In terms of foreign investors, examples are Thailand's Siam Cement, which is investing in building a US$142 million paper

Table 5.9 Paper production and consumption (Ton)

Unit	2008	2009	09/08 (+/−) (%)
Production	1,114,416	1,133,831	1.74
Newsprint	59,816	20,531	−65.68
Printing/writing paper	254,100	262,500	3.31
Kraft, packaging	642,300	736,000	14.59
Tissue	73,000	68,300	−6.44
Export joss paper	85,200	46,500	−45.42

Source Vietnam pulp and paper association, 2009

mill to produce 220,000 mt/year of packaging paper by 2009, as well as a joint venture at the An Hoa Pulp & Paper mill between Hanoi's General Export and Import Co and India's Ballarpur Industries; and Thailand's Martin Group is building a pulp mill in the northern province of Tuyen Quang worth US$200 million that will start operating in 2009. Eucalyptus and acacia plantations have already been developed to feed these mills.[20] In terms of direct investment, Vietnam will invest more than USD 1140 million in implementing 15 projects to produce pulp and paper, including USD 9 million to build a 15,000 ton tissue paper factory in Cau Doung.[21]

How to overcome paper consumption to production gaps?

By the end of 2009, there were indications that some big economies were showing signs of recovery. Vietnam's economy has overcome difficulties and seems to be recovering, with a GDP growth rate of more than 5.8%. An export, which was in a negative growth rate, is starting to move back towards a positive growth rate, even though foreign direct investments are still low. At the time of the global economic downturn, Vietnam's paper industry had restructured its operation and items. Investment in the paper industry remains continuous, products are sold out and large scale investment projects are ahead of schedule. However, according to Vietnam Pulp and Paper Association, the competitiveness of Vietnam paper products has not significantly improved, and the products are still of lower quality and higher cost than in other regional countries. As a result, Vietnam still needs to import pulp and paper products from neighbors, such as Indonesia and Thailand.

The government, represented by the Ministry of Finance, announced a policy to reduce the import tax on paper by between 7 and 12% in September 2008, depending on the type of paper.[22] Several newspapers reported that the tax cuts were a result of proposals by the Pulp and Paper Association; however, Vu Ngoc

[20]See "Vietnam Paper Industry: Flying under the radar", (http://www.glgroup.com/News/Vietnam-Paper-Industry).

[21]See "Paper Industry in Vietnam", News from Embassy of the Socialist Republic of Vietnam in the USA, February 4, 2001.

[22]See "Paper shortage leads to import tax cut", Vietnam News, 5 September 2008.

Bao, Secretary General of the Association, told the Vietnam News Agency that the "reduction would seriously affect local paper producers, who were having difficulties reducing production costs in face of rising material costs. Foreign giants such as Japan, China, the US and South Korea challenge the competitive capacity of local producers".[23]

Provision of Raw Material

Due to the effects of a rapid increase in woodchip exports in the past few years, as well as the appearance of new pulp and paper mills, treeless forest and hills are being covered by acacia and eucalyptus and the government has eased this growth with easy access for forest plantation concessions to private companies. This has greatly contributed to the rapid development of the pulp industry in Vietnam. As a result of the development of forestry industries, wood demand is increasing. Vietnamese hardwood imports have increased from 1 million m^3 annually in 2003 to almost 4 million m3 in 2008 and 2009.[24] At least 80% of this imported volume is re-exported as finished and semi-finished products such as furniture.

Factories must use FSC-certified wood and attain FSC chain of custody certification if their products are to carry the standard's logo. Because wood product buyers in Europe, the US and Japan have increasingly demanded FSC-certified wood products, so Vietnam has a relatively high proportion of FSC chain of custody certified factories. Bai Bang Company is eager to obtain FSC certification for its forest plantations through practicing sustainable forest management, under the guidance of SFMI.[25]

Another approach to increasing production is by recycling paper, accounting for 67% (2008) of the total amount of raw material used for paper production (except for joss paper, made by semi-chemical pulp), mainly used to produce containerboards, boxboards and tissue paper. 35% of waste paper is also used for newsprint. According to a Pulp and Paper Association officer, the recovery rate of used paper in Vietnam for recycling remains low, only 26%, accounting for 67% of total waste paper recycled (the remainder is imported) (Interview, March 5, 2010).

In February 2007, the MARD started a program to establish 2.4 million hectares of plantations over the next five years in the northern mountainous region. The plantations, according to Hua Duc Nhi, the Vice Minister of MARD, are intended to provide raw material for the pulp industry, which will annually produce 700,000 tons of pulp.[26] There is also the Five Million Hectares Reforestation Program

[23]See "Paper projects kick off to cool down shortage", Vietnam News, 11 September 2008.

[24]See "*Vietnam: A Forestry Investment Opportunity*", (*Google: Forestry industry in Vietnam*).

[25]SFMI is NGOs which actively involve in socialization to obtain FSC-certified forest. It is an abbreviation of Sustainable Forest Management and forest Certification under the guidance of Dr. Vu Nham, Former chief cooperation and scientific Department of Vietnam Forest University. Interview with Vu Nham, on March 4, 2010 in Hanoi.

[26]See "Vietnam: What is happening in the pulp and paper sector"? in World Rainforest Movement (http://www.wrm.org.uy/bulletin/115/Vietnam.Html).

(5MHRP), intending to plant one million hectares of industrial tree plantations to feed the pulp and paper industry.

Other actors also support the creation of plantations, such as the German government, which is establishing plantation projects in five northern provinces of Vietnam. The Asian Development Bank (ADB) has approved a US$45 million loan for an afforestation project in the central highlands, and the World Bank is funding a Forest Sector Development Project in four central coastal provinces. These projects intend to plant 66,000 ha of plantations.

Concluding Remarks

Plantation forestry in Vietnam has developed rapidly since the 1990s and 2000s. 600,000 ha of plantation forests in 1990 increased to 1.7 million hectares in 2000. This has had a positive impact on domestic and foreign investment, providing cash to boost growth in the domestic economy and strengthening the socio-economic capacity of farmers in district and rural areas. Domestic paper production achieved 900,000 m^3 tons in 2005, 1,114,416 m^3 tons in 2008 and finally 1,133,831 m^3 tons in 2009. The rationale for this success is as follows:

First, the Vietnam government launched an economic reform, shifting from a centrally planned system to a market oriented economy. The main issue in the forestry sector has been the application of a policy of "land distribution" and "timber concession" areas to stakeholders, such as households, individuals, the cooperative sector, private companies, state forest enterprises, etc.

The impact on the livelihoods of local farmers and income generation increased rapidly and the government now obtains annual foreign exchange earnings of US $2.5 billion from forestry industry products.

Second, the government has encouraged domestic and foreign investors to also be involved in a wide range of timber plantation activities and provides incentives, such as easy accessibility to timber concession areas, financial institutions, tax relief on import of machineries and more. This policy has attracted investors, cooperative sectors, farmers, NGOs, donor agencies from international banks, such as the World Bank, ADB and foreign agencies such as CIDA, SIDA, JICA, GTZ, UNSAID, CIFOR, CIAR, etc., to actively engage to tree planting in Vietnam. The impact of this policy is that today, eucalyptus urophilla and acacia mangium plantations have increased, with a positive response from stakeholders, participating in the planting of trees on a massive scale. In spite of criticism from NGOs concerning the ecological damage from a wide range of fast growing trees, such as acacia and eucalyptus, affecting soils, leading to floods and soil erosion during the rainy season, the government and other stakeholders have been positively engaged in tree planting in conservation and special use forests.

Part III
Japan

Chapter 6
Japan: The Initial and Development of Pulp and Paper Industry

Introduction

The paper industries are rapidly developed in Japan. Currently, Japan is placed at the third rank after US and China, which produced 30.8 million tons paper among major producers of pulp and paper in the world (2005). This phenomenon, pulp and paper is categorized number thirteen (13) among the largest manufacturing in Japan, which total amount reached Y 6.8 trillion and absorbed 34,839 employees. From this perspective, the role of government as a direct actor, especially the Meiji government is so significant to discuss. This government launched 'affirmative action' on political stability and economic development. In the case of economic development, Meiji launched policy by establishing facilities, such as banking, infrastructure, inviting foreign experts to build factories, and even sent scholars to study western civilization. This paper focuses on the role of government to launch policy and regulation for paper industries in Japan.

The Role of the Government

The role of the government in Japanese economic development has been regarded as very significant. Initially, the group of leaders who brought about the Meiji Restoration formed "the government," and laid the foundations for economic development at the same time the Restoration happened. Earlier successes provided the basis for subsequent economic policy. From this perspective, one might hold that economic development in Japan is wholly attributed to the strong guidance given by the government.

The first steps in modern economic development were possible when the country was united under the very strong and farsighted centralized Meiji government, which aimed at 'the enrichment and strength of the nation' through military and

H. Hidayat, *Sustainable Plantation Forestry*,
https://doi.org/10.1007/978-981-10-7653-4_6

economic development. Particularly important is the fact that the Meiji government was active not only in framing political and economic policies, but also in implementing social changes. The government maintained political stability, built a mass education network, an industrial order, and achieved political strength and empire. By the end of the Meiji Period (1868–1912), Japan was one of the world's great powers, the envy of some neighbors, and master of others.[1]

These achievements, the role of the government in Japanese political stability and economic development (industrialization) have been regarded as very significant. The government led the industrial development. They modernized the old factories set up by the shoguns and lords. They built paper industries (1872), railways from Tokyo to Yokohama in 1872, improved the roads, post office (1871), ships for a navy and merchant marine, Tokyo Electric Light (1886), Tokyo Gas (1885), textile mills before 1890, and later, iron and steel works, engineering plants and shipbuilding yards, especially in north Kyushu and in the Osaka and Tokyo areas (Dempster 1969: 175–176). In 1877, the government sponsored a great exhibition in Tokyo to spread knowledge of new farming and industrial methods. The government also sold the new factories cheaply and many passed to ex-samurai families, who gradually gained more and more influence as they built up trade, industry and banking concerns, and became the great financial groups, or Zaibatsu,[2] of which Mitsui, Mitsubishi, Sumitomo and Yasuda were the largest. Almost every industry in Japan today had its beginnings in those established by the Meiji Government.

A steady flow of capital was essential for the continuance of this rapid industrialization. Little came from abroad, but at home, although at first it came from taxes and samurai bonds, soon industry made a vital contribution from its profits. As they paid little to their workers, who were used to living frugally, most of the profits could be ploughed back into industry and the rate of capital formation was very high. Therefore, in many cases, private enterprises went along with government policies in building the infrastructure for industrialization, and then went their own sweet way once that foundation had been built. The fact that the Meiji government was dictatorial, but not 'totalitarian', implies that there was room for other economic forces to operate. Moreover, such freedom was admitted in a wide range of economic activities. This is one reason why the Japanese economy could enjoy some advantages associated with competitive markets without falling victim to monopoly capital or extreme protectionism. The importance of a government in economic development is usually judged by the great number of policies it has adopted. In Japan, there is considerable discrepancy between the declared policies of the government and the way in which those policies were in fact administered. However, it is generally believed that economic development of Japan can be explained largely by the role of the government.

---

[1]See Jansen (1968).

[2]Zaibatsu means the big traders that eventually became a large group of financial and private companies in the Meiji era, such as the Mitsubishi, Mitsui, Sumitomo, Yasuda, etc.

Besides, the role of group of Japanese modernizers such as Fukuzawa Yukichi, Shibusawa Eiichi, Tokugawa Yoshinobu, Okuma Shigenobu, Ito Hirobumi, Soejima Taneomi, Mori Arinori, etc., who contributed to the Meiji Restoration formed "the government," and the government laid the foundation of economic development at the same time the Restoration occurred.[3] Nakayama told that the Meiji government's policy on modernization of industrialization and military programs were very significant steps in the early of her government. First, the government sought to maintain the country's independence. Second, it sought to ensure that the social order was maintained domestically, and strove to promote political unification as the basis for economic activity. Third, it removed the legal restrictions on changing employment, and thoroughly reformed existing institutions. Fourth, it established a foundation for greater productive efficiency by making sweeping revisions in the legal, educational, tax, and currency systems (Nakayama 1975: 121–122). From this perspective, considering the role of the Meiji government, it was not necessarily totalitarian, despite its politically dictatorial character. This is of some importance for an overall evaluation of the government's role in Japan. The great emphasis placed on the role of the government Meiji in laying the foundations for development beside the response of the role private sector. However, these foundations alone did not sustain Japan's economic growth. There must have been other vital elements with which the government worked to promote further progress after the basic foundations had been laid. The major issue, then, is the relationship between the government and these other vital elements such as the role of private sectors.

Government and Private Sector

The success of the government in laying the basic groundwork during the early years of Meiji, along with the support which it enlisted from the public, was ultimately due to the close coordination of its efforts with those in the private sectors. This meant that the government had responsibility to the private sectors and the encouragement of creative private initiative, enterprise and implementing business decisions. Then, any assessment of government's contribution to economic progress necessitates a closer examination of the role played by private enterprise, and consideration of the relationship between the government and private sectors. Therefore, the variety of institutions through which the government either directly or indirectly influences the decisions of private sectors must be focused on. The tax system, tariff policy and general industrial policy, including the relative weight given to agriculture and the small or medium-sized enterprises all deserve mention (Nakayama 1975: 124–125).

---

[3]They were young boys who had brilliant intellects and were sent to abroad (United States and West Europe) to study Western sciences (industrialization) at the end of Tokugawa era and eventually became pioneers of 'modernization' in transition to the Meiji government; for further information, see Harootunian (1970).

One of the first steps of the Meiji leadership was the concentration of the major productive facilities in the hands of the government. So, the government forged ahead with national policies and production targets that were largely non-economic in nature. Its initial intentions were to firmly assert its leadership over the private economy. Its policies for manufacturing were similar to those pursued by the German government, which from the beginning sought to establish government control to some extent. The necessities of guidance by the state were accepted as a basic premise, primarily with an eye on developing the productive power of the nation. Obviously, from this basic premise emerged the close contact between the state and the large enterprises; the intimate relationship which later developed between the major banks and the industrial giants; and the appearance of various semi-official organizations which sought to promote cooperation and autonomous control in all private industries.

By way of contrast, it is interesting to note that the Meiji government made no attempt to interfere with or control the smaller enterprises. The only exceptions were in the silk and tea industries which provided the stable source of exports during the initial years. In these two industries, which were characterized by a vast number of small enterprises, the government also sought to set up a system of 'autonomous' control similar to those found among the large enterprises. In those cases, in terms of among the large firms, the government not only promoted cooperation and autonomous control, but even positively promoted business mergers in some industries which were in need of large amounts of capital and technological renovation. Thus, the relationship between the government and the large enterprises rapidly deepened. This in turn accelerated the emergence of a firmly implanted plutocracy in the long run, particularly the zaibatsu, which stood quite apart from the government so that the entrepreneur or respectable businessmen could not openly occupy the top position in the political arena. The entrepreneurs in the zaibatsu, which had close contact with the government, came from the samurai class in most cases. They shared with the political elite the same social standing and cultural background. Nonetheless, samurai class could not occupy authoritative political positions while still being entrepreneurs. This basic fact suggests that the relationship between the government and the big enterprises was much more narrowly defined than might have been expected from limited consideration of the government's political objectives alone. In this case, the scope for independent initiative by those in private industry, as well as their ability to compete among themselves and with the government, had not been greatly limited (Nakayama 1975: 126–127).

In many cases, private enterprises went along with government policies in building the infrastructure for industrialization, and then went their own sweet way once that foundation had been built. They were, in order words, quite opportunistic. Thus, while the controls of the government were considerable, in fact only a few industries were subject to the direct control of the government, at least on paper, because it was very strategic industries. Paradoxically, it might rather be said that Japanese enterprises enjoyed comparatively greater competitive freedom in the prewar period despite the seeming omnipresence of government controls.

Moreover, it might also be pointed out that the government devoted great efforts to creating favorable conditions for free enterprise, and eagerly awaited the fruits of such competition.

After the establishment of many new industries in Meiji times, the period from 1880 to 1931 was one of political stability and expansion of industry in scope and in scale. For instance, the Sino-Japanese war of 1894–1895 provided an impetus to industrial development, while the unexpected success of the Japanese encouraged them to even greater patriotic fervor and brought the country to the notice of the great powers as a new force to be reckoned with the Far East (Dempster 1969: 176). Japan extended her frontiers, using her new military power, to build up an empire in the east. She gained control Formosa (Taiwan) in 1895. In 1905, she defeated Russia and acquired Southern Sakhalin and in 1910 moved into Korea. This brought Japan's growing power to the notice of the West.

Development of Pulp and Paper Industry

Introduction

The role of academics such as Eiichi Shibusawa, which his brilliant ideas could encourage conditional spheres to establish entrepreneurship among Japanese. Among the entrepreneurship ideas namely to establish Oji Paper Company, as the pioneer of pulp and paper industry in Japan. This paper discusses the development and significance of paper industry, raw material, its impact on environmental problem.

The Pioneer of Pulp and Paper

As previously discussed, the role of Japanese modernizers who formerly studied in West Europe and United States were very significant to launch the modernization of Japan. In this case, one of the great figures who built the Meiji economy, Shibusawa Eiichi (1840–1931) was unique. He started his post-Restoration Career in the Finance Ministry, but soon left it to become a banker-entrepreneur.[4] He founded one new company after another; in all, he is associated with, more than five hundred enterprises. Yet he created no industrial empire. Among the companies he helped to found were the Oji Paper Company (1872), Tokyo Marine Insurance (1879), Tokyo Gas (1885), Tokyo Electric Light (1886), Tokyo Chemical Fertilizer (1887), Shinagawa Glass (1888), and Ishikawajima Shipyard (1893). The two spearhead industries of Meiji economic development were cotton spinning and railroads.

[4]For further information about the role of Shibusawa Eiichi in Business, see Craig (1994).

Table 6.1 Pulp and paper among the largest manufacturing in Japan

Manufactures	Total amount Y (trillion)
1. Transport machinery	49.5
2. General machinery	24.8
3. Chemical	23.1
4. Foods	21.9
5. Electric appliances	17.5
6. Electronic parts, devices	17.3
7. Information communication machinery	12.6
8. Metal	11.9
9. Steel	11.6
10. Beverage, tobacco, feed	10.
11. Oil products, coal products	9.6
12. Plastic products	9.6
13. Pulp, paper, paper converted products	6.8
14. Ceramics	6.7
15. Printing and allied industries	6.6

Source Census of Manufactures Report by Industries (METI 2003); See also Japan Pulp and Paper Co., Ltd. (2005, p. 5)

Shibusawa was active in both. Shibusawa was also instrumental in establishing the Tokyo Stock Exchange (1878) and the Tokyo Chamber of Commerce (1878).

Focusing on the development of pulp and paper industries, it is better to refer to a valuable speech of Eiichi Shibusawa as one of (the founders of the modern Japan). He said," Western civilized nations have attained their full development in every field. This is because they have devoted themselves to the cultural and scientific development by promoting the spread of higher education. In Japan we have to strive for cultural and scientific development. In order to attain Japan's development as a modern nation, we have to do a lot of things. The important thing is to promote printing industry in order to publish a great number of newspaper and books which are conveniently available to everyone."[5]

These words had great influence to encourage private companies and Japan's government to launch modernization programs one of them was to further develop pulp and paper industries. Currently, these industries obtained very strategic positions and were subsequently categorized number 13 among the 15 largest manufacturing industries in Japan, which total amount Y 6.8 trillion and absorbed 34,839 employees (Table 6.1). The largest manufacturing shows transport machinery which total amount Y 49.5 trillion, general machinery Y 24.8 trillion in the second, and chemical Y 23.1 trillion in the third rank.

[5]For further information, see The History of Oji Paper, published by Oji Paper Co., Ltd., 2004, p. 163.

Table 6.2 Pulp imports by origins (2005) (in metric tons)

Origins	Grand total
Canada	820,470
U.S.A	585,524
Brazil	322,861
New Zealand	229,396
Indonesia	132,835
Chile	95,426
Russia	57,658
Sweden	48,370
Finland	15,548
South Africa	10,960
China	6294
Other	34,714
Total	2,360,056

Source Ministry of Finance (2005); Pulp and Paper Statistics (2006, p. 11)

Based on Pulp and Paper Statistics 2006, development of this industry reached 21 units for pulp manufacturers and its production of about 10.6 million metric tons in 2004 and rapidly increased to become 10.7 million in 2005. Although, at the same time, Japan imported pulp from many countries of about 2.5 million metric tons in 2004 and decreased to be 2.3 million in 2005 (Table 6.2).

Meanwhile, paper and paperboard manufacturers rapidly developed to 419 units and production of 31.3 million m^3 tons in 2005 (Table 6.3). Japan also, at the same time, exported recovered paper to other countries (Table 6.4).

It was obvious that pulp and paper industries in Japan are rapidly developing because of three rationales. First, the government provides a good environment for growing these industries through incentive credits from the bank, tax law and provides good infrastructure (port and road). Second, the sustainability for procuring raw materials (wood trade) in overseas and domestic as well by huge their plantation trees as raw materials. Third, research and development respective pulp and paper industries well managed. There are top 12 Paper Companies financial results for the March 2005 (Table 6.5).

Table 6.3 Recovered paper products (in metric tons)

Year	Total net paper and paperboard supply for domestic consumption
2001	31,071,938
2002	30,666,872
2003	30,929,580
2004	31,383,951
2005	31,380,357

Source Ministry of Economy, Trade and Industry; Ministry of Finance (2005); Pulp and Paper Statistics (2006, p. 12)

Table 6.4 Recovered Paper Exports by Destination 2005

Countries	Amount (in metric tons)
China	3,108,449
S. Korea	177,430
Taiwan	169,819
Thailand	165,181
Vietnam	53,272
Philippines	22,982
Indonesia	7656
Other	5693
Total	3,710,482

Source Ministry of Finance (2005); Pulp and Paper Statistics (2006, p. 12)

Table 6.5 Top 12 companies' financial results for March 2005 (billion yen)

Company	Sales	Profit after tax	Total assets	Employees	Profit to sales ration
Oji Paper Co	1,185.1	43.3	1,606.2	18,634	2.7
Nippon Paper	1,179.7	24.4	1,530.0	13,774	1.6
Daio Paper	395.4	4.3	638.1	7,974	0.7
Rengo Co	391.2	10.9	424.7	9,385	2.8
Mitsubishi Paper	234.7	24.7	319.5	4,902	–
Hokuetsu Paper	151.2	7.0	221.4	2,782	4.6
Chuetsu Pulp and Paper	110.6	1.5	156.5	1,916	1.4
Kishu Paper	50.9	0.7	63.4	1,470	–
Tokai Pulp Co	54.1	0.9	72.4	1,134	1.7
Tomoegawa Paper	44.0	0.6	41.9	1,172	1.4
Tokushu Paper	22.1	1.4	53.5	543	6.3
Mishima Paper	36.1	0.9	39.2	861	2.5

Source Pulp and Paper Statistics (2006) (Japan Paper Association), p. 18

Raw Materials (Wood Trade)

Japan's total wood demand was 105,382,000 m^3 in 1987. Only 31% of this was produced domestically; the other 69% was imported (Tables 6.6, 6.7 and 6.8; Fig. 6.1). Table 3.6 shows changes in the volume of Japanese wood imports (1960–1999). In 1960, Japan's wood demand reached 71,303,000 m^3, but kept increasing until it exceeded 110,497,000 m^3 in 1972. Because of the economic recession caused by the first oil shock of the 1970s, the demand for wood then decreased until the mid-1980s, before increasing again in the late 1980s. Since then, the self sustenance rate has fallen steadily, and by 1996, it had dropped to 21%. If Japan imports such a large amount of wood, what is her position in the global trade of

Table 6.6 Changes on total wood demand in Japan 1960–2005 (1000 m^3)

Year	Total wood demand	Wood import volume	Self-sustenance rate (%)
1960	71,303	7541	89
1966	82,470	25,041	70
1972	110,497	63,354	43
1978	106,344	71,363	33
1981	94,586	60,603	36
1984	93,963	58,772	37
1987	105,382	72,212	31
1997*	109,905		
1999	97,812		
2000	99,263		
2003	87,191		
2004	89,799		
2005**	86,305		

Note Volume calculated in terms of round wood volume
Source Forestry Agency (various years), Ringyo tokei-yoran; See Also Iwai (2004), Forestry and the Forest Industry in Japan, p. 246; See also Forestry Agency (Rinya-Cho) (from * to **)

Table 6.7 Changes in the volume of Japanese wood imports, 1960–1999 (1000 m^3)

Year	Round wood	Sawn wood	Chips	Wood-based panels
1960	6223	156	–	–
1969	33,741	2066	4114	63
1975	35,650	2612	11,340	165
1978	42,653	3857	13,116	85
1981	29,220	3898	12,508	191
1987	32,292	7397	14,026	2444
1990	28,999	9082	19,043	4066
1996	21,336	12,281	26,445	7463
1999	16,551	9740	25,295	6034

Source Forestry Agency (various years), Ringyo tokei-yoran; Food and Agriculture Organization (FAO), FAO Yearbook, Forest Products; Yoshiya Iwai and Kiyoshi Yukutake, Loc Cit

forest products? Table 3.7 shows the five leading wood-importing countries for three different types of wood materials in 1998. Japan ranked first in importation of round wood (log) and chip and particles, and was second to the United States in importation of sawn wood. Obviously, when the volumes for the three materials are totaled, Japan ranks first, outstripping the US.[6]

MAFF (Ministry of Agriculture, Forestry and Fisheries) divides timber demand into two broad categories: (1) industrial use or youzai; and (2) 'others'. The later

[6]For further information, see Iwai and Yakutake (2002).

Table 6.8 Leading wood-importing countries, 1998 (volume, 1000 m^3)

Rank	Round wood		Sawn wood		Chips and particle	
	Country	Import (Vol)	Country	Import (Vol)	Country	Import (Vol)
1	Japan	15,190	USA	44,940	Japan	22,610
2	Finland	9328	Japan	7765	Canada	1749
3	Sweden	9300	Italy	7295	China	1699
4	China	7150	UK	7031	Italy	1134
5	Canada	6278	Germany	5822	S. Korea	885
World total		89,329		115,760		37,883

Source Food and Agriculture Organization (FAO) (1998), FAO Yearbook, Forest Products

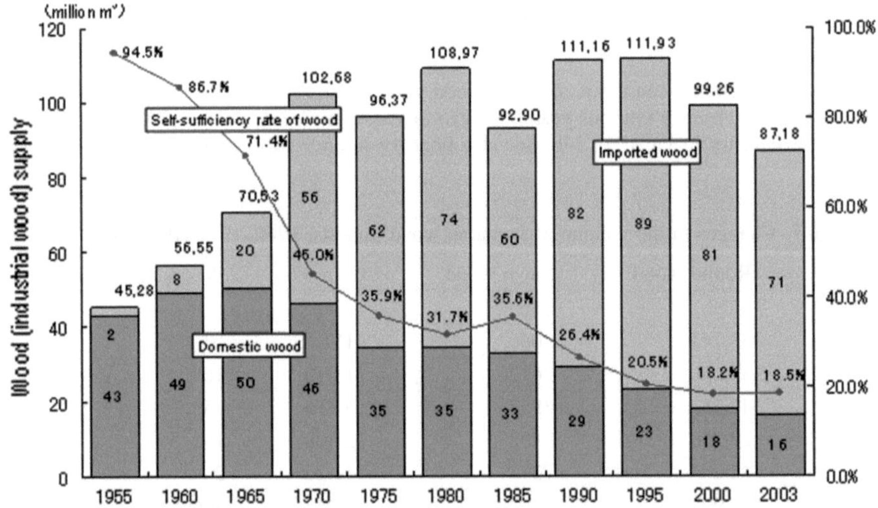

Fig. 6.1 Trend of supply/demand and self-sufficiency rate of wood in Japan. *Source* www.rinya.maff.go.jp; see Lumber Supply and Demand Chart

contains firewood and logs for the production of Shiitake mushrooms. The demand for these products will usually be satisfied in local markets so the level demand is not small but also great significance for international timber trade. The former category of youzai is further divided into seizaiyouzai, or saw logs and finished and semi-finished solid wood products, pulp and chips, veneer logs and 'other uses'. Other uses include transmission poles, pit props, timber piling and scaffold poles. These figures for 1996 are shown in Table 6.9.

Table 6.9 The demand for timber in Japan and variant timber uses

Demand (000 m^3 RWE)	Domestic production (%)	Variant
Industrial wood (Youzai)	112,324	20.0
Saw logs and lumber (Seizaiyouzai)	49,758	32.5
Veneer logs and plywood	15,726	1.4
Pulp logs and chips	43,822	12.9
Other	3018	14.2
Other uses		
Firewood	749	
Shiitake bed logs		967
Total	114,040	21.2

Source Forestry While Paper (1996); See Blandon, P.R, Japan and world timber markets, UK: CA International Press (1999, p. 28)

Wood Chips

At present, the paper manufactured in Japan is made from 60% used paper and 40% wood pulp. In addition to using timber harvested from planted forests, wood pulp can be made from low–quality timber from natural forests and lumber residuals. In Japan, 71.1% (2005) of paper is recycled; this is one of the highest recovery rates in the world.[7] Moreover, as used paper is recycled again and again, the fibers rub against each other and, as a result, their cohesive strength gradually deteriorates. For this reason, it is estimated that paper can only be recycled three to five times. In modern Japan, the demand of paper has risen at the same pace as GDP; in particular it grew rapidly from the 1980s to the 1990s. In recent years, it had been assumed that the increasingly widespread use of Information Technology (IT) equipment and the accompanying shift toward the paperless office would lead to a fall in demand for paper, and yet this has not happened; demand continues to be at a high level (for example in 2004) consumption of paper reached 31,383 million metric ton and in 2005 reached 31,380 million.

Apparently, to overcome the demand for these raw materials, Japan's paper-making companies are working to ensure the efficient use of timber resources through such steps as increasing the ratio of used paper in their products and using lumber scrap. At the same time, based on the principle of "Create what you use", for some time now they have been involved in afforestation[8] projects principally aimed at securing a stable supply of high-quality wood chip as raw material. These

[7]See Overseas Plantations for a Sustainable Society of the Future: Forest Plantations and Pulp and Paper, published by JOPP (Japan Overseas Plantation Center for Pulpwood), 2005, pp. 4–9.

[8]Afforestation means conducting reforestation program (planting trees such as Eucalyptus globulus, Acacia mangium, Pinus radiate, etc.) for wood chips as raw materials in previous forest areas which currently became a critical land or covered by bushes.

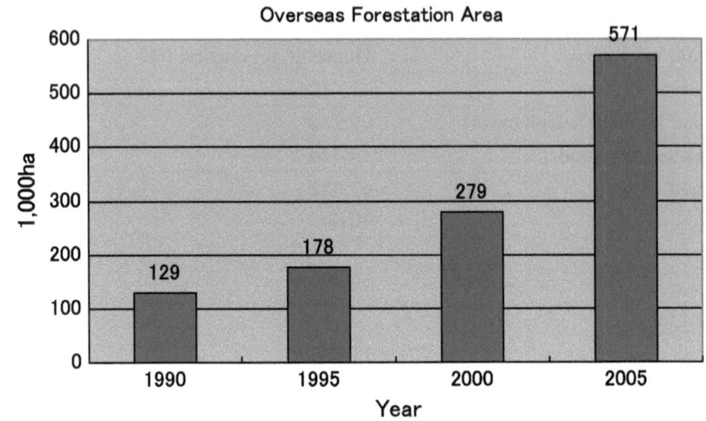

Fig. 6.2 Overseas afforestation areas. *Source* JOPP (Japan Overseas Plantation Center for Pulpwood) (2005, p. 9)

plantations projects are not only being conducted in Japan, but also overseas. Based on a JOPP report, at the end of 2005, there were 33 forest plantation projects conducted overseas by Japanese companies connected with paper making. These were located in 9 different countries, mainly in the southern hemisphere, particularly Australia, Southeast Asia, South America, and South Africa. The total area of these plantations was 570,000 ha (Fig. 6.2).

The list of companies involved in these projects includes not only papermaking related firms but also trading companies (Itochu Corp, Mistui, Marubeni, Sumitomo, Mitsubishi, etc.), electric power companies (Tohoku Electric Power, Osaka Gas, Tokyo Electric Power, Chubu Electric Power, the Chugoku, Shikoku, etc.), publishers (Kodansha Ltd.) and printers (Toppan printing), automobile manufacturers (Toyota Motor Corp, Honda), manufacturers of office automation equipment (Fuji Xerox), and mail-order retailers. An increasing number of companies that do not directly use timber as a raw material for paper are joining these projects, and this reflects the greater priority being given to the role of forests as absorbers of CO_2 and thus as a means to prevent global warming.

The Significance of the Pulp and Paper Industry

It is clearly shown that Pulp and Paper industries are very significant in modern history of Japan. As an architect of this industry, Shibusawa Eiichi advocated "Dototku Keizai Goitsu Setsu", the theory of Harmony and Economy, all his life. And it is known that the number of business enterprises which Eiichi was involved in founding and fostering exceeds 500. Currently, Japan is the third largest world's paper and paper board production with about 30.8 million ton in 2004, and per capita consumption of 247 kg.

The production systems of Japan's dominant paper manufacturers are integrated from pulp through to paper and paperboard production.[9] In 1995 the producing companies' integrated mills consumed 86% of total pulp production. Some paperboard manufacturers who use wastepaper, however, produce very little pulp at their mills. No Japanese firm specialized in pulp production only because of two underlying factors: Firstly, Japanese companies are reluctant to carry it out, because of the lack of land for planting trees as raw materials. Secondly, the advantages of this business are anticipated to decrease in modern times. This is due to shortages of a variant product which will eventually affect the lack of income generation of the firms.

The paper manufacturers' financial condition has greatly deteriorated in recent years due to a downturn in this cycle caused by the recession after the collapse of the "bubble economy" (in the decades of 1990s-early 2000s). Apparently some paper manufacturers are once again trying to strengthen their management through merger and acquisition. Japanese paper manufacturers began reorganizing through two successive large-scale mergers in (1993–1997). For instance, a merger happened between Oji paper and Kanzaki paper to form New Oji (1993), Oji Paper with Nippon Pulp industries (1979) and also between Oji Paper and Honshu Paper (1997) to become Oji Paper. In 1995, New Oji Paper produced 3.6 million tons of paper and paperboard (12.0% production share) and Nippon Paper 3.0 million tons (10.1% share). Oji Paper's production rapidly increased and reached 6.0 million tons in 1997 (19.4% production share) and sales were about Y 980 billion. Currently, Oji Paper is the largest company in the Japanese pulp and paper industry in Japan.[10] The world paper market is continuing to grow, both in terms of production and consumption, and the pulp and paper industry is said to be a growth industry. Currently, the center of production and consumption is shifting from North America and Western Europe to Asia. The Pulp and Paper Groups, which have its base in Japan (such as Oji and Nippon Paper) are favorably positioned in close proximity to the Asian market, which will see significant growth. Therefore, the Nippon Paper Group and the Oji Paper's vision in 2015 is to become two of the top five Pulp and Paper Companies in the world although they are currently positioned number seven and eight among top 10 major global pulp and paper companies in 2003 (Table 6.10).

[9]Grade and Classification of paper and paperboard is newsprint, printings, and communication (copy paper, thermal paper, forms, etc.).

[10]See Noda (2002).

Table 6.10 Results of major global pulp and paper companies in 2003 (in trillion yen)

Company	Sales of P&P business	Consolidated sales	Operating profit
1. International Paper (USA)	2.5	2.9	129 (billion yen)
2. Stora Enso (Finland)	1.3	1.6	63.4
3. George Pacific (USA)	1.3	2.3	135.5
4. Svenska Cellulosa (Sweden)	1.2	1.2	111.6
5. UPM-Kymmene (Finland)	1.1	1.3	102.7
6. Kimberly-Clark (USA)	1.1	1.7	279.8
7. Oji Paper (Japan)	1.0	1.2	73.9
8. Nippon Paper (Japan)	1.0	1.2	55.7
9. Weyerhaeuser (USA)	0.8	–	–
10. Smurfit-Stone Container (USA)	0.7	–	–

Source Nippon Paper Group (http://www.np-g.com/e/aboutvision/index.html)

A Brief History of Paper

The invention of paper in China dates from ancient times: the early Qianhan Dynasty has been found in mounds in Ganshu Province in China (176–141 BC) and Baqiao paper predating the Qianhan Dynasty has been discovered. Today, the familiar name of Ts'ai Lun is generally remembered as a great figure who improved the method of papermaking. In fact, Ts'ai Lung worked for Emperor Chang (A.D. 77) in a higher position with honors and he appointed him Chief of the Imperial Supply Department. He was entrusted with the important function of making swords, implements and furniture for the imperial household. It was during this period of his success that he found a way of increasing production of paper.[11] This invention was so significant, in retrospect, for writing and writing materials of ancient China. A country in which learning was revered, China early on developed, possibly in the twelfth century before Christ, an ideographic system of notation based on figures used in divination. The great impact of paper invention in China was to encourage its use in cultural and civilizations throughout the world. The Japanese transformed it into Washi[12] and the Europeans into Western-style paper.

There are some versions of historical events regarding paper transmission into Japan. Based on a source, it transmitted into Japan in the late second or third century. But, according to Kojiki, Japan's oldest extant chronicle (712), Wani Kishi

[11]See Narita (1954).

[12]Washi is an original Japanese paper. It made from kozo (majority), mitsumata and ganpi tree. Currently only 400 of household make Washi in Japan, but in the Meiji era (1868–1912) reached 70,000 units of households. This Interview held with Fujiwara, staff of Paper Museum, Oji, Kita-ku, Tokyo on October 8, 2006. See also document's exhibition in this Paper Museum.

(Wani) of Paekche arrived in Japan in the 16th year in the reign of the Emperor Ojin (285) carrying 10 scrolls of the Analects of Confucius and one of the "Thousand Character Classic". The Kojiki also says that the Japanese began making paper in the 18th year of the reign of Empress Suiko (610) during the Asuka period. The role of paper eventually affect very significant in Japan. As seen, the main production areas were: Shuzenji, Mino, Kamakura, Sugihara (Banshu), Nara, Yoshino, Koya, Minato (Senshu), Nishijima (Kai), Nishinouchi (Hitachi), and Mizoguchi (Chikugo). The products had an exhaustive list of uses: Kamiya-gami (shuku-shi, rinshi- gami, usuzumi-gami), shiki-shi, Kara-kami (bamboo paper for sliding doors and folding screens), hana-gami, shoji-gami, seicho (used for documents), yakutai-shi (paper for medicine pouches) and ro-gami (wax paper). Over the past 300 years, both the quality and variety of Japanese paper has increased, along with its utility and beauty, and it has become a fully commercial product.[13]

During the Yedo period (1603–1867), all daimyos or feudal lords encouraged paper-making in their own provinces. Each daimyo had a specially designated paper factory for his own use, and granted the monopoly of sales of paper to a paper-dealer in his province. In those days, rice, paper and lacquer constituted the chief articles of tributes to the daimyos.[14] Among the representative species of paper of this period, clay-contained torinoko or Settsu, nishi-no-uchi and hodo-mura of Hitachi and Shemozuke, and hosho of Echizen can be cited. Of these, however, the most popular paper for daily use was what has been called hanshi, manufactured in nearly every province, as shown in documents belonging to this period. The most well-known hanshi were Yanagagawa hanshi of the province of Chikugo, Yamashiro hanshi of Nagato, Tokuji hanshi of Suo, Sekishu hanshi of Iwami and Ozu hanshi of Iyo.

From the Meiji Era to the Second World War

The role of Meiji Emperor in launching modern Japan was very significant. The Japanese government sent missions to Europe and America to learn advanced Western technologies. Mission members clearly realized that "paper" manufacturing and the printing industry were necessary to modernize Japan. Meanwhile, there was some talk about the possibility of starting papermaking business. At that time, Shibusawa was a high official at the Ministry of Finance. He advised development of paper-manufacturing business. His views in regard to this matter were summarized as follows:

> ...The high development evident in the West is largely due to the high level of culture among the people. Many factors may contribute to the spread of culture, but the foremost is undoubtedly printing. By means of printing, a great number of books, newspaper, and

[13]The History of Oji Paper, Loc Cit, pp. 162–163.

[14]Kiyofusa Narita, Loc Cit, p. 27.

periodicals can be made available; and the material for these publications would be necessarily be paper. It is obvious then that papermaking and printing are, so to speak, the fountainhead of civilization.[15]

In those days, there were three large business companies in Japan: the Mitsui, the Ono and the Shimada, which ran the financial organization of the Meiji government. Shibusawa persuaded these companies to render real service to the country by establishing a papermaking industry. Eventually, the efforts of Shibusawa were fruitful to organize with a little capital as possible, about Y 150,000. In November 1872, these three companies agreed to apply for a permit to start a company under the name "Paper Company." In the sixth year of Meiji (1873), the demand of machine-made paper was limited to governmental use, and at about this time, the Ministry of Finance began to issue paper money and public bonds. However, the demand for printing paper gradually increased as newspapers, books, and magazines were being printed by the Western method.

Relating to the paper company industry, Oji Paper and Nippon Paper Company were founded at this time. The existence of paper industries led to a newspaper publication boom and many translated books were published in the end of 1870s. Because much Western-style paper consumed in Japan at this time was imported, and the demand for such paper was increasing, Japanese paper companies began producing it domestically.[16] In 1877, domestic production of Western-style paper was 547 tons, compared with imports of 771 tons. In 1875, the Oji Mill, which introduced the Western method of paper manufacturing, began operations. In 1933, big pulp and paper manufacturers, such as Oji Paper, merged with Fuji Paper and Karafuto constructed new mills on the northern island of Hokkaido where they started the mass production of paper, particularly newsprint. This company virtually monopolized the Japanese market, producing 80% of the paper and 95 of the pulp. Under this phenomenon, paper imports peaked at 90,000 tons in 1936; Japanese paper production reached its prewar peak of 1,540,000 tons in 1940. Apparently, in terms of raw materials, in the northern territory these was abundantly available in the form of yezo spruce (picea jezoensis) and todo fir (Sakhalin fir, Abies sachalinensis). The manufacturers also built mills in Sakhalin, where more of the same pulpwood could be obtained, enabling mass production of paper during the 1910s. From this point, domestic self-sufficiency in Western-style paper was established.

Unfortunately, when Japan defeated in World War Two (1945), it eventually had a negative affect on the development of the Japanese pulp and paper industry, First, because Japan had lost overseas territories like Sakhalin, the industry lost abundant fiber resources and much factory equipment. As a consequence, paper production decreased to 210,000 tons in 1946. Second, in 1949, Oji paper was divided into several companies by the US occupation government. Many smaller pulp and paper

[15]Ibid, pp. 94–95.

[16]Hideshi Noda, Loc Cit, p. 215.

manufacturers emerged and competition resumed. The prewar monopolistic pulp and paper market was transformed virtually overnight into a competitive industry.[17]

In contrast, demand for paper expanded greatly in the rapidly growing postwar Japanese economy. Paper began to be used for various products. Cardboard boxes, for instance, replaced wooden boxes and were widely used as containers for transporting and protecting goods. Thus, paperboard demand also increased rapidly. Nationwide, pulp and paper manufacturers expanded production, investing in facilities and equipment and promoting technological innovation.

Impact on the Environment

The rapid development of industrial sector especially pulp and paper industries had an impacted on environmental pollution. There has been considerable debate about human development and the environment. The human activities include industrialization development in advanced countries, such as Japan in recent years that has caused local pollution problems to be compounded by environmental destruction on a global scale.[18] Although, it also led people to realize that economic development aimed at making life more affluent can, if not properly managed, ultimately jeopardize humanity's survival. Then, it is necessary to think about the concept of development from environmental perspective. But we must also recognize that human existence inevitably has some effect on the environment. The best we can realize is to minimize the negative impact our actions have on environmental conditions now and in the future.

During the 1950s and 1960s, Japan obtained the reputation of being a showcase of environmental pollution. Some images of victims of mercury and cadmium poisoning were published all over the world. Smaller cities, such as Minamata (Kumamoto Prefecture), Yokkaichi (Mie Prefecture), Fuji city (Shizuoka Prefecture), Tokyo bay, Niigata, etc., became domestically and internationally famous for their pollution-related diseases. Probably no other country had come to feel the consequences of unrestrained industrial growth as early and as painfully as Japan.[19] It happened, in the beginning, during the latter part of the 1960s. Government and private companies gave up their ecologically ignorant attitude. This critical condition of pollution encouraged more Japanese who were affected by pollution to start to protest loudly, supported by NGOs activities against industrial and governmental activities that harmed the environment and impaired human health. They were supported by environmentalist activists such as researchers, journalist, judges and NGOs. The roles of these groups were very significant to encourage about changes by imposing revolutionary sentences on polluting

[17]Ibid, p. 216.

[18]See Fujisaki (1995).

[19]See Weidner (1989).

enterprises. This effort was fruitful for people's consciousness on sensitivity to environmental problems spread among the public. All this eventually led to a fundamental about-face in Japan's environmental policy. The government supported implementation of strict environmental regulations on these industries and enforced them, which was rewarded with quite spectacular success in individual environmental areas. Therefore, other countries now looked to Japan in the hope of finding solutions to their own environmental problems. Japan did not only shed her image as "pollution nation"; in some respects she now stood out as a paragon of effective anti-pollution policy.

The next paper discusses the role of two of the largest pulp and paper industries in Japan in terms of its development, raw materials, production, distribution, revenues, research and development. In contrast, the consequences of its pulp and paper industrialization, which were ecologically ignorant of the effect on environmental pollution. NGOs criticizes on pollution, ecological damage, support livelihood of local community and so forth.

Profile of a Pulp and Paper Company

Introduction

The roles of private sectors as direct actors to encourage pulp and paper industries are very significant. The existence of pulp and paper industries in the Meiji government was rapidly developed. There are many fascinating players of this sector, such as Oji Paper, Nippon, Daio, Rengo, Mitsubishi, Hokuetsu, Chuetsu, Kishu, Tokai, Tomoegaea, Tokushu, Mishima, etc. These, among 12 actors, comprised the largest paper companies in Japan, from the viewpoint of sales, profit, total asset and employees. This paper focuses on the two biggest actors, namely Oji Paper and Nippon paper, which placed number seven and eight rank among ten (10) as major global pulp and paper companies.

The Oji Paper Company

Oji Paper Group is the largest business among pulp and paper companies in Japan with annual sales of Y 1185 trillion (March 2005), that rapidly became Y 1430 trillion in 2016. Net income reached Y 36.56 billion in 2016. The group's core company Oji Paper, is a publicly listed company on the Tokyo Stock Exchange and obtained profit after tax of about Y 43.3 billion and more than

18,600 employees in 2005.[20] For more than 132 years, (1873–2006) Oji Paper has contributed to cultural progress and the improvement of people's lives as the leading company in Japan's paper industry. This company, in 2003, produced about 8.9 million tons of paper and paperboards (Japan Pulp and Paper 2005).

History of Its Development

Oji Paper Company, which is familiarly known as "Soshi Kaisha" was founded on the 12th of February in 1873 in the village of Oji in Tokyo Prefecture by Eiichi Sibusawa. Obviously, Shibusawa intended to promote the modernization of Japan through paper manufacturing. The company, capitalized at Y 150,000, was the first joint-stock company in Japan.[21] The early construction of Paper Mill in 1874, under Kojima-Gumi the supervision of the Western engineer by Frank Cheethmen from British, Bottomley and Walsh from the United States.[22] Kojima-Gumi Company construct an entirely Western style-building was quite a new experience for them. The construction of the mill was completed by June 1875, and when it was equipped, paper-making was started. Imported from England were a Fourdrinier paper machine, seventy-eight inches wide, and the largest of those were installed in the six mills established in this country between 1872 and 1879.

In 1875, the Oji Mill, which introduced the Western method of paper manufacturing, began operations. In 1889, Keta Mill, Japan's first wood pulp mill, began operation on the bank of the Keta River, in Shizuoka Prefecture. In the first stage of Western paper manufacturing, the company positively developed new technology and new products in addition to the introduction of the newest facilities. So, we can see the company had already sought to create its corporate culture, characterized by a positive attitude toward better technology, which has been handed down to the present Oji Paper. And its development later (1893), the company named changed to 'Oji Paper Co. Ltd.' It constructed a large mill in 1910 with the newest facilities at the village of Tomakomai in Hokkaido in order to meet the increasing demand of paper for newspaper, books and magazines. Therefore, in order to look for raw material resources, the company directed its attention to a vast field, located on the Chitose River originating in Lake Shikotsu.

This rapid development of Oji Paper occurred while Ginjiro Fujiwara appointed in 1911 to be president. He had a good leadership and rehabilitated the company in many sectors. He devoted himself to expand its paper manufacturing business into Manchuria and the Korean Peninsula until he retired from its presidency in 1940.

---

[20]The second largest is Nippon Paper Group Y 1179.7 billion, Daio Paper Corporation Y 395.4 billion, Rengo Y 391.2 billion and Mitsubishi Paper Mills Y 234.7 billion. See "Top 12 Companies Financial Results on Pulp and Paper Industries" in Japan for the March 2005; Pulp and Paper Statistics 2006: Japan Paper Association, 2006, p. 18.

[21]For further information, see The History of Oji Paper, Tokyo: Bunkado Printing, 2004, p. 163.

[22]Kiyofusa Narita, Loc Cit, p. 102–104.

It was remarkable that Fujiwara developed its paper pulp business in Sakhalin and merged the company with Fuji Paper and Kurafuto Kogyo in 1933, creating the biggest paper company in Japan. In this case, Fujiwara has been considered the father of the Oji restoration. As known, the key to success in management of the company, Fujiwara embraced the revered Eiichi Shibusawa's teachings: "paper manufacturing industry should perform its duty for society." Shibusawa set forth his principles of business management that emphasized the need to combine good business practice with good ethics. So Fujiwara consistently managed the company considering the prospects for the future and education to cultivate the abilities of staff members.

Following Japan's defeat in World War II, the Excess Economic Powers Decentralization Act was implemented, effectively dividing the company into three components: Tomakomai Paper, Jujo Paper, and Honshu Paper. The newly founded Tomakomai Paper Co., Ltd. strove to reconstruct the Tomakomai Mill through improvement and perfection of equipment. As a result, in 1951, the company could produce 498 tons a day, which surpassed the highest prewar production record of 475 tons. One thing that was given very strategic attention by Oji Company is the significant institute of research laboratory. To parallel this, meanwhile, Central Research Laboratory was founded at Shinonome district in Tokyo (in 1957) in order to promote the improvement of the quality of products and to develop new products. Another, Institute for Forest Tree Improvement was established at Kuriyama in Hokkaido (in 1956) and also in Kameyama, Forest Tree Improvement Institute opened in Mie Prefecture (in 1957). They began to carry out research on how to improve the breeding of trees for the purpose of propagating tree planting.

As a result, under this strategic development, Oji Paper was rapidly setting up the production of paper, proceeding to carry out the plan for the modernization of mills and technical innovations along with the promotion of setting up larger plants. Eventually, the company merged with Kita Nippon paper (1970), Nippon Pulp Industries (1979), and Toyo Pulp (1989). In 1993, Oji Paper merged with Kanzaki Paper to become Oji Paper, and furthermore, in 1996, New Oji Paper and Honshu Paper merged to become Oji paper (Fig. 4.1). Currently, Oji is one of the largest and most influential paper manufacturers in Japan as well in the world.[23] Obviously, the portrait of management is based on group structure (Figs. 6.3, 6.4, 6.5, 6.6, 6.7, 6.8 and 6.9).

Wood Raw Material Policy

Oji Company's policy is to procure raw material throughout a sustainable forest management. There are some procurement guidelines: (1) expand procurement of wood from certified forests; (2) increase the use of plantation trees; (3) utilize

[23]For further information, see: www.ojipaper.co.jp/English/group/corporate/history.html.

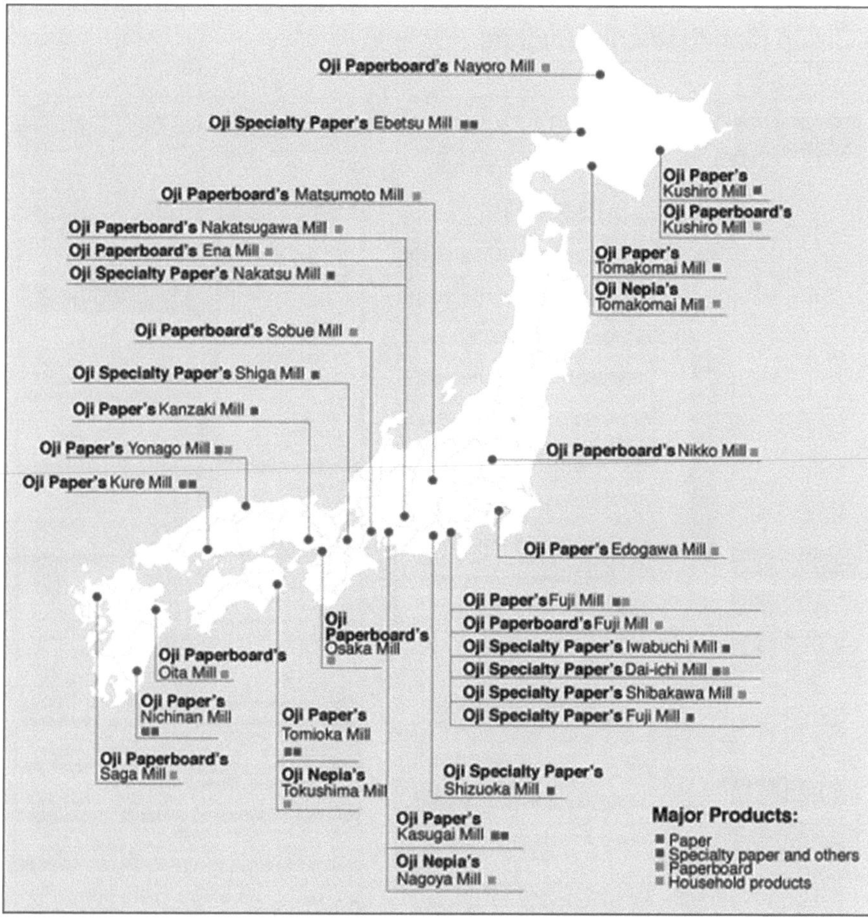

Fig. 6.3 Holding and its branches of Oji Company. *Source* See about Oji Paper Group, ojipaper.co.jp/English

unused wood effectively; and (4) ensure raw material traceability. Oji paper will work to trace the origin of wood raw material and to confirm that it was from well-managed forests. Oji Paper will be particularly vigilant about not purchasing from illegal logging.

The objective of sustainable forest management is to achieve environmentally, socio- economically sound forest management. The criteria of environmental sustainability: preserve biodiversity, ecological processes. Meanwhile, social sustainability means sustain human society, which relies on forests. And economic sustainability means sustain the viability of companies and communities on forests. Therefore, objective criteria and indicators have been established for evaluating sustainable forest management for Japan and regions with similar natural conditions and social backgrounds. On the other hand, forest certification entails a defined

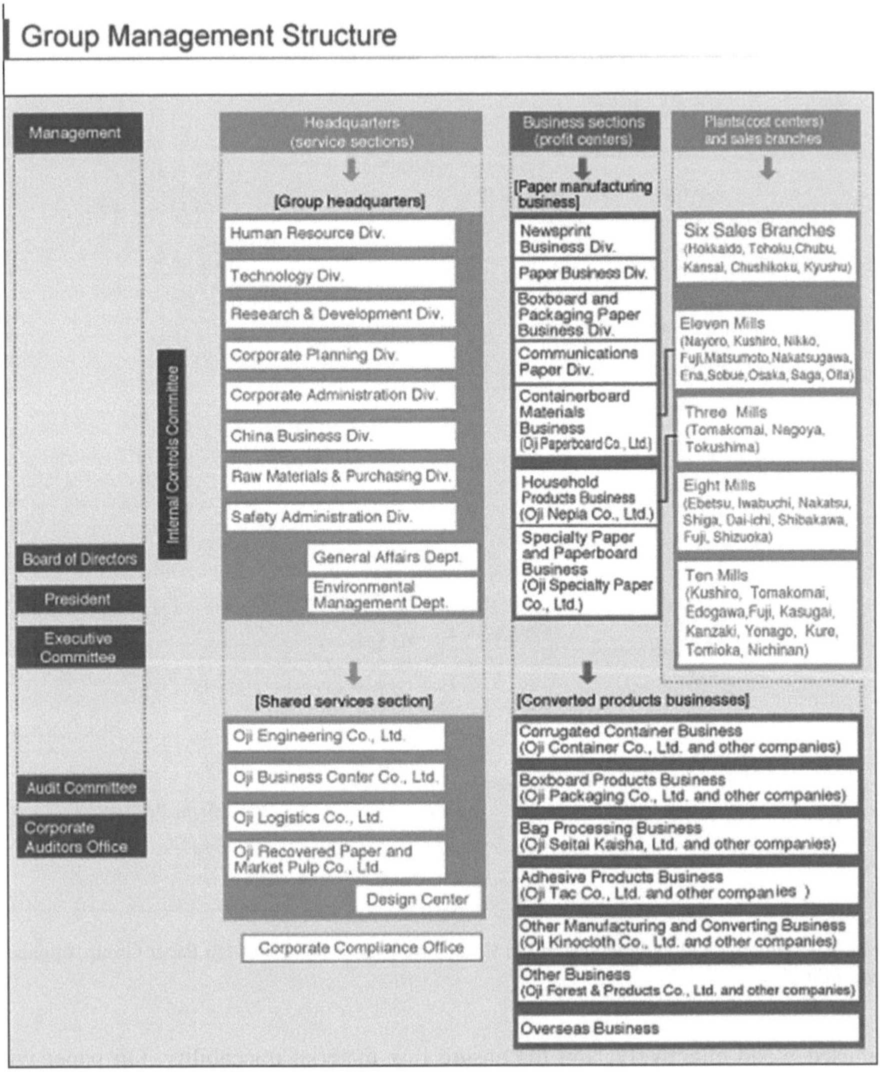

Fig. 6.4 Structure of Oji's management. *Source* See ojipaper.co.jp/English (Group structure)

FY	Material		
FY2004	Imported hardwood 58%	Imported Softwood 15%	Domestic Softwood 24%
FY2011	Imported hardwood 66%	Imported Softwood 12%	Domestic Softwood 19%

Fig. 6.5 Composition of wood chips procurement. *Source* ojipaper.co.jp Ibid. *Note* The Fiscal Year (FY) 2011 plan includes wood raw material procured for a new plant

FY04	Oji Paper Plantation 8%	Plantation trees 72%	Sawmill residue 9%	Low-grade, natural trees 13%
FY011	Oji Paper Plantation 16%	Plantation trees 81%	Sawmill residue 6%	Low Grade, natural trees 13%

Fig. 6.6 Increase plan in plantation trees and certified wood for import. *Source* ojipaper.co.jp Ibid

Fiscal Year	Plantation
Fiscal Year 2004	140,000
Fiscal Year 2011	300,000

Fig. 6.7 Expansion plan of Oji's Paper overseas plantation (ha). *Source* Ibid

process of evaluation and certification by in-depth third-party organizations that a forest is well managed according to sustainable forest management criteria. Internationally recognized forest certification schemes include the Forest Stewardship Council (FSC), a Sustainable Forestry Initiative (SFI). The Sustainable Green Ecosystem Council (SGEC) is the locally recognized forest certification scheme in Japan.

Oji paper's effort to procure raw materials has to plant industrial timber plantation from many countries. For example, in China, Thailand, Vietnam, Laos, New Zealand, Australia (Tasmania, Adelaide, Brisbane, Victoria, Albany, etc.) and Canada.[24] These representatives in some countries inspect shipment of wood chips, advice on quality, and meet with suppliers. The ship cargo is prepared by Oji Company. These annual shipments number about 200. In contrast, Oji Forests & Products has representatives in major cities across Japan to meet with suppliers.

Research and Development (R&D)

The progress of a company depends on utilizing Research and Development (R&D) as a strategic function to step head into the future. In the case of Oji, as the largest paper company in Japan, because of an optimal utilization of R&D for driving progress, the company makes an important contribution to society through R&D activities targeted toward the discovery of new possibilities in paper products and the forests and trees from which it is made. Apparently, new technologies, products and services are a driving force for continuing growth and success. On the other hand, Oji Company focuses on accelerating the creation of the new advances by increasing its research and development throughout:

[24]See Pulp and Paper Statistics 2006, Loc Cit, p. 24.

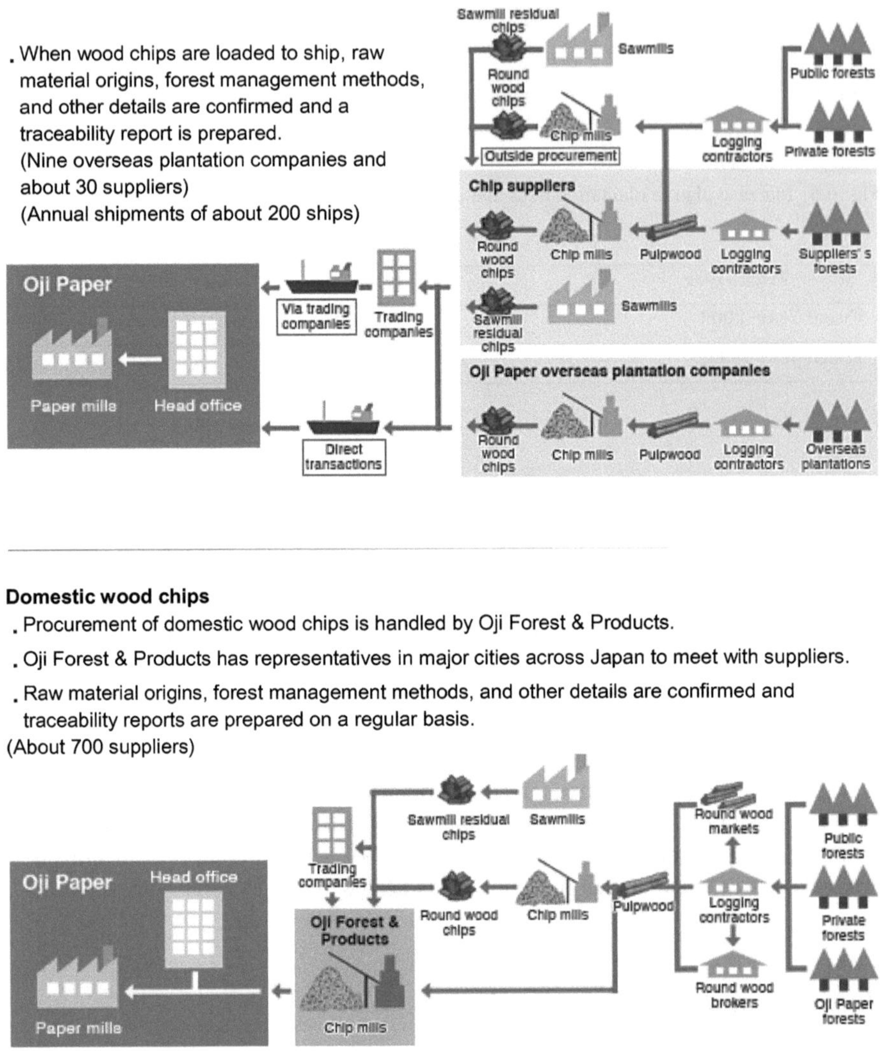

Fig. 6.8 The process of raw materials and shipping to the head office. *Source* ojipaper.co.jp/English

(1) Pulp and paper research laboratory; In charge of strengthening core business and supporting technically development of new products; (2) Imaging media development laboratory; With the rapid advance of information age, this company is developing strategies to meet customers' need for paper as a role of information communication media; (3) Fundamental technology research laboratory; Conducting research in peripheral fields of paper, development of new materials, new manufacturing methods and systems, and biotechnology applications; (4) Forestry research institute; In view of environmental technology on a world

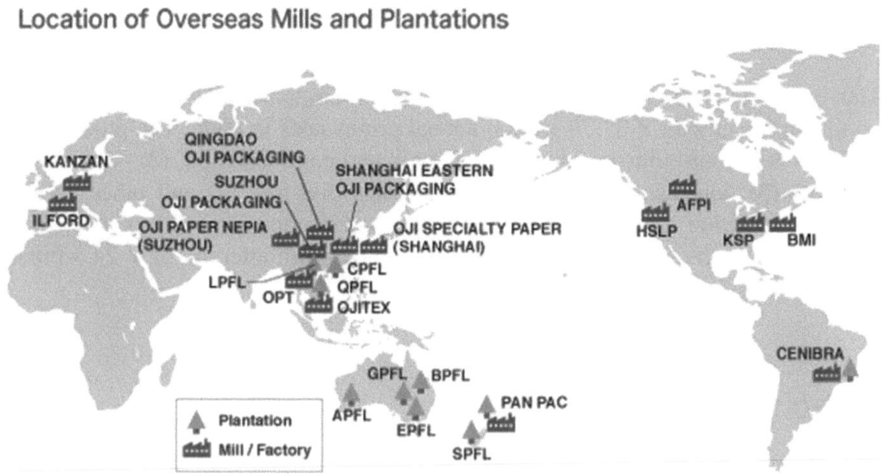

Fig. 6.9 Tree plants in overseas

scale, developing efficiency of afforestation abroad and new species which are suitable for afforestation area and paper manufacturing material; (5) Material analysis center; High analysis and evaluation technology are indispensable to keeping and improving high quality products and development of new products; and (6) Intellectual property department and R&D management development; It is supporting promotion of reserving rights, such as new technology and maintenance of Company's rights.

Corporate Code of Conduct

As mentioned above, a key success of Oji is an optimal utilization of R&D Besides, Oji also strongly adheres to a "Corporate Code of Conduct". This code of conduct as guiding principles for corporate activities is based on an awareness of the company's responsibilities as a corporate citizen and on high ethical principles appropriate for an organization that enjoys the society. The implementation of a code of conduct involves significant steps for Oji Company, such as compliance with the law; harmony with the environment; supply of sale, useful products and services; communication with society; coexistence with the international community; contribution through manufacturing; and achievement of employee satisfaction.

Nippon Paper Group

This company is the second largest after the Oji Company among paper business groups in Japan with Y 1.179 trillion annual group sales in 2005. The group's core business is in four industries, namely: (1) pulp and paper division; (2) paper related division; (3) housing and construction materials division; (4) and other division (Fig. 10). Nippon Company is a publicly listed company in the Tokyo Stock Exchange with the total assets Y 1.530 trillion and profit after tax of Y 24.4 billion and number of employees 13.774.[25]

History of Its Development

The initial stage of Nippon paper began with a merger of pulp and paper groups. For instance, between Fuji Seisho which built Kushiro Mill in 1920 and Kyushu Seisho, Yatsuhiro Mill started operation in 1924. They agreed to a merger in 1933 and built Hokkai Kogyo. On the other hand, this company also built Sanyo Pulp Kogyo in 1937 and established Iwakuni Mill in 1939 and in 1938, Kokusaku Pulp Kogyo established Asahikawa Mill in 1940. It agreed to a merger in 1945 between Asahikawa Mill and Yufutsu Mill under Dai Nippon Saisei-Seishi.

After these companies' merger, Nippon Paper Industry was established as paper manufacturer on August 1, 1949, with capital 104,873 million yen. Under the leadership of Masatomo Nakamura who rapidly developed these industries for many branches and established promotion of structure and launched philosophy. The Nippon Company emphasizes his philosophy "as member of society, we shall proudly promote activities that contribute toward social development". There are consists three principles: namely, (1) that contribute to cultural heritage and development; (2) that contribute to conservation and improvement of the environment; (3) and that contribute to the development of communities.[26] It is obvious that in all of its operations, the Nippon Company has been consistently committed to working in harmony with the environment, sustainable business development, fairness, and risk management as a member of society will be of critical importance to its success.

Radical steps were taken again on March 30, 2001, when Nippon Paper Group was formed as Nippon Unipac Holding, which was a pure holding company of Nippon Paper Industries Co., Ltd. and Daishowa Paper Manufacturing Co., Ltd., to facilitate the companies' integration of their operations. Nippon Unipac Holding changed its name to Nippon Paper Group, Inc., on the first October 2001. On April 1, 2003, Nippon Paper Group introduced a new organizational structure under

[25]See Pulp and Paper Statistics (2006, p. 18).

[26]See "Philosophy and Social Contribution Subcommittee," in http://www.np-g.com/e/csr/ideology/social_commission.html.

◎ Pulp and Paper Division

Nippon Paper Industries Co.,Ltd.
Kitakami Paper Co.,Ltd.
KOYO PAPER MFG. CO., LTD.
Daishowa North America Corporation
Nippon Paper Industries USA Co., Ltd.
Nippon Daishowa PaperboardCo., Ltd.
Nippon Daishowa Paperboard Tohoku Co., Ltd.
Nippon Daishowa Paperboard Kanto Co., Ltd.
Nippon Daishowa Paperboard Yoshinaga Co., Ltd.
Nippon Daishowa Paperboard Nishi Nippon Co.,Ltd.
Nippon Paper Crecia Co., Ltd.
NP Trading Co., Ltd.
HAGA Paper Trading CO., LTD.
Kokuei Paper Co., Ltd.

◎ Paper-Related Division

NIPPON PAPER-PAK CO., LTD.
Nippon Seitai Corporation
Nippon Paper Chemicals Co., Ltd.
FLOWRIC CO., LTD.
Nippon Tokan Package Co., Ltd.
Sakurai Co., Ltd.

◎ Housing and Construction Materials Division

Nippon Paper Lumber Co.,Ltd.
South East Fibre Exports Pty. Ltd.
NIPPON PAPER UNITEC CO., LTD.
Daishowa Unitec Co.,Ltd.
Kokusaku Kiko Co., Ltd.
PAL CO., LTD.
N & E CO., LTD.
Daishowa Uniboard Co., Ltd.
Kunimoku House Co., Ltd.

◎ Other Division

NIPPON PAPER DEVELOPMENT CO., LTD.
SHIKOKU COCA•COLA BOTTLING CO., LTD.
NIPPON PAPER LOGISTICS CO., LTD.
IWAKUNI-KAIUN CO., LTD.
Kyokushin Transport Co., Ltd.
NANKO UNYU CO., LTD.
Hotoku Co.,Ltd.
Daishowa Logistics Co., Ltd.
Asahikawa Grand Hotel Co ., Ltd.
Graphic Arts Communication

Fig. 6.10 Related companies

which the paper business and paperboard business "the Group's two core operations" were consolidated and reorganized under the Nippon Paper Industries Co., Ltd. and the Nippon Daishowa Paperboard Co., Ltd., respectively. Currently, The Nippon Paper Group, consists of Nippon Paper Group, Inc., Nippon Paper Industries Co., Ltd., Nippon Daishowa Paperboard Co., Ltd. and their 145 subsidiary companies and 49 affiliated companies, that are actively involved in the manufacture and sales of pulp and paper (Fig. 6.11).

Wood Raw Material

The Nippon Paper Group was very smart in procurement in terms of 'raw materials' by establishing plantation areas through global supply chain management, such as from New South Wales, Australia, Victoria, Burnbury, Green Triangle, Western Australia, Victoria, Chile, South Africa, Quesnel, B.C., Canada, Peace River, Alta., Canada, Port Angels, USA, Longview, Washington, USA, and China.[27] The production of raw material (Eucalyptus trees) to become wood chips and shipped with its own cargo ship owned by Nippon Company directly to address in his own company.[28] The Nippon Paper's basic policy when procuring raw materials, is namely, (1) environmentally friendly raw materials procurement; (2) socially aware raw materials procurement; (3) promotion of dialogues with stakeholders; Tree Farm Initiative afforestation activities in overseas countries to create sustainable sources of hardwood chips. The company set a target of more than 100,000 ha of afforestation areas, more than one million dry tons of wood chips supplied annually from afforested areas, achieved by fiscal 2008. This company acquired forest certification and total imported hardwood chip to 100% by 2008.

Besides, the Nippon Company also established a philosophy and basic policy on 'product safety' in October 2004. This company considers it to be its corporate responsibility to provide safe products and services. The activities of the entire group are aimed at gaining high levels of trust from customers and making an increased contribution to society. The philosophy integrated into the company "to work to improve safety at every stage of the life cycle of our products, from design to manufacture, supply and disposal, and to provide products and services that the public can trust."[29]

---

[27]For information "Overseas Ventures" in terms Nippon Paper's raw materials, see Pulp and Paper Statistics (2006, p. 24).

[28]The three largest pulp and paper companies (are Oji, Nippon and Daio), they own their cargo ships for shipping wood chips from overseas at least 91 units. Interview was carried out with Fujiwara, staff of Paper Museum, in Oji, Tokyo, on October 8, 2006.

[29]See http://www.np-g.com/e/csr/ideology/products.html.

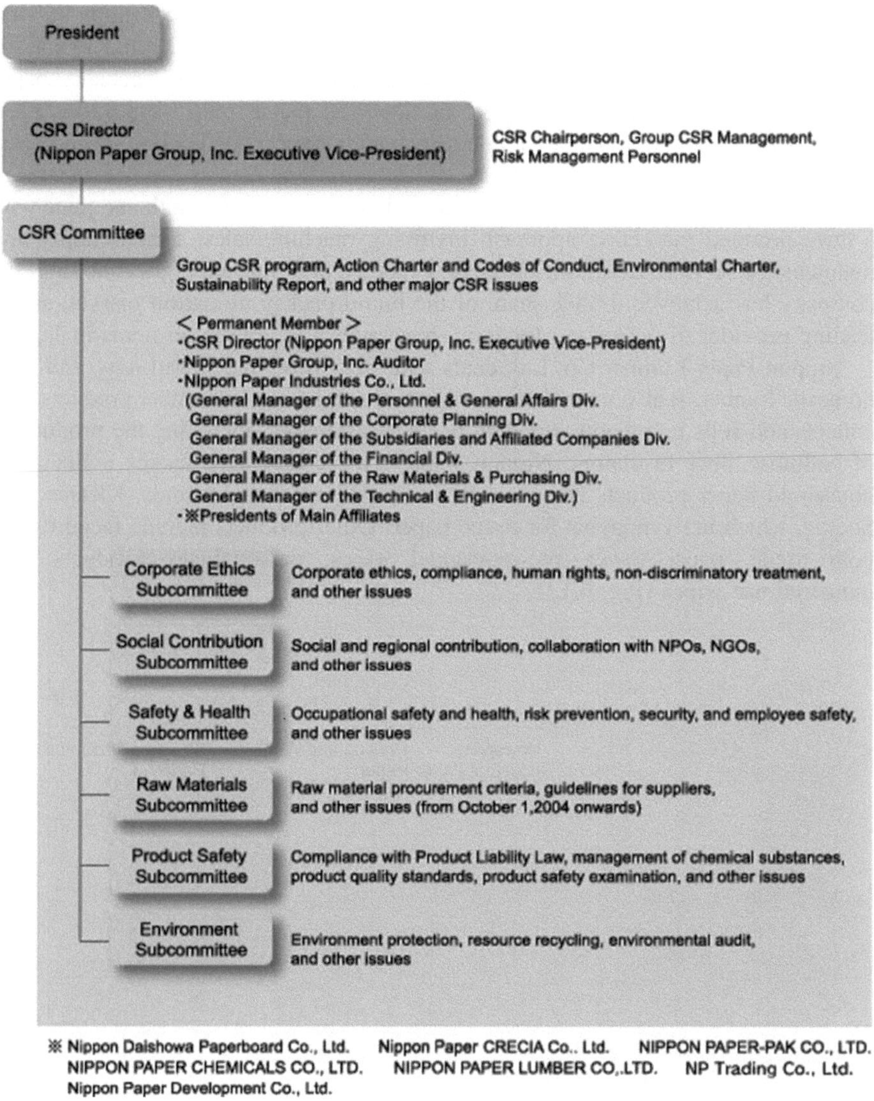

Fig. 6.11 Nippon Paper Group's CSR promotion structure (from October 1, 2003 onwards)

Products

This company in 2003 produced 8.7 million tons of paper and paperboards (Japan Pulp and Paper 2005). The product quality of the company products is based on the optimal utilization of Research and development (R&D) in many divisions (Fig. 6.11). In the case of Nippon Paper Group, namely Nippon Daishowa

Paperboard Co. Ltd. operates a number of paperboard business. Paperboard has diverse everyday applications as low-cost and light weight packaging materials, typified by corrugated paperboard and paper boxes. Meanwhile, specialty paper products namely, Teabag filter paper, air filter bag paper, pulp wrap paper, meat casing paper, insulation paper, wiper paper, tracing paper, oiled paper, facial tissue, etc.

Nippon Paper Co. Ltd. provides liquid-packaging cartons for milk and juice with a three-pronged marketing approach involving machine sales, carton supply and maintenance service. Based on abundant expertise and outstanding technology, the company has achieved a 33% share of the liquid-packaging carton market, as the leading provider of packaging for food, beverages and household items in Japan.

Nippon Paper Lumber Co. Ltd. deals with raw wood and sawed logs, and sells domestic lumber. It also manufactures plywood and laminated lumber products, and collects and sells pulpwood, wood ships and wastepaper, including the production of cellulose fiber insulators. Nippon Paper Crecia Co., Ltd. makes a lineup of household paper products that include two world-renowned brands, Kleenex and Scottie, which are synonyms for tissue paper. Other products include facial tissue, both tissue, paper towel, pre-moistened wipes, personal-care products, and industrial-use wipes (Fig. 6.12).

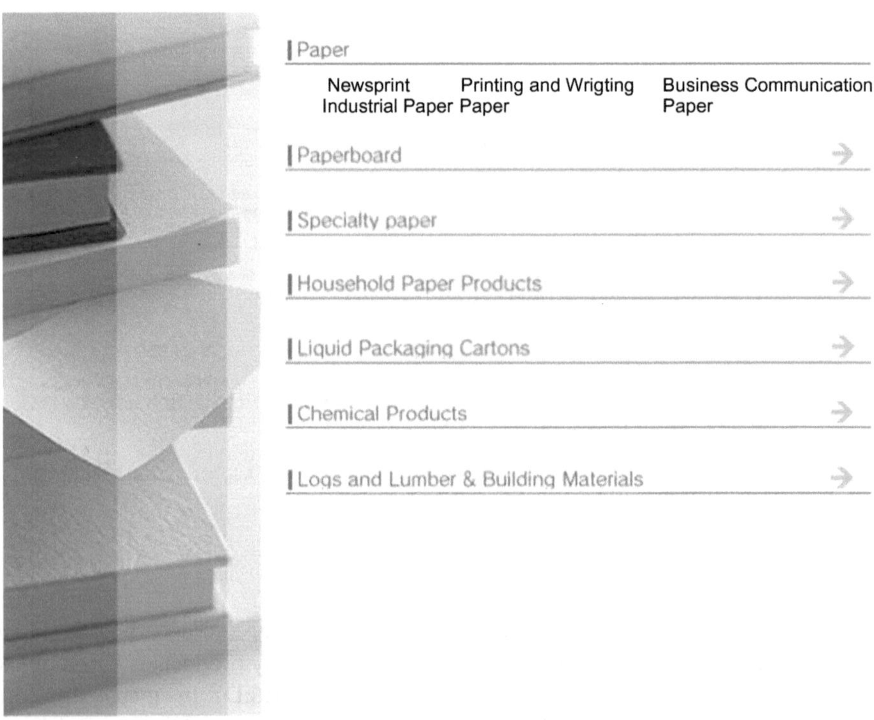

Fig. 6.12 Products

Research and Development (R&D)

The Nippon Paper Group's utilization of R&D is very significant. It considers its strategic vision to produce a high quality of paper products in the future. The company emphasizes its corporate activities in sustainable development that is in harmony with nature and seeks to help achieve a recycling-based society and worldwide environmental protection; the company is constantly developing new technologies and products through fresh innovation. The Company's R&D policies are "Responsiveness to user needs," Augmentation of our global competitiveness," and "Regard for the improvement," and it conducts comprehensive R&D activities from research fiber resources through the improvement of production technology efficiency and the development of high-quality products with innovative features. Therefore, to implement efficient and comprehensive R&D from raw materials through finished products, Nippon Paper Industries has consolidated its research functions in Tokyo's Kita ward, where it undertakes all of its paper and pulp R&D activities.

The Research and Development organization consists of five laboratories: (1) the Forestry Science Research Laboratory, which takes charge of forestry and biotechnologies; (2) the Pulp and Paper Research Laboratory, which develops pulp and paper technologies; (3) the Product Development Research Laboratory, which manages the development of new products to meet emerging needs; (4) the Intellectual Property Department, which manages the intellectual property generated by our R&D activities; (5) the R&D Planning and Administration Department, which handles the general affairs of the division. Thus, by optimizing these R&D functions, leading products area developed constantly with an eye towards the future.

One of the R&D findings, Nippon Paper Industries has developed two tissue culture technologies known as "photoautotrophic culture technology and low-temperature storage techniques," with which it has succeeded in mass clone propagation of Eucalyptus globulus and other products (Fig. 6.13).

Nippon Paper Group Makes Acquisition in Hokuetsu Paper Mill

Generally speaking, a company who makes an acquisition of another company has two prerequisites: (1) the healthy managerial condition of its company and ability get profit; (2) the maximal utilization of R&D that by making acquisition to another company will get more integration. Obviously, the two prerequisites above have been fulfilled by Nippon Paper Group to make acquisition of shares in Hokuetsu Paper Mills (the sixth largest paper mill in Japan). Then, Nippon Paper purchased 13,356,000 Hokuetsu Shares (8.49%) on July 28, 2006, making it the second largest shareholder in the papermaker after Mitsubishi, the trading company, which

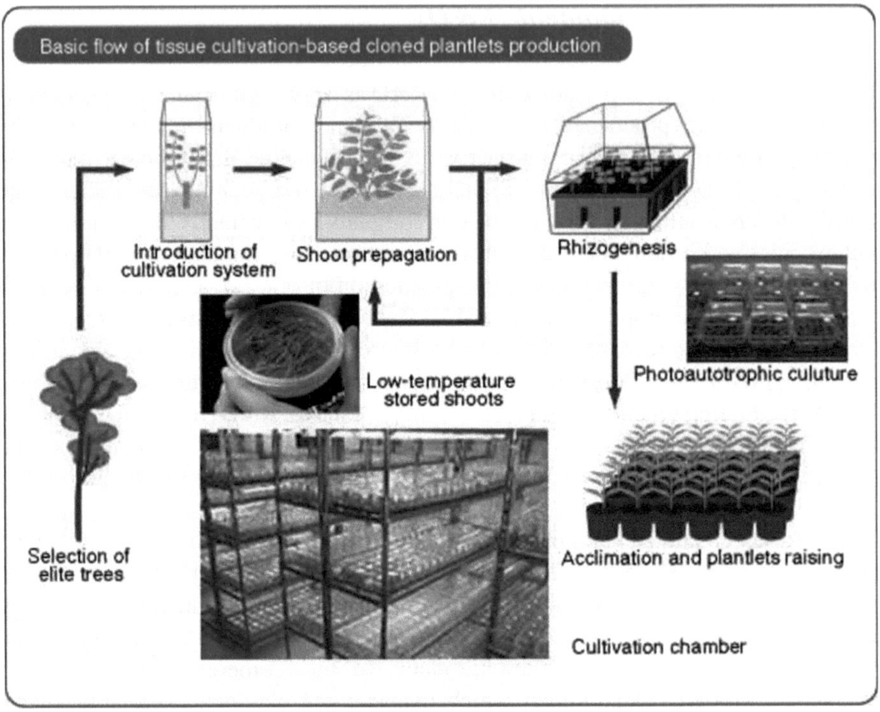

Fig. 6.13 Basic flow of tissue cultivation-based cloned plantlets production

had a 24.4% stake on August 7 (The Japan Times, August 25, 2006). In a report submitted to the government the same day, Nippon Paper Group said the shares were purchased by Nippon, a fully owned subsidiary, for a total of Y 10.7 billion, or Y 806 per share, on average (The Japan Times, August 5, 2006).

The objective of the share acquisition, according to Nippon Paper, is not intended for a controlling stake of Hokuetsu Paper, but aimed at potential business integration, etc. On the other hand, Nippon Paper has been strongly aware of the importance of the strengthening of its business foundation and operational rationalization by way of consolidation and business integration, and has, itself, taken a lead in the industry with this respect.

Why does Hokuetsu paper refuse a takeover of her share (40%) by Oji Paper? But this Hokuetsu prefers to work with Mitsubishi and Nippon Paper. There are some arguments. Firstly, an Oji takeover would damage shareholder value. The company said a merger with Oji would cut its pre-tax profit in the fiscal 2009 year by between Y 13.4 billion and Y 17.9 billion. Secondly, its tie-up with Mitsubishi would generate at least Y 3 billion in additional profits, as it would include procurement and distribution benefits. Thirdly, most of Hokuetsu Paper's employees were against Oji's Paper's acquisition. Fourthly, Oji's takeover plan has drawn fire from local government officials and business leaders in Nagaoka and Niigata

Prefecture. For example, regional lenders Hokuetsu Bank and Daishi Bank, each have a 2 percent stake in Hokuetsu, and have also refused to sell to Oji (The Japan Times, August 25, 2006). Mr. Masaaki Miwa (Hokuetsu President) said: "We are fully confident that we can block Oji's attempt." The Japan Times, August 10, 2006. Herewith, this paper highlights the business strategy of Oji and Nippon companies (Table 6.11).

Environmental Problems

Introduction

The pollution from pulp and paper industries, which subsequently can turn into environmental catastrophe, should be well managed. Otherwise, it can seriously affect environmental issues and human beings, such as Minamata and Itai-Itai diseases. In this case, the role of indirect actors, such as NGOs (Fishery and Farmers Association), local people and academics in criticizing pollution caused by these industries was significant enough to be explored. This paper discusses how NGOs, academics and local people launched critiques of Honshu Paper and other companies due to pollution of the Edogawa River in 1958 and Tagonoura harbor in the 1980s.

High Level Economic Development

Japan launched her industrialization after World War Two to produce goods for civil markets in 1950s. When food, clothing, chemistry and other goods became plentiful after the war, a production and marketing system naturally developed, and this led to Japan's high- economic growth period. Electricity and electronic related appliances became more common and motorization (cars, motorcycles, etc.) began to change people's everyday lives. At the end of the 1950s, the production of consumer goods, such as washing machines, refrigerators, transistor radios, television sets, tape-recorders, motorcycles, cars, and the like increased with great rapidity. Because of the intense competition between companies, there was a great deal of investment in infrastructure to increase the efficiency of mass production systems.[30] This high-growth era also brought a revolution in consumption. For instance, parents were driven to buy television when their children started to visit a neighbor's house to watch television and failed to return home for dinner. Refrigerators and washing machines became part of Japanese family life.

[30]See Hoshino (1992).

Table 6.11 Business strategy of Oji and Nippon

Issues	Oji	Nippon
Raw material	Certified forest; 140,000 ha (2004); 300,000 ha (2011)	Environmentally friendly; 100,000 ha (afforestation); Hardwood import 100% by 2008
	Overseas 80% (New Zealand, Australia, Chile, China, Vietnam, Laos)	Overseas 70% (S. Africa, Australia, Brazil)
	Domestic 20% (Japan)	Domestic 30%
Product scope and annual sales and profit, total asset	8.9 million tons of paper and paperboards; Y 1.185 trillion sales; Y 43.3 billion profit; 1.6 trillion asset	8.7 million tons of paper and paper specialty; Y 1.179 trillion sales; Y 24.4 billion profit; 1.5 trillion asset
Product market	Aggressive expansion Overseas (16 mills): China, EU. Full vertical integration	Aggressive expansion; Overseas (17 mills): China, EU, USA. Full integration
Production facility	Japan (32 mills): R&D priority	Japan (22 mills); R&D priority
Management and institutionalization	Professional managers; Merger other companies Listed in Tokyo SX: actively selling equity share and find out liability from bank	Professional managers; Merger other companies Listed in Tokyo SX: actively selling equity share and liability from bank

Source Processing data by writer based on information

These symbols were later replaced by the three Cs: car, cooler, and color television.[31] The concentration of the factories were located in designated new areas in the Pacific Cost belt that starts in the Kanto plain areas (Tokyo, Yokohama, Yokkaichi, Nagoya, Kobe, Osaka, etc.,) and runs all the way down to Kitakyushu, passing the northern rim of the Setonaika Sea. By 1974, the belt area alone accounted for 84.5% of Japan's entire industrial production.

This growth in industry brought many young people from the rural areas into the industrial cities such as Tokyo, Osaka, Nagoya, Yokkaichi and Kobe. Urbanization rapidly increased, and by 1970, 44% of the population of Japan was concentrated in these urban centers. In contrast, the farming areas, which had experienced an oversupply of labor for a very long time, now faced labor shortages, and this resulted in the rapid mechanization of agriculture.

[31]Shigeaki Fujisaki, "Development and the Environment," Op Cit, p. 15.

Pollution Issues

During the first phase of high economic growth in the 1950s–1960s, the greatest environmental pollution problem was caused by dust, smoke and other airborne particulate matter. The main source of energy at that time was coal. Dust collectors and other methods of particulate matter control were not provided, and all of the chimneys belched forth black smoke. The effluents from sulphuric-acid based pulp processing industries are the same in this regard, that is, they are not directly and immediately apparent as the cause of environmental problems. Paper Companies, such as Honshu in Tokyo, automotive vehicles gas emission, a major iron and steel complex in Yahata, northern Kyushu, was pouring 27 tons of particulate matter per day into the city's air, and in Kawasaki City, situated in the Tokyo Bay industrial area, the amount was 23 tons.

Also, especially in the period from 1970 to 1974, great damage was caused by photochemical smog and other effects resulting from atmospheric pollution attributable to auto emissions. For instance, the number of victims affected by photochemical smog totaled 200,000, counting only those cases that had been reported.[32] In early 1970s, almost a million persons were either killed or injured in road accidents annually. As a result, the number of school and young children who were orphaned in this way totaled more than 60,000, about 40% of whom were from families already receiving social relief at the time. These accidents caused by environmental disruption were brought about by automotive vehicles.

Along with the black smoke, there was also a great amount of red smoke that spread over the sky and polluted the water in the river and sea. This smoke, dust, contaminated water and particulate matter caused pollution issues. From the end of the 1960s to the beginning of the 1970s, the so-called major pollution cases (the Kumamoto—Minamata disease, Edogawa River pollution, the Niigata Minamata disease, the Itai-Itai disease, Yokkaichi Astma, Tagonoura's Beach Pollution, etc.) became forums for open discussion of polluter responsibility.[33] In the early 1970s, Tokyo's pollution problems were so severe that protesters began using the slogan "No More Tokyo." Although many problems remain, including nitrogen oxide levels that still exceed environmental standards, Tokyo's air is now incomparably cleaner than in those days, and fish have returned to the city's rivers.

It was obvious that Minamata disease was the most massive pollution problem caused by Nippon Chiso's manufacturing to strike Japan in the 1950s. The village of Minamata located on the west coast of southern Kyushu, was traditionally supported by rice farming and by a cove in the port which allowed the production of salt. In general, Minamata disease was characterized by symptoms typical of methyl mercury poisoning, and it was described and recognized in patients who developed

[32]See Tokue Shibata, "The Influence of Big Industries on Environmental Policies: The Case of Car Exhaust Standard," in Shigeto Tsuru & Helmut Weidner (eds.), *Op Cit*, pp. 100–101.

[33]See Jun Ui, "Anti-Pollution Movements and Other Grass-Roots Organizations," in Shigeto Tsuru and Helmut Weidner (eds.), Op Cit, p. 113.

these same severe symptoms in 1953–1960. In the initial stage, it was discovered that 17 people in all had died after showing the same symptoms and characterized by a profound sense of shock at the high death-rate.[34] Surveys were conducted after the second outbreak of Minamata disease by the Medical Department of Kumamoto University in 1971. The result of the survey indicated that 158 Minamata disease patients had been discovered in a sample of 50,000 persons. However, 114 of the 158 had already been designated Minamata disease patients, and later surveys indicated that even greater numbers than reported in the survey had also been designated as victims.[35]

What was the reaction from Minamata fishermen? During the summer and autumn of 1959, the Minamata Fishermen's Association and Fushimi Sea Fishermen's Association demanded compensation from the company for damage perpetrated by the chemical complex. The company refused to make any payments on the grounds that the cause of disease was not understood to be related to the operation of the chemical complex. However, the company did decide to pay a small amount of sympathy money. The social tensions in and around Minamata City rose to a crescendo, and on November 2, 1959, 4000 Fushimi Seas Fishermen congregated to demand a National Diet members' inspection of the polluted area. On the way home from this rally they broke into the factory and destroyed office equipment. This event subsequently magnified the issue in the national news media and as a whole, some three and half years after it was discovered. The labor union of this factory (Nippon Chiso) protested and was supported by the Japan Socialist Party and they criticized the fishermen's riot. Apparently, if this action had not been taken by the fishermen, the Minamata disease would never have become national news in Japan.

This paper would like to focus on discussion on the effect of Pulp and Paper Industries in Japan which eventually affect water pollution in Edogawa River in Tokyo and Tagonoura, in Fuji City, Shizuoka Prefecture.

Edogawa River Pollution

It initially happened when Honshu Paper,[36] which was located about 8.8 km upstream from the mouth of Edogawa River, bought a new semi chemical pulp machinery (SCP) costing Y 1.1 billion from United States in April 1958. This machine could cut wood from logging and directly processed it, with mix with

[34]Jun Ui, "Minamata Disease," in Jun Ui (eds.), Op Cit, pp. 104–110.

[35]Ibid, p. 124.

[36]Honshu Paper was originally extended company from Oji Paper Company, one of Shogo Shosha. When Japan was defeated in the war by USA and its allies in 1945, the USA's policy launched economic reform in Japan, especially that Shogo Shosha (big companies) must be divided into branches and open their shares for public trading.

Fig. 6.14 Edogawa river in recent year (Chiba Prefecture). *Source* Picture was been taken by the writer, in December 2006

water and ammonium sulfate. Although this machine could efficiently produce a huge quantity of pulp, the water waste subsequently affected water pollution (Fig. 6.14).[37]

This factory dumped pitch black wastewater into Edogawa River and caused great damage to the downstream fishing grounds. Based on the Fishermen Association report, this Edogawa water pollution affected 90% of shell species (Asari, Hamaguri) near sea dead; and 60% shell species far sea dead, and Ayu fish could not bear their eggs. Then, the loss of money from Hamaguri production of about Y 32,570,000 and Nori (sea weeds) of about Y 22,964,000. In that time, Urayasu fishermen consisted of 75% of the total population of 16,471 million. The remaining 25% were categorized as related to the fishing industries.[38] Table 6.12 shows the lost money suffered by fishermen in Japan affected by industrial water pollution.

Twenty-three Urayasu fishermen on April 23, 1958 in downstream areas protested this pollution against the company and demanded negotiation. They made

[37]Discussed with Dr. Yuri Sato on October 18 and November 6, 2006, Kawashima, Takenori, 1958, "Urayasu Gyomin Sodo no Hoshakaigaku-teki Kosatsu," (Legal Sociologistic Consideration on Urayasu Fishermen Disturbance), in Juristo, No. 159, August 1, pp. 2–4.

[38]Discussed with Yoko Kawano, on November 3, 2006, Wakabayashi (2000).

Table 6.12 Water pollution of industries and fishermen victims in Japan

Year	Number of cases	Amount of loss due to pollution 1000/kan@	Value of loss of money 1000/yen
1950	295	7229	65,340
1951	333	5494	104,990
1954	706	2454	379,054
1956	478	2800	778,812
1957	810	–	1,342,508
1958	749	–	5,231,133

Source Department of Fishery (1958). 1 kan is 3.75 kg (Japanese weight)

two requests to the company: First, the company had to pay compensation to the Fishermen Association which totaled an amount of Y 9,000,000 for their loss of income. Second, the compensation money must be paid after the installation of a new machine for cleaning water in front of the Chiba Prefecture officer, head of the company and head of the Fishermen Association.[39] Unfortunately, this protest was not well responded to by the company. Another protest was held by Urayasu Fishing Cooperative Association on May 14, 1958 to the Urayasu Local government, Chiba and Tokyo government officer. The warning protest from Tokyo officers to the head of Honshu Paper Company could not stop throwing water waste to the river. Thus, the greatest protest from fishermen and Fishermen Cooperative that was managed by Okajima Tasaburo[40] was held on June 10, 1958, when ten buses and 1000 people came to Diet (parliament) and the Tokyo government office. The head of group and other members met and discussed with Shojoro Kawashima, a member of Diet from Urayasu. Unfortunately, both parties left the discussion unsatisfied, particularly on fishermen side. On the return home, in the afternoon, the head of the group and other Fishermen Association members requested the bus drivers to go to the company location. Although, police men had totally forbidden protesters to enter this company, they eventually entered the factory and cut the pipe line which brought water- waste to the river.

This protest turned into an incident among fishermen, factory workers and police men. The amount of heavy accidents suffered included 35 men and 108 light ones. The total injured numbered 108 people. Among them were 37 persons injured from policemen, 3 from journalist and 3 persons from the company, and subsequent arrests of 81 men. As a result of the Urayasu fishermen disturbance, the Tokyo Governor on June 13, 1958 required stopping operations of this factory. Besides,

[39]Keiko Wakabayashi, Op. Cit, p. 328.

[40]In fact, Okajima Tasaburo was not a fisherman, he was the owner of the shop, but he was a powerful man from Yakuza group, as seen some tattoos in his body. He eventually worked and colluded with Oriental Lando Company Group (Mitsui Real Estate, Kessei Dentetsu Railway and Asahi Tochi) to provide land in Urayasu for the Disneyland area in 1960s. Okajima eventually became a rich man and mayor of Urayasu City. Because of his criminal action, Okajima was subsequently caught by police and put in jail.

Diet institute also proposed to investigate the Urayasu incident and required the factory to stop operating and paid "compensation" to the fishermen. The response of this company was to subsequently pay compensation about Y 5,100,000 to the Urayasu Fishing Cooperative Association.[41]

This Urayasu fishermen incident invited academician critiques, one by lawyer Takenoro Kawashima who said: "this incident totally reflected that the law condition could not realize in the field. Even its implementation showed discriminately to local people, which they have no political bargaining due to environmental pollution protested by local people. Then, other stakeholders (local government, businessmen, Diet and police) just gave response that eventually affect incident."[42]

Other scientists commented regarding Edogawa River Pollution such as Kenji Yamazaki from Meiji University and Tomoko Yamazaki from Iwate University that the movement to revive Tokyo Bay, however, was initiated by local inhabitants and activities of fishermen. When we open the history on the 'environmental pollution,' it becomes clear that "the fishermen have been sensors for pollution and subjects of experimentation of some cases such as Minamata in Kumamoto, the second Minamata (Niigata), Yokkaichi, Tagonoura, etc. Furthermore, they said that this Edogawa pollution incident triggered the legislation of two laws governing water and air quality.[43]

From a positive perspective of this Urayasu incident, the central government proposed drafting a new law to Diet (National Parliament) on making "Environmental Law". This was an initial step on 'environmental law' in Japan on December 25, 1958. This law consists of two emphasize materials namely: first, a law on protection of air quality for factory management; and second, law on water-waste from factories. It was actually, two shortages on realization in the field of these laws. Firstly, the harmony between industrial holders and People's lives still could not be reached, yet. Secondly, registration for water areas system was difficult to realize in reality. It happened, because of negotiations still not being made clear, especially regarding the definition of 'registration' on water area systems among stakeholders and also registered versus unregistered water areas systems.

Observing this pollution impact, Diet released "Anti Pollution Act" in August 1968 for seven leading issues namely: air, water, soil, voice, quake, land slide, and contamination. The purpose of this Act is to protect Japan from pollution of various elements, which subsequently affect Japanese people's lives and maintain Japan's earth quality.

---

[41]Keiko Wakabayashi, Op Cit, p. 334.

[42]Kawashima Takenori, Op Cit, p. 3.

[43]For further information see paper of Yamasaki and Yamazaki (2000).

Tagonoura Harbor Pollution

Tagonoura water pollution in 1980s was caused by waste liquor of paper and pulp manufactories. The pollutants were found to contain such ingredients as fiber, lignin, resin, fatty acid salt, inorganic fillers, messes of microorganisms, and others produced from manufactories. When the polluted water is mixed with seawater, SS becomes more likely to be flocculent and cohesive. Finally, SS that flows into Tagonoura harbor has a concentration of 30–40 ppm, which means approximately 20,000 tons annum In Shizuoka Prefecture and Fuji district, there were 142 paper-pulp factories, of which five are owned by major manufacturers (such as Tokai Pulp and Paper, Tokushu Paper, Kida Mill, Oji Paper has four factories, Nippon Paper Crecia, etc.). Those factories utilized abundant water from the Fuji Mountain System and used to pump up 1.25 million m^3 of water per day from numerous wells and discharge the water in the rivers, i.e. the Numagawa, Wadagawa, Egawa, Kouruigawa, Akafuchigawa, Takigawa and Uruigawa, etc., and streams in the estuary Tagonoura Harbor.[44] At present, 81% of the industrial effluent is being discharged in Togonoura Harbor through the industrial drainage channels that were constructed subsequently (Fig. 5.2). The quantity of the above waste water amounts to 1.2–1.4 million m^3 per day. The rest, namely 19%, is discharged into the rivers flowing in Tagonoura Harbor. Therefore, the sludge accumulating in the harbor contains not only masses of fibrous fine organic flocks, but also earth and sand that have flowed from the rivers. The earth and sand from black mud have an offensive odor from H2S developed within; when exposed to the air, its color changes to dark gray (Fig. 6.15).[45]

Although the contaminants have been reduced, the volume of pollutants discharged from the industrial drainage channels has too large a load compared with the self-purification capacity of the waters around Togonoura Harbor. In the harbor, SS continues to accumulate, and the sediments are turning into sparopel rotten mud. As a consequence, virtually no fish or shellfish are living and breeding in the harbor. This pollution subsequently affects fishermen's loss of their income.

The reaction and hard protest from anti-pollution movements by the local residents, fishermen and general public were very strong for pollutants to be reduced. An agreement on pollution control was concluded between Fuji City and the Paper Companies concerned, thereby a regulation was established on the effluent concentration and companies eventually set up antipollution facilities. Then, the problem arises of how to control the load of pollutants discharged into the Tagonoura Harbor water area and reduce it to within the area's self- purification capacity? The solution to overcome this critical problem is to realize a stricter application of the environmental quality standards. Namely, those standards should be applied strictly to the relevant rivers rather than to the seawater area and raise the applicable water quality standard to a level at which fish can live. The next step to

[44]For further information see Ogushi et al. (1998).

[45]Ibid, p. 136.

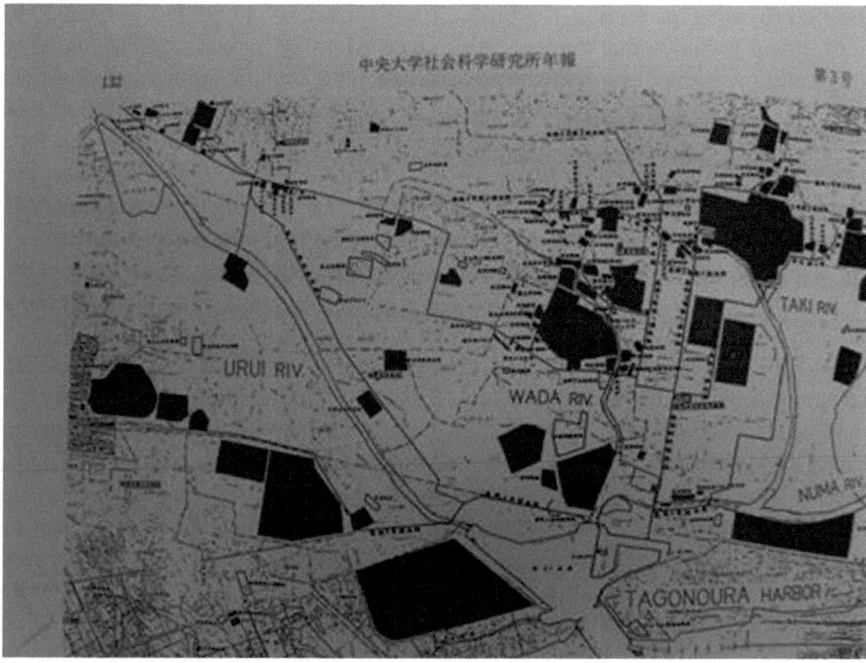

Fig. 6.15 Water pollution in Tagonoura harbor affected by Pulp-Paper Factory Mills. *Source* Ogushi et al. (1998). Water Pollution in Fuji District (paper)

take for that purpose is to build industrial effluent treatment facilities where waste water from factories will be purified to a quality level at least more desirable than the water flowing out of Tagonoura Harbor at present, and the water thus purified will be discharged directly to seawater areas outside the harbor without passing through the rivers running into the harbor and through the harbor itself.

Stakeholders' Reaction to Environmental Destruction Analysts Comment

Jun Ui, Professor of Economics in Institute of Regional Studies, University of Okinawa, represented as having an academic background, launched a critical comment on environmental destruction. He said that critical pollutions occurred in Japan due to the political attitudes, and the misguided supervision provided by national and local governments functioning in collusion with business organizations (private companies) greatly exacerbated environmental problems. In this case, pollution victim-based citizens' movements were suppressed on the basis of national security. In the post-Second World War period, the national government did not change these basic structures, and government policy has consistently been one of providing protection for business organizations. In pollution-related disputes, it was common that the legislature stood as proxy for corporations, and even anti- pollution agreements concluded between municipalities and enterprises was used as a shield to protect both the enterprises and municipal authorities from victims' actions. The symbiotic relationships between Japanese politics and business are now so solidified on an organizational level that they represent true state monopoly capitalism. Under the aegis of the conservative party (Liberal Democratic Party/LDP) that has long ruled Japan, officials in the highly organized government bureaucracies take management positions in large corporations upon their retirement from government service, or become conservative politicians. This symbiosis between business and government is to be found in all industrial sectors.[46]

Another analyst, Tokue Shibata, former Director of the Tokyo Metropolitan Research Institute for Environmental Protection, an institute founded by the Metropolitan Government in 1968. Since that time, the Japanese government has placed heavy emphasis on nurturing the automobile industry (such as Toyota, Honda and Nissan, etc.), offering various incentives to manufacturers and wide-ranging facilities. Firstly, a tariff-barrier was set up in a move to protect the industry against powerful western competitors. Secondly, special tax benefits were granted to reduce the tax burden and encourage capital accumulation. Thirdly, investment in plant and equipment was facilitated by extending long- term, low-interest loans through government sponsored financial institutions. Fourthly, to establish limits to the volume of automobile exhaust gases in October 1972. The guidelines provided for a reduction of up to one tenth of the current concentration of carbon monoxide (CO) and hydrocarbons (HC) by the fiscal year 1975; and for nitrogen oxides (NOx) by the fiscal year 1976. This regulation for automobiles in Japan, especially for Tokyo Metropolitan City, is based on the guidelines from Environmental Agency of Japan on 'Environmental Pollution Control' with regard to establishing a standard for the control of auto emissions.[47]

In fact, in 1950s, prefectures such as Tokyo, Osaka, and Kanagawa already enacted ordinances for pollution control. Even so, these merely regulated the procedure for approving the installation of those factories likely to cause air and water pollution. Thus, local governments had to revise their ordinances so that more proper and effective regulation could be enforced with regard to the new types of industrial pollution. In this course of this process, arguments as to the compatibility of national laws and local ordinances arose and must be properly adopted.

---

[46]Jun Ui, "Overview," in Jun Ui (eds.), Industrial Pollution in Japan, Op Cit, pp. 2–3.

[47]Tokue Shibata, "The Influence of Big Industries on Environmental Policies: The Case of Car Exhaust Standards," in Shigeto Tsuru & Helmut Weidner (eds.), Op Cit, pp. 44–102.

The Role of Nongovernmental Actors

By the end of the 1960s, anti-pollution movements had become strong nationwide. The diligent work done by supporters of Minamata disease victims, as well as movements resisted corporations that continued to poison the biosphere by refusing to process their wastes. In 1964, citizen's movements were able to stop the planned construction of large petrochemical complexes in the cities of Mishima and Numazu on the Pacific coast. Also, Narita's farmers who launched resistance movement against the construction of Narita Airport, provided a great deal of encouragement to local citizens' movements all over Japan.

The specific character of Japanese nature conservancy and anti-pollution movements derives from the fact that they started as "federations" of lower middle-class persons, such as school-teachers, local government workers, small shop-keepers and landlords, as well as fishermen and farmers, who are mostly outside the market economy, with the central core nevertheless still among the lower-middle and poorest classes.[48] There were some common tendencies among many of the local movements during the 1970s. The target of this local movement was frequently the strengthening of local self-governing bodies, since those development projects that are targeted for resistance are mostly prepared, or supported, by central government. Local movements are basically populist in character.

The pattern of organization derives from a local organization, already existing, becoming the basis of the movement. In this case, most organizations and formal groups retain a pyramidal hierarchy of leadership over its members. In large-scale projects, such as the construction of oil, coal, and pollution from pulp and paper, mining, chemistry, etc., concerned fishermen's and farmers' cooperatives frequently began to resist; although these groups were rather susceptible to pressure from government and industry holders, sometimes being forced in the end to accept monetary compensation for the loss of fishing or faming rights. A new tactic towards the end of the 1960s was the court actions, which proved quite useful. The victims of pollution (such as Minamata disease in Kumamoto and Niigata, Itai-Itai disease, Yokkaichi Atsma) decided to go to court because there were no other forums for open discussion of cause and effect relationship and polluter responsibility.

The Response from Central Government and Locals

The attitude of central government to local movements has consistently been to regard them as a problem of 'national security,' reacting with hostility and oppression under a policy that is designed to protect industry. In 1970, this

[48]Jun Ui, "Anti-Pollution Movements and Other Grass-Roots Organizations," in Shigeto Tsuru & Helmut Weidner (eds.), Op Cit, pp. 112–114.

government policy was brought under explosive criticism on the part of public opinion, so that the government proposed a dozen pollution control laws in the National Diet and decided, and eventually established an "Environmental Agency" in 1971. The main task of this Agency is responsibility for environmental problems. This represented some progress in the actual control of pollution, when compared to the former legal system. The Environmental Agency is not a fully-fledged ministry, but a special department (gaikyoku) of the Prime Minister's Office (PMO). The legal status given to the heads of ministerial agencies involves assisting the Prime Minister in his role as head of the Prime Minister's Office in the administrative activities under his supervision. If necessary, they are able to resort to the Prime Minister as a final measure in order to break any deadlock caused by severe inter-ministerial conflicts. According to one conventional definition, supplied by Prof. Isao Sato of Sophia University,

> …in some cases, they are not sufficient in size to be an independent ministry and, in other cases, their administrative affairs relate to be business of several ministries so that it is considered inappropriate for any one ministry to assume overall responsibility for them.[49]

Further, three research and training institutes are attached to the Environmental Agency. Among them, the National Institute for Environmental Pollution Research, which began functioning in 1974, is located on the site of the Tsukuba National Academic Center and has a staff of about 250, more than 27% of the Agency's total staff of 914 in 1985.

In realization of 'environmental law', Central government has transferred the implementation of certain tasks to local government officials, and a number of national laws prescribe the duties of local government. Local government ordinances, enacted for the protection and improvement of the environment, can largely be classified into the following categories:

(1) ordinances for pollution control;
(2) ordinances relating to general environmental policy;
(3) ordinances for nature conservation;
(4) other ordinances for specific objectives, e.g., for environmental impact assessment.[50]

Ordinances for pollution control provide the regulatory schemes for pollution in the respective areas of jurisdiction. They prescribe the regulations standards, facilities and substances subject to the regulation, the method of inspection and monitoring, etc. All prefectures have enacted such as ordinances for pollution control and, in addition, 499 municipalities have enacted similar ordinances. Some local governments have also adopted administrative guidelines, for preliminary consultation and granting permission to site and install factories and plants,

---

[49] See Tsunao Imamura, "Environmental Responsibilities at the National Level: The Environment Agency," in Shigeto Tsuru & Helmut Weidner (eds.), Ibid, pp. 43–44.

[50] Hidefumi Imura, "Administration of Pollution Control at Local Level," in Shigeto Tsuru & Helmut Weidner (eds.), Ibid, p. 62.

as designated by prefecture governors or mayors. For example, in Japan, the metropolitan government of Tokyo enforced a revolutionary ordinance for the prevention of pollution in 1969 centered on a declaration of the basic human rights of Tokyo's citizens, speaking specifically of a 'comfortable life' for all her citizens.[51] Under strong pressure from citizen groups and progressive local authorities, the Japanese automobile industry had to seek to improve car engines so that they would consume less fuel. This was their basic attitude towards solving the problem of air pollution caused by gas emissions. They were more or less pressured into doing this by the grass-root campaigns. As a result, remarkably efficient and economical engines came onto the Japanese market from about 1975 onwards.

Concluding Remarks

Japan as rank three exporter of paper reached 30.8 million tons in 2004 after United States and China has been properly managed pulp and paper industries from upstream until downstream. Japan is very aware that she does not own rich resources; therefore, the pulp and paper company optimized functioning R&D (Research and Development) for many strategic developments. Firstly, how the company (Oji and Nippon) can invent the best quality seeds (Eucalyptus globulus) and plant afforestation programs for getting productive timber as raw materials in overseas. This so called-raw material procurement is categorized as sustainable forest management; Japanese companies properly managed their raw materials (timber) as strategic keywords of "upstream level" in their industries. From this perspective, Oji and Nippon Companies fully understand how to minimize the effects of ecological damage downstream such as flood, soil erosion, forest fires, etc. In this case, Japan's forest and its environment have been properly managed. The consequence, it was very rare to see flooding of rivers and soil erosion in some districts in Japan as long as sustainable forest management is well managed. Interesting to note that pulp and paper industries of Japan to procure raw materials, most Japanese pulp and paper companies prefer to plant industrial timber plantation (HTI) as raw materials procurement, which highlights on environmentally friendly in overseas (80%) from total of wood demand compared with domestic orientation just reach 20% of its plantation.

These companies realize that if they are guilty of ecological damage such as utilizing illegal timber for their industrial raw material, Japan consumers could boycott their products and criticize the company for sourcing their raw materials through unfriendly environmental or unsustainable forest management.

Secondly, R&D targeted to innovative technology in terms of water treatment equipment to minimize water pollution. This strategic invention taken by Oji and Nippon after the tragedy of Honshu Paper (that eventually merged with Oji Company) and other paper industries seriously affected water pollution in Edogawa

[51]Tokue Shibata, Op Cit, pp. 107–108.

River in 1958 and Tagonoura Harbor pollution in the 1980s. Yet, the Urayasu Incident, because of Edogawa River pollution, positively affected the adoption of 'environmental law" in Japan by the Central government, which eventually proposed the drafting of a law to Diet (National Parliament) on December 25, 1958. This environmental law aims for national regulation to prevent pollution, which emphasizes protection of air and water quality from many industries holders. As a result, environmental pollution has been relatively reduced in many districts in Japan, as company holders especially pulp and paper industries have to obey these environmental regulations. From the viewpoint of implementation, this clearly shows indiscriminately in action. If stakeholders, especially industry holders, are found guilty, they must be punished.

Thirdly, R&D activities have been targeted toward the discovery of new possibilities in excellent paper and paperboard products (newspapers, journal papers, etc.) As we know, the Japanese papers consumer market is very large and usually they request high quality paper for many printing publications. Japan also exports these products to Southeast Asian countries, such as Indonesia 7656 tons, Vietnam 53,272 tons, Thailand 165,181 tons, Philippines 22,982 tons and also to China 3.1 million tons, S. Korea 177,430 tons and Taiwan 169,819 tons (Pulp and Paper Statistics 2006).

Part IV
Environmental Services

Chapter 7
Yakushima-Japan: Sustainable Forest Management

Introduction

Japan's land territory covers 380,000 km^2; her position is categorized as number 61 in the world. But Japan's territorial waters and EEZ combined are 12 times as large (4,470,000 km^2) as its land area, placing it number six in the world (Kuwahara 2013).[1] The National Parks, which registered 30 units, covers 2,091,163 million ha, or occupying 5.5% of the country's total land, play a significant role for forest conservation and also a crucial role in protecting the Japan's wealthy ecosystems, such as forests, waters, wetlands, seashores and coral reefs and wildlife therein and preserving them for future generations. The National Park System aims at protecting the scenery that represents the Japan's natural beauty. Its existence also aims at providing visitors opportunities for the experience, enjoyment and better understanding of the country's natural environment by (National Parks of Japan, Ministry of the Environment, 2012).[2] One of the National Parks is Yakushima, located 60 km south—southwest of Kyushu mainland, Kagoshima Prefecture (Fig. 7.1). Kagoshima Prefecture owns 605 islands, including 28 inhabited ones. As a specific Yakushima Natural Park and World Heritage, it has a very unique ecosystem and collection of forest trees especially Japanese Cedar (*sugi Cedar*) in Yakusugi Land (Fig. 7.2). One of the biggest trees is *Jomon Sugi*, with a circumference of 16.4 m, a height of 25.3 m and estimated to be from 2600 to 7200 years old.

The condition of National Park for instance Yakushima is very significant as World Heritage which was recognized by United Nations Educational Scientific

[1]The first is the US, Australia, Indonesia, New Zealand, Canada and Japan. For further discussion, see Kuwahara Sueo, "Culture and Society in the Islands of Kagoshima," in Kawai Kei, *The Islands of Kagoshima: Culture, Society, Industry and Nature*. Kagoshima University Research Center for the Pacific Islands (KURCPI). 2013, p. 2. Or refer to Japan Institute of Construction Engineering: http://www.jice.or.jp/quiz/kaisetsu_04.html/02.
[2]For further information open website (http://www.env.go.jp/park).

H. Hidayat, *Sustainable Plantation Forestry*,
https://doi.org/10.1007/978-981-10-7653-4_7

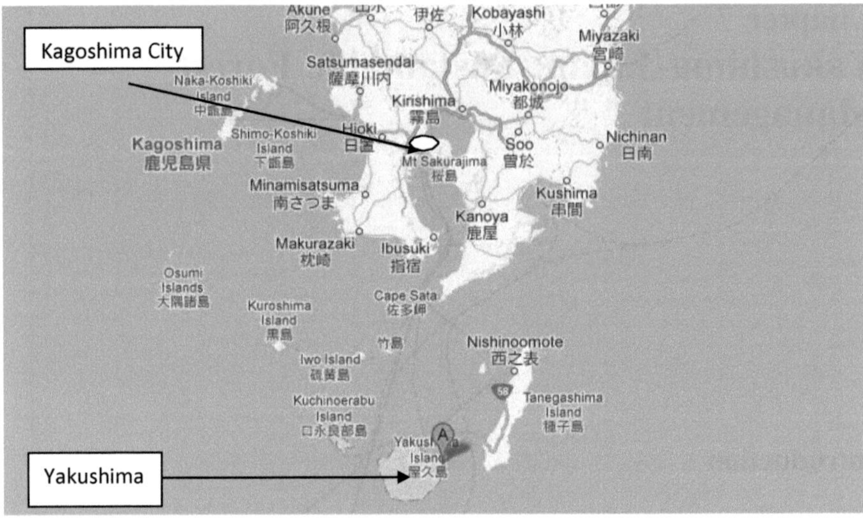

Fig. 7.1 Kagoshima Prefecture and Yakushima Island. *Source* Google (2013)

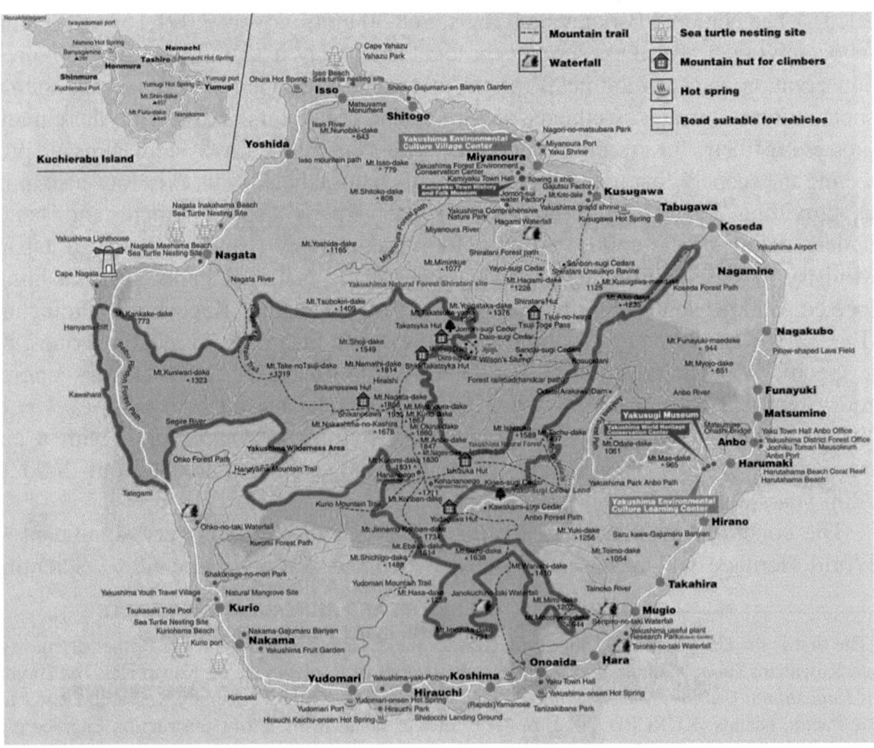

Fig. 7.2 Field study of Yakushima National Park and World Heritage Area. *Source* Yakushima environmental learning map (2010)

Organization (UNESCO) Committees of a significant on-going ecological and biological processes in the evolution and development of terrestrial, fresh water, coastal and marine ecosystems, and communities of plants, trees and animals, Kagoshima Prefecture Office and local communities had a positive response. From an economic and social viewpoint, the recognition of Yakushima as a World Heritage site rapidly developed a 'tourist industry' in Kagoshima City and especially in Yakushima Island. It provided the development of infrastructure (roads ports and Airport), transportation (ferry, Toppy Jet Ship, planes), and an increasing number of tourists visited the island who needed accommodations (restaurant, hotel, *Minshuku* (a private house for rent), *Ryokan* (Japanese style hotel), guest house, Youth Hostel, etc.

This paper focuses examines the 'partnership' management model among stakeholders to manage Yakushima National Park and World Heritage from the issues to be discussed, such as a landscape protection, protecting animal and plants, patrolling to control protected and special zones, implementation of regulations to bring portables for every visitor to visit *Jomon sugi*, and to conduct research in *wilderness area* for academic purposes. The impact of partnership management eventually maintained the ecosystem, water supply and positively contributed to 'environmental' services (*Yakushima Denko*) and tourist industry in Yakushima. Therefore, for this purpose the interview among stakeholders is very necessary in relation to how they manage the park and the impact on Yakushima's designation as a National Park and World Heritage site, such as Ministry of Environment officer, Forestry Agency, Yakushima Town Office, Tourist Agency Association, Farmer, the owner of Minshuku, etc. (Fig. 7.3).

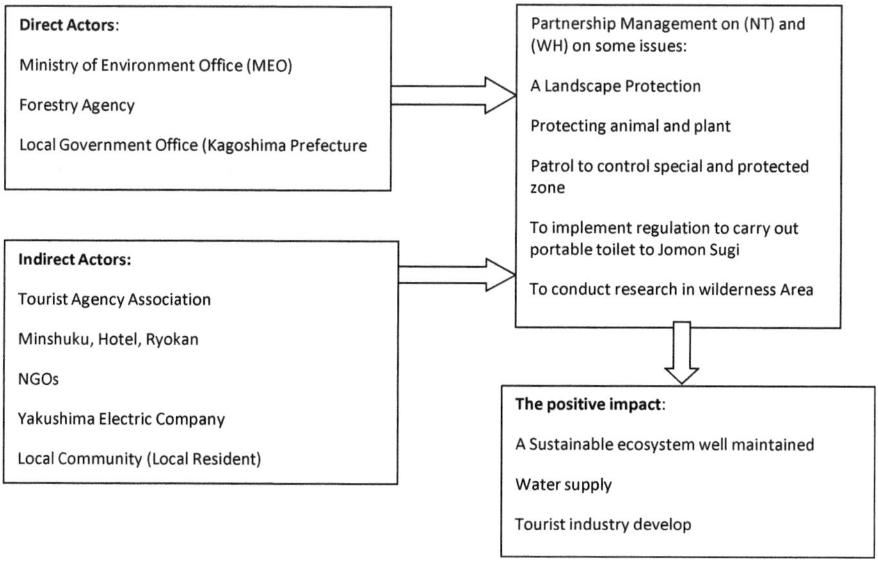

Fig. 7.3 Stakeholders analysis in managing Yakushima National Park (NT) and World Heritage (WH). *Source* Based on the theoretical framework of Blaikie and Brookfield (1987)

Review on Natural Park

The aim for the establishment of a Natural Park is to preserve beautiful scenic areas and their ecosystems and to contribute to the health, recreation and culture of citizens. It was enacted by law in 1957. Japanese natural parks system provides three kinds of parks: national parks, quasi-national parks and prefectural natural parks.[3] Recently natural parks are registered 401 units in Japan. They consist of 30 units of National Parks; 56 units Quasi-National Parks and 315 units of Prefectural Natural Parks (Table 7.1; Fig. 7.4). The Natural Environment Bureau of the Ministry of Environment (MOE) is responsible for the management of national parks. Quasi-national parks are designated by MOE in response to request from local governments. Meanwhile prefectural natural parks are designated and managed by local governments. The budgets for National Parks have responsibility for managing national parks. The Ministry of Environment (MOE) has established various budgets for Natural Park Facility Development Program, the Green Worker Program, night soil disposal improvement, and the auxiliary ranger program.

The mission of the natural parks is conserving natural resources as well as providing the public with access to nature for their enjoyment. Therefore, to achieve these goals, the natural parks provide recreational facilities, such as natural recreation forests (Arakawa and Shiratani area) in Yakushima, trails and campgrounds. Historically, national parks were established in Japan starting in 1911. Parliament members discussed the establishment of national parks based on a petition. And its development in 1920 was discussed among member of Diet to prepare legislation. The Diet (Japanese Parliament) eventually agreed on a National Park Law in 1931. The first group was decided in March 1934 namely three national parks: Setonaikai (i.e., the island Sea), Unzen and Kirishima. In 1957, the Natural Parks Law was enacted to replace the 1931 legislation (MOE 2011).

Positive developments occurred from 1960s to the 1970s–1990s by indication of the increasing visitors to national parks. Actually, visitors enjoy visiting national parks in various ways: mountain climbing, hiking, and recreation in the natural forest, to view the fascinating landscape in the valley, river side, and coastal sea are among the very popular activities in national parks. Those activities enable visitors to interact with nature, ecosystems, forests and marine resources more deeply. As registered in the Japan Tourist Agency 2004, visitors to national parks reached 3.5 million; 2.9 million visited quasi-national parks and 2.6 million visited prefectural natural parks (The MOE 2010). The increasing number of tourist visiting national park at the end of the 1990s has been debatable; according to the findings of Thomas Edward Jones (2012), that "in Japan and other developed countries, visitation at Nature-Based Tourism (NTB) destinations such as national parks rose

[3]Definition of Natural Park: places of excellent scenic beauty and important ecosystems and worthy of designation as national scenic sites. Quasi-national Park: Places of natural scenic beauty almost equal to that of the national parks. And Prefectural Natural Park: places of local significance as designated by the local government (MOE 2011).

Table 7.1 Table of area figures for natural parks (ha)

Classification	Number of parks	Area of park	Percentage of total land area
National parks	30	2,091,163	5.533
Quasi-National parks	56	1,362,613	3.606
Prefectural natural Parks	315	1,977,528	5.233
Total	401	5,431,304[a]	14.372

Source The Ministry of Environment (MOE) (2010).
[a]The total land area of Japan is 37,790,697 ha
Source Area Surveys for Japan by Prefecture and Municipality as conducted by the Geographical Survey Institute (2004).

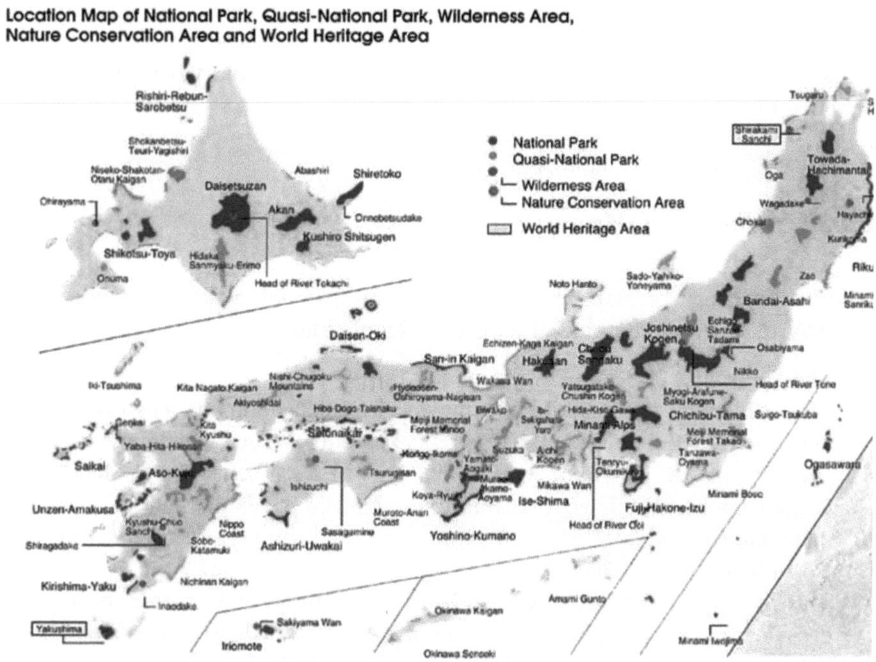

Fig. 7.4 National Parks, Quasi and Prefectural National Parks of Japan. *Source* Ministry of Environment Japan (MOE) (2010)

steeply in the postwar Japan, but has since peaked and is now in a state of decline."[4] In contrast, as amwriter, I do agree with Jones's opinion, that visitors to national parks still increased in number, as showed in findings of Yakushima National Park and World Heritage from 1991 to 1992 just reached 25,049, but rapidly developed

[4]For further discussion, see the paper entitled: "Changing Demographics in Japan's National Parks: Towards a Targeted Marketing Strategy for Nature-Based Tourists." Please refer to the Journal *Tourism and Hospitality Management*, Vol. 18, No. 1, pp. 95–109, 2012.

to become 40,975 in 1995–1996 and in 2008–2009 was registered 50,379 (Yakushima Town Office Statistic 2012).

In line with the development, environmental pollution happens in urban areas as excessive development activities subsequently contribute to social problems. To solve the problem, the government established the Environmental Agency in 1971 and the Ministry of the Environment (referred to as **MOE**). In 2002, a new law was issued to regulate activities in special zones, creation of regulated utilization areas and preparation of new systems for scenic landscape protection agreement.

Park land ownership is complexly mingled in Japan, incorporating private land, state land and local government land. Therefore a *characteristic* of Japanese natural parks highlights that various land owners cooperatively maintain the landscape of a park. The rationale is that Japan is small and densely populated with a long history of private land ownership. From this viewpoint, the Japanese government had to create natural parks not necessarily where it owned land, but where it recognized the need to preserve nature.

Conservation and Management

In order to properly manage and conserve of national parks, according to the Ministry of Environment, based on certain regulations:

(1) The landscape Protection Agreement
The agreement aims to lighten the burden placed on the landowners. The agreement activity involves some stakeholders, such as landowners, the Ministry of the Environment (MOE), local governments, Park Management Organizations that have the ability to manage lands in national park and quasi-national parks.

(2) Beautification Activities
This beautification activities, such as maintaining the scenic beauty and ecosystem is very significant in natural parks. In this case, the cooperation is between local governments, NGOs as well as MOE, for cleaning parks. MOE designated the first Sunday of August as "Natural Park Cleanup Day". On that day, MOE encourages a nationwide natural park cleanup campaign in cooperation with local government and volunteers.

(3) The Green Worker Program
Under this program, the Ministry employs local residents knowledgeable about the local nature and social institutions to perform various activities including conservation of the fauna and flora, clearing alien species and cleaning locations of difficult access.

Management System

The existence of a management system is very significant, especially how to manage and harmonize coexistence nature with ecosystem and biodiversity in national park. There are four institutions such as: headquarters; regional environmental affairs office; park rangers and partnership with local government that have a significant role in management system (MOE 2010).

(1) Headquarters
It is necessary to ensure coexistence nature, ecosystem and biodiversity in National Parks. In this case, the Nature Conservation Bureau of Ministry of Environment (MOE) is responsible for managing it. Three sections mainly involve in managing national parks namely: division of conservation and utilization within national parks; division of park facilities and conservation technology; and division of nature appreciation undertake various activities to promote visitors' appreciation of the natural world.
(2) Regional Environmental Affairs Offices
There are seven Regional Offices to carry out regional environmental administration, such as Hokkaido, Tohoku, Kanto, Chubu, Kinki, Chugoku Shikoku and Kyushu (e.g. Aso and Yakushima). These offices have responsibility for national park management.
(3) Park Rangers and Auxiliary Rangers
Park rangers' task is to carry out patrols, conduct research, implement interpretation programs and facilitate coordination among volunteers to manage park resources by examining development activities and conserving rare wildlife in parks. Actually, the number of rangers had been registered at 46 persons in 1959 and rapidly developed to become 246 in 2005. Rangers are assisted by an assistant ranger or auxiliary ranger for managing the park. In 2005, 60 people were employed as auxiliary rangers in 47 districts.
(4) Partnership with Local Governments
It is very significant to have partnerships between MOE and local governments and forestry agency for national park management in Japan. For instance, the MOE office in Kyushu has the responsibility to manage core zones, special zones and to control ecosystems in national park. Prefectural governments (local governments) have the responsibility to establish roads, bridges, picnic sites and administering some licensing procedures. And the Forestry Agency in Yakushima has to establish feet step for recreation forest areas, the signs of trees age, to minimize the number of deer's and monkeys in cooperation with hunter associations.

Yakushima

A National Park and World Heritage Site

As mentioned, these national parks occupy 5% of Japan's land and preserve the land for future generations. Yakushima Island, which is categorized number 29 among 30 national parks in Japan is located 60 km south-southwest of the Kyushu mainland, Kagoshima prefecture. It is a roughly circular island with a circumference of about 130 km. The island's central location is dominated by a series of peak mountains over 1800 m in height, for instance, Mt. Miyanoura is 1935 m, the highest mountain in Kyushu island. From this viewpoint, Yakushima is well-known as the "Alps on the Ocean" (Diagram Yakushima 2000). This specific topography has created a range of climates on the island, from the subtropical in the coastal areas to cold in the mountainous areas and even fallen 'snow' in winter time, for instance, January and February.

The total area of Yakushima is registered 50,000 ha, and mostly covered by national forest (40,000 ha, 76% from all island) (Forestry Agency 2012). In 1964, Yakusugi forest within 18,738 ha became national park land. It aims at providing visitors chances for the experience, enjoyment and a better understanding of Yakushima's island and natural environment (forest, waters, trees, ecosystem, and marine resources). A National Park Special Area, (Kirishima-Yaku National Park) was gazetted in 1964 under the National Parks Law, comprising land on Yaku Island and Kirishima National Park on Kyushu mainland. A Forest Ecosystem Reserve was established in 1992, comprising the nominated area and various adjacent blocks of land. Its development as the Centre of Yakushima Island, and parts of the island's southern and western coastal lowlands were accepted as a Biosphere Reserve in 1980 (Yakushima World Heritage Area Management Plan 2010).

In December 1993, part of Yakushima Island was inscribed as the World Heritage List, in the Committee Meeting in Columbia. The total area of the Heritage Area is 10,747 ha, and contains approximately 21% of the total area of the island as a World Heritage site. The location stretches from the area centered on Miyanoura-dake in the central mountainous area; the Heritage Area continues west to the ocean over Kuniwari-dake, south to Motchom-dake, and east to Aiko-dake. The Heritage area contains mountains and hills, and ranges from latitude 30° 15′ to 30° 23′ north, and 130° 23′ to 130° 38′ degrees east.

Actually, the history of conservation in this area is long. A series of measures in were implemented to establish conservation measures in the Area, beginning with the designation of an Academic Reference Forest Reserve in the national forest. At its development in 1924, the Area was designated as the Yakushima Old Growth Japanese Cedar Forest Natural Monument (the designation was changed to Special Natural Monument in 1954), and was incorporated into a national park in 1964.

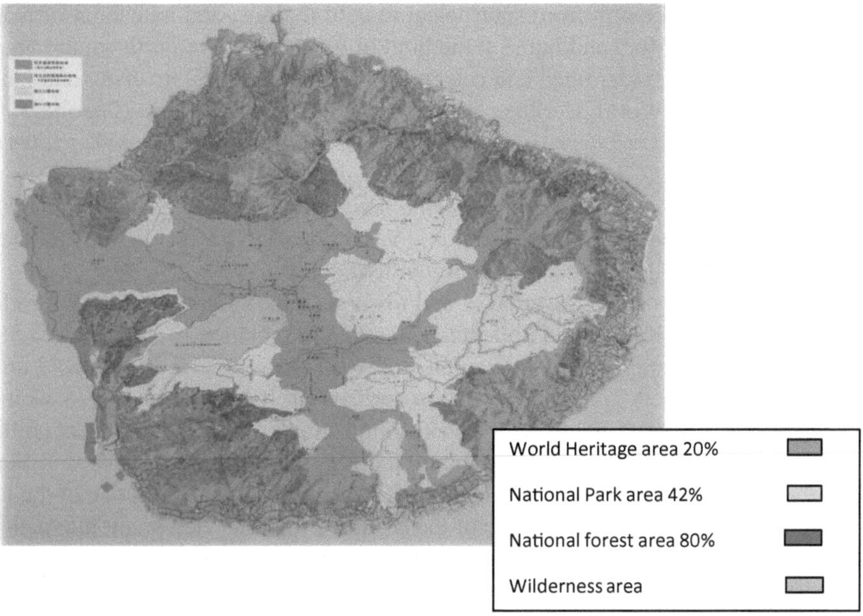

Fig. 7.5 Protected areas and its history in Yakushima. *Source* Yakushima Protected Areas Map (2010)

A decade later, in 1975, the area was further designated as *Wilderness Area*[5] and was expanded in 1980 to be a Biosphere Reserve site; in 1993, it was appointed a World Heritage Area (including Yakushima National Park Area) and in 2005 Nagata Beach was designated as a Ramsar Site to protect turtle inhabitants (Yakushima World Heritage Area Management Plan 2010) (Fig. 7.5).

Accordingly, based on the World Heritage Convention, the area was deemed, "an outstanding example representing significant ongoing ecological and biological processes in the evolution and development of terrestrial, fresh water, coastal and marine ecosystems, and communities of plants and animals." It was also considered to "contain superlative natural phenomenon or areas of exceptional natural beauty and aesthetic importance." (Yakushima World Heritage Area-Management Plan 2010). Therefore, the rationale of Yakushima's selection as a World Heritage site include its unique forest flora, including Yaku-sugi cedar several thousand years of age, such as Jomon sugi, Kigensugi, Hahakosugi, etc. Besides, its vertical

[5]Is defined as an area that fulfills the following conditions: that it has not been affected by the activities of man; that has preserved the primitive conditions, that it is of a considerable area in size, and also has its natural environment to be conserved. This definition taken by the Director-General of the Environment Agency, according to the Nature Conservation Law (MOE document 2004).

distribution of vegetation from subtropical to cold regions; and its habitat of many indigenous plants, flora and fauna at the limit of their respective north-south ranges.

These species are identified and categorized as the biological diversity and fauna contained within World Heritage properties such as: *Abies firma/Japanese fir, Apodemus speciousus/Large Japanese Field Mouse, Ardisia crenata/Coralberry, Camellia sasanqua/Camellia, Castanopsis cuspidate/Chinquapin, Cervus Nippon/ Sika Deer, Chloranthus glaber, Columba janthina/Japanese Wood Pigeon and Cryptomeria japonica/Japanese cedar* (Yakushima-UNESCO World Heritage Centre 2010) (see http://whc.unesco.org/en/list).

Yakushima consists of 24 villages and town such as Yaku-cho, Kamiyaku-cho, and Kumage-gun, Kagoshima Prefecture. The total population is registered 13,589. It consists of 6641 men and 6948 women in 2010. The productive age from 15 up to 64 years old reached 7779 or 57.3% that consists of 4005 men and 3774 women (National Census 2010, published by Yakushima Office 2011). In more recent times, the population mostly work in primary industry (agriculture, forestry, fishery, environmental services, and individual hotel/cottage) which has been one of the key factors supporting the economic life of the islands. From this viewpoint, interactions have taken place between the islanders and outsiders. As a consequence, harmonious coexistence between nature and people should be maintained in the island.

Flora, Fauna, Plants and Vegetation

The flora is variety species, comprising more than 1900 species and subspecies. Of these, 94 are endemic, mostly concentrated in the central high mountains. They range from sub-tropical plants, such as the banyan tree (Ficus superb var.japonica) near the seashore (Fig. 7.6); Gajamaru tree (Fig. 7.7), Machilus thunbergii,

Fig. 7.6 Banyan tree near the Isso seashore. *Source* Individual archive (January, 2013)

Fig. 7.7 Gajamaru tree near Isso seashore. *Source* Individual archive (January, 2013)

(a) (b)

Fig. 7.8 **a** Macaque (monkey of Yakushima profile). *Source* Google (2007); **b** Yakusika (Deer of Yakushima profile). *Source* Individual Archive (January, 2013)

chinquapin (Castanopsis cuspidate) and evergreen oaks. And the remaining 200 species are at the southern limit of their natural distribution. A distinctive characteristic of the vegetation is the exuberance of foliage, particularly at higher elevations.

The fauna consists of 16 mammal species. Four mammal subspecies, *including macaque* (Fig. 7.8a) and *sika deer* (Fig. 7.8b) are categorized as endemic to the island. Among the 150 bird species present, four, including Ryukyu robin, Izu Island (Turdidae subspecies), White-eye (Zosteropidae) and Japanese wood pigeon, have been designated as Natural Monuments. Yakushima insect species include Yakushima Wonderful Green Hairstreak (Lycaenidae), Great Orange Tip (Pieridae), Restricted Demon (Hesperiidae), Tibicen esaki (Cicadidae), Rhipidolestes aculeate yakushimensis (Megapodagrionidae), Common Map (Nyaphalidae), Chinese Windmill (Papilionidae), Prismognathus anglaris tokui (Lucanidae), etc. (Diagram Yakushima 2000: 48–49). The number of Yakushima's endemic species and subspecies is small compared with those on the main islands of

Okinawa and Amami Oshima, but is remarkably large when compared with those on the Izu archipelago.

Several plants can be found in Yakushima, such as *Yahushimashakunage (Rhododenron spp.), Sakuratsutsuji (Rhododenron spp.), Marubatsutsuji (Rhododenron spp.), Yakushimaebine (Calanthe spp.), and Yakushimakanran (Cymbldium spp.)*.

Vegetation is significantly different from the mainland. For instance, subtropical vegetation near the coastline is different from vertical vegetation in upper land. Cool temperate zone grows coniferous forest. Meanwhile, warm temperate broad leaved forest previously covered extensive areas of south Japan.

Yakusugi-Beginning of Yakusugi Logging

As mentioned previously, the scientific name of sugi is Cryptomeria japonica, a type of cedar native to Japan. Sugi that are found growing naturally on the mountains of Yakushima Island are known as Yakusugi. Traditionally, the Island Mountains have been considered to have a spiritual value and the 'Yakusugi' were revered as sacred trees. Yakusugi or so called Yakushima cedar has both properties, and was the most valuable products of the island in the past. When the Shimadzu family, as the head of Satsuma clan governed in power, Yakusugi began to be harvested in 1457 to build special buildings, for instance, the Osumi-sho-hachiman-gu shrine (now the Kagoshima shrine). The products of Yakusugi trees were sent for Hidoyoshi as Shogun to build Hoko-ji temple in Kyoto in 1588. In 1595, the Shimadzu Family established the Yakushima-okime (Yakushima Regulation) and control over forest products (Yakusugi Museum Magazine 2000).

Logging in the Edo Period

When the Tokugawa Shogunate was established in 1603, all heads of regional clans (*Han*) in Japan, including the Shimadzu family in Kagoshima were required to show loyalty to the Tokugawa Shogun. One of the manifestations of loyalty was to pay 'tribute', for instance hiragi[6] to Shogun. In 1612, the Shimadzu clan promoted the Yakushima Island retainer to the position of local magistrate, and Yakushima became under their direct control. In realization of their control in 1624, the Shimadzu established official ship office and at major ports and checkpoints in order

[6]*Hiragi* was used as roof made of from Yakusugi timber. It is approximately 50 cm in length and 10 cm in width, that have the same function as roofing tiles. Hiragi was used to pay land taxes as well as to acquire daily commodities during the Edo period. According to the ledger (Yakushima-tegatasho-kibo-cho) in 1728, that for land taxation purposes, one bag of rice was equal to 2310 pieces of *Hiragi* (Yakusugi Museum Magazine, 2000: 21).

be able to monopolize trade in Yakusugi. As was suggested by Tomari Jochiku[7] as Buddhist scholar, full-scale harvesting of Yakusugi timber began in 1640. As a result, Hiragi formally was used for standard for land tax payment for the local people in 1728. And its development in 1843 was issued 'Memorandum on the Encouragement of Mountain Resource Use' was enacted to secure Yakusugi.

In the Meiji Era and After

The Meiji Era (1868–1912) launched a program so called 'restoration' in many sectors to modernize Japan. During Meiji, the administration government proposed a forestry sector movement towards 'public ownership'. In 1879, following the land tax reforms of 1873, Yakushima Island's forests were divided into government and privately owned zones. As a consequence, a large part of the forest in Yakushima Island was established as a national forest at the end of the Taisho-era, by judicial process "Kokuyurin-keieino-taiko" (Outline for Managing National Forests) in 1921, which was sometimes so called the "Yakushima Kenpo" (Yakushima Constitution), was established, and eventually came into full force (Yakusugi Museum Magazine 2000: 49). The government constructed a trolley rail from Anbo to Kosuidani about 6 km and eventually opened forest for logging in 1923. Kosugidani village was located upstream of Anbo River between 600 and 1500 m above sea level, surrounded by high mountain ridges and almost in the central part of Yakushima Island. According to Kuwahara, the village started in 1923 as the resident area for workers with the establishment of logging camp under "Development Work of National Forest". Besides, Ishizuka Village was a residential quarter which was opened 4 km upstream of Anbo River from Kusugidani Village. According to Miyuki Matsumoto, who had stayed at Ishizuka Village from 1951 to 1967, that there were about 25 workers during the time, some of whom were from Kumamoto and Miyazaki prefecture (Kuwahara et al. 2010: 2–3). Regarding the population of Kosugidani, there were 100 people in 1923, which rapidly increased to 300 people from 1932 to 1943 and decreased to become 100 people with 30 households in 1945, but in its peak time reached 540 people with 133 households in 1960. The population decreased to 442 people with 167 households in 1965 and 4 people with 2 households in 1970. There was a snowfall in winter time, and the temperature from December to February was 7 °C. In 1963, there was a heavy snowfall of more than 3 m in Ishizuka area and 1 m in Kosugidani. Thus the traffic was closed for more than 40 days and foodstuffs were transported by helicopter (Ibid: 2–3).

[7]Jochiku Tomari, the Buddhist scholar man who told the villagers to make 'use' of Yaku-sugi cedar. He was born in Yakushima's Anbo village in 1570. As a child his name was Ichibei, son of a ship builder. Ichibei was sent to Honbutsu-ji temple in Anbo, where he began to study the Hokke sect of Buddhism under the name of Nissho. When he grew to be an adult he went to Honno-ji temple in Kyoto for study and training. (Diagram Yakushima, 2000: 55).

The Kosugidani project office was opened in 1923 and about 2900 ha of Yakusugi primeval forest were appointed as a target of cutting and some of them are still kept as natural monuments, for instance, *Kigensugi and Jomon sugi*. During its development in 1950s, the trend moved from restoration to development and harvesting began to be conducted on a full scale. At this time, forest resources were strongly sought after to speed Japanese economic growth. The massive logging of Yakusugi began in 1955, while chainsaws were introduced and used for cutting trees for a decade from 1960 to 1970s and subsequently effects of forest logging peaked until the log operation closed, because of environmental damage such as flood and soil erosion in 1970.

As an illustration in 1960s (Show a period) while logging activities (Fig. 7.9) were in peak time, the logging workers family registered about 133 households and 540 people stayed in Kosugidani and another Ishizuka Village. There were two public schools: elementary and junior high schools for the children. But its logging cutting resettlement were decreased; in terms of numbers, people reached 442 in 1965 and stayed 167 households and in 1970 just left 2 households and 4 people

Fig. 7.9 Troli brought log of Sugi tree after cutting in Yakushima forest area. *Source* Kuwahara et al. (2010)

while logging activities totally stopped.[8] The rationale behind closing of logging activities were highlighted two factors. First, an internal factor that occurred was a protest movement throughout Japan by local people on environmental damage such as deforestation (Yakushima, Amami, Saetama, etc.) and sea pollution (Minamata diseases) in Kumamoto, Yokkaichi, Niigata, Shizuoka, etc.) (Interview, March 5, 2013). Second, external factors were highlighted that the price of log imports was cheaper than in Japan. In line with this timber trading, most Japanese timber industries imported log from Southeast Asian countries such as from Philippines and Malaysia (Sarawak and Sabah) in 1960s and from Indonesia (Kalimantan and Sumatra) in 1970s (Hidayat 2004).

In the case of Yakushima forest, a excessive logging in 1960s without implementation of sustainable forest management eventually had an effect on forest degradation. The impact of deforestation seriously affected soil erosion, flood, lack of water in drought time. The local community surrounding Anbo and other villages against large scale logging increased and the movement toward 'conservation' program was launched. As a result, in the end of 1960s, a logging ban was issued by the government and subsequently closed Kosugidani project office in August 1970, after nearly five decades in operation.

It is a tradition told in Yakushima that only cedars which are older than thousand years were called 'Yakusugi'. Although there is no difference when comparing the trees whose age is under thousand years from aspect of biological view. Some oldest Yakusugi trees stood near the trails. But unfortunately, we have no chance to see them because of the difficulty entering the deep forests. For visitors who would like to see Yakusugi, two approaches have been provided for seeing some of them. One is Shiratani—Unsuikyo Park that is located in the upstream of the river Miyanoura 600–1300 m above sea level and can be reached from Miyanoura by using a car. In Shiratani areas, cedar trees such as Yayoisugi cedar can be seen, one of 100 largest trees of Japan, *Nidai Ohsugi cedar, and Bugyosugi cedar* (Fig. 7.10). Another called Yakusugi-Land is located on the upstream of Anbo river about 1000–1300 m above sea level. Covering an area of 270.33 hectares the nature park has four hiking courses. The park is covered by a number of huge ancient cedar trees such as *Kigensugi (Fig.* 7.11*) Buddhasugi (Fig.* 7.12*), Hahakosugi, Odasugi*, etc. Besides both recreational forest locations the most popular tourist destination is

[8]The development of resettlement for logging workers sometimes increased and decreased in Arakawa district in Kosugidani village during 1923–1970s. After the logging company was closed in 1970, most workers returned home and get a normal job. The former logging workers, some of them still live and are around 80 s years old and stayed in Chiran, Kagoshima Prefecture. Chiran district was very popular place in Japan, while some young pilot airplane bomber recruitments for suicide (*kamikaze*) in World War two (1942–1945) originally came from it. For further information see Kuwahara, Sueo et al., *Yakushima no Kosugidani to Ishizuka Shuraku no Seikatsushi* (The Monograph of Life in Kosugidani and Ishizuka Village in Yakushima Island), Kagoshima University. 2010, p. 3.

Fig. 7.10 Bugyosugi in Shiratani Unsuikyo area. *Source* Individual achieve (March, 2013)

Fig. 7.11 Kigensugi, circumference 8.1 m, aged 3000 years old. *Source* Author archive (March, 2013)

to visit the *Jomon-sugi cedar* (Fig. 7.13). The greatest of the Yakusugi cedars is registered, with a circumference of 16.4 m, with a height of 25.3 m. It is variously estimated to be from 2600 to 7200 years old.

Another species of trees, such as *Tsuga (Tsuga spp.: hemlock) (Fig.* 7.14*), Momi (Abies spp.:fir), Himeshara (Stewartia spp: Stewartia), Harigiri (Kalopanax spp.) and Shirodamo (Neolitsea spp*.). The location can be reached by the Anbo forest road, which was built for forestry operation on the inner areas (Yakusugi Land 2010). This road is familiar as the road for tourism.

Fig. 7.12 Buddasugi, with circumference 8.0 m, approximately 1800 years old. *Source* Individual archive (March, 2013)

Fig. 7.13 Jomon Sugi is the greatest Yaku-sugi cedar with circumference reaching 16.4 m. *Source* Diagram Yakushima (2000)

Institutions, such as the Ministry of Environment, Forestry Agency in Kyushu region and Kagoshima local government are responsible for managing, controlling and giving sanction on a National Park and World Heritage Area (Table 7.2).

Fig. 7.14 Tsuga tree (red color) in Yakusugi Land. *Source* Individual archive (March, 2013) (Color figure online)

Table 7.2 Major Yakusugi Cedars in Yakusugi Land

Name	Circumference	Height	Age
Kigensugi	8.1	19.5	3000
Hahakosugi (*Mother* and *Child* Cedar)	9.0	31.1	2600
Odasugi	8.2	28.6	2600
Buddhasugi	8.0	21.5	1800
Tenchusugi	8.2	33.8	1500
Mitsunesugi	9.3	26.1	1100

Source Yakusugi Land published by Yakushima Island Natural Recreation Forest (2011)

Findings in the Field

The Management of Yakushima

The focus of the discussion on 'partnership' management covers some issues to be observed, such as a landscape protection, protecting animal and plant, patrolling to control special and protected zone, to bring portable toilet, and to conduct research in wilderness areas. Besides, environmental services on water utilization on Dam and Waterfall for electricity power supply and use by *Yakushima Denko* (Yakushima Electricity Company). The impact of partnership management on a sustainable ecosystem, water supply and tourist boom to visit Yakushima by actively involved to *Minshuku* (Private house owned by individual).

As mentioned, those three institutions such as The Ministry of Environment (MOE) Office, Forestry Agency of Kyushu, Branch in Yakushima, and local government (Yakushima Town government Office) are responsible for making 'partnership' to manage and to monitor national park and world heritage.

The first discussion highlights in terms of 'a Landscape Protection Agreement' is one of the objectives on a Natural Park Law. The related actors, such as Forestry Agency, local government, the MOE, park management organizations and other stakeholders (local community and local authorities) have to be aware that it is restrictions to establish on the installation of buildings or structures that may potentially alter the natural environment or landscape. Besides, the felling of trees or other plants, the extraction of soils or rocks, capturing animals of plants and other actions in the special and core zones are strictly forbidden.

In this context, its operations could be concluded on necessary 'partnership' between landowner (Forestry Agency), the Ministry of the Environment (MOE), local government, Park Management Organizations that have the ability to manage lands in national and quasi-national parks. Therefore, based on the agreement, the park management Organizations manages and protects the land on behalf of the landowner. In this sense, park management organization has a duty to conduct vegetation restoration, maintenance of park facilities, dissemination of information, etc. The task of MOE office (Yakushima Environmental Culture Learning Center and Yakushima World Heritage Conservation Center) near Anbo also hold poster exhibition about national parks and the necessary on the performance on World Heritage on the dissemination of information and training on awareness to pupils, visitors and local communities on keeping environment and to protect forest ecosystem (Interview on January 23, 2013).

The office holds training and academic scientific committee discussion on environment and ecosystem of nature to local communities, pupils, visitors, students, NGOs officer, at least twice annually. Besides, the Forestry Agency has the task to make the signs in board regarding the profile and the age of trees in natural recreation forest, plants, animals and vegetation. Another task of the Forestry Agency is to construct the feet step in natural recreation forests in Shiratani Unsuikyo, Jomo Sugi Cedar area, natural recreation forest in Yakusugi Land, Arakawa district, and hiking in the mountain side. The budget of making boards, feet steps, forest operation on reforestation, salary of workers, and maintenance forest trees derives from the selling of old trees which have reached 30 years old provided for tree cutting in national forest areas by the private company. As an illustration, the price of sugi reaches Y 15,000/m^3 and Hinoki costs Y 18,000/m^3. Meanwhile the local government (*Kagoshima Prefecture*) has the task to construct and maintain the transport infrastructure, such as roads and bridges towards recreation facilities from Anbo to Yakusugi Land and from Miyanoura to Shiratani Unsuikyo and its surrounding.

The second discussion is regarding the necessity of 'protecting animals and plants'. The mission of National Park and other Natural Parks, among others, is to protect exceptional natural landscapes (forest landscapes, coral reefs, fauna and flora). In this sense, animals and plants that cannot be captured or collected inside

Special Zones of National park and Quasi-National Parks without permission are designated to ensure the diversity of the ecosystem inside Natural Parks. According to an interview with Forestry Agency officer in Anbo, because of the raising numbers of monkey and deer, approximately 20,000 and 15,000 respectively (it is more than local inhabitant of Yakushima 13,589 persons), and these animals subsequently are a threat to forest rehabilitation (young tree leaves) and agricultural farming (orange and vegetable) of local people. To make fences to protect young trees from monkeys and deer, Forestry Agency has the task of cover fence on forest rehabilitation in the special zones.

In this context, to prevent increasing number of deer, Forestry Agency invited a "hunter Association" from Kagoshima City to hunt limited numbers of deer. As illustrated in 2012, 340–400 deer were shot and subsequently buried in the soil. The price of one head of deer costs Y 5000 for hunter.

Third, to establish 'patrol' conducted control in the special and protected zone. To realize the aim of patrolling the areas, making 'partnership' between Forestry Agency, the MOF office and local staff. Furthermore, ranger staff under the authority of Forestry Agency has been increased from 10 to become 30 persons and supports with volunteers.

The forth discussion is planning to carry out 'portable toilet' to Jomon Sugi and pay Y 1000 for tax entrance for every visitor. These both issues will be provided by Yakushima town Office and invited other stakeholders such as The MOE Office Brach, Forestry Agency, The Tourist Association, and local community. The Yakushima town Office is responsible for handling the portable toilets in certain places and making a 'recycle system' for waste (Interview, on January 23, 2013). The aim of the planning regarding reducing number of tourists to Jomon Sugi, especially in the peak session that reached 300,000 visitors. These huge numbers eventually had an affect on environmental damage and pollution. If this regulation is finally accepted, it will be implemented in the mid-year 2013.

In the peak seasons of summer, July and August, and golden week in May, tourists poured into Yakushima for climbing and hiking to mountain and visit natural recreation forest in Jomon Sugi, Shiratani Unsuikyo and Yakusugi Land. The role of Park Management Officer, which is supported by local government Office (police institution) strictly ban using individual cars or vehicles to visit certain recreational places, because of limited park land. For this reason, visitors should use public transport (bus) from Miyanoura and Anbo to certain destination in recreation natural forest. This prohibition of individual cars is to prevent damage caused to vegetation, habitation and breeding environments of wild animals and plants by the use increasingly road vehicles on parklands.

Fifth is the discussion to conduct research in "Wilderness Area". The study on ecological research should invite professional researcher partnerships, such as Universities, Yakushima World Heritage Conservation Center and the Yakushima Forest Environment Conservation Center. The aims of the ecological research to gain a better understanding of the current natural environment in terms of collecting a new plants and ecosystem. The result of the research should produce academic reports, collect documents and disseminate information to public users (Interview, January 23, 2013).

Non-Profit Organizations (NPOs)

One of the NPOs is so called "Yakushima Biodiversity". The aim of the NPO is to protect endangered species and to make basic data of plants, such as *Gentiana, Yakushima pinus amamiana, pinus armandi*, etc., in special zones. Kenshi Tetsuka is one of the leaders of Yakushima Biodiversity who manages 160 members and carry out field work on endanger species in summer session. Actually, there are 1900 plants and 90 units of endemic species registered in Yakushima. In operation for field work during 3–4 days, NPO established camp in the forest and utilized a GPS system for mapping and monitoring the condition. The NPO's fund derives from Yakushima Town Municipality Office and Yakushima Environmental Culture Foundation.

The task of the NPO is to collect basic data in terms of endangered species to give to the Yakushima Town Office and to disseminate information to educational institutions, such as elementary school and high school students in Yakushima about how they can keep maintaining the nature and ecosystem (forest trees and plants) by learning experiences and also to publish basic data for the socialization to the next generation.

The significance of protecting "*Umigami*" (big Turtle sea) is very urgent in Nagata Inakahama Beach area. Besides, based on Japanese belief that turtle (*Tsuru Kame*) is the symbol of life, it can live up to 100 years old (Interview, February 6, 2013). The turtle is categorized as one of endangered species based on *CITES* (*Convention on International Trade in Endangered Species of Wild Flora and Fauna*) Convention, and the Japan government has made signature to protect it. Most turtles visited in Inaka-hama beach is red and blue color (Fig. 7.15).

The local government made partnership with Yakushima Umigame-Kan, one of *NPO* (*Non-Profit Organization*) to protect turtle wildlife. Local government provides offices for NPO in Nagata area, near the Inaka-hama Beach (Fig. 7.16) by renting of houses for NPO volunteers and provides infrastructure for turtle life exhibition. The ticket costs Y 800 for watching video, listening to lectures on regulation campaign, such as not making a light in beach and not passing through certain sand areas and a guide to the beach for every visitor. On the other hand, NPO officer also campaigned and actively involved the participation for local people and pupils of elementary up to secondary school, such as do not eat eggs of turtles, protect landscapes, and clean the beach twice annually. The variety of turtle species come is red and blue color to the beach. Both species must be kept according to CITES convention. As a result of NPO's campaign, the local

Fig. 7.15 The red and blue turtle in Inaka-hama Beach, Nagata. *Source* Individual achieve (February, 2013)

Fig. 7.16 The Nagata Inaka-hama beach while turtle landing and breeding eggs. *Source* Individual archive (2013)

communities have increased awareness to protect turtle inhabitants in the Nagata Inaka-hama Beach and also the behavior of visitors are conducive pro eco-friendly (Interview February 6, 2013). As an illustration, in 1987, turtle eggs could be breeding 200 units and eventually increased to become 1500 units in 1989–2000. According to a report from NPO Office in Nagata, the breeding of eggs could be increased 4000 units in 2000–2013. Based on field survey (NPO office 2011) estimates that every egg from range 1500–2000 units just live to be children

Table 7.3 A range of time of blue turtle breeding in Inakahama beach area

Month	No. of landing	No. of breeding	Normal	Not-normal	% of Frequency breeding
4	9	4	339	0	44.4
5	660	418	6704	59	63.3
6	1845	107171	9638	11	58
7	2344	1140	13,563	24	48.3
8	141	78	0	0	55.7
9	0	0	0	0	50
12	0	0	0	0	100
Total	4999	2711	30,217	94	54.1

Source *Yakushima no Okeru Umigame Setai Chosa Hokoku* (Research Report on Turtle Life in Yakushima), published by NPO Yakushima Umigame-Kan (2011)
No: number

100 and among of them could to be parents reached 3–10. It is frequently often that turtle almost the range time 60 years do not give 'eggs' because of lack their meal and happened environmental pollution. The rationale behind decreasing number of turtle children is that most of them could be swimming as far away as Canada and USA beaches and most of them die mid trip or are eaten by big fishes such as shark and whale. Some of them also died because they caught in the nets of fisherman. Meanwhile the parents just make swimming neat, for instance, surrounding South Korea and China Sea. It means there is a positive impact of a campaign to local people and pupils to maintain turtle inhabitants. In June and July while the peak of turtle came to have breeding eggs in Inaka-hama Beach sand, several visitors come to this area to watch the fascinating scenes of turtle life (Table 7.3).

The NPO office in Nagata also collaborates with Fukuoka and Miyazaki NPOs to keep turtle life in their respective areas. The sharing experiences among NPO for turtle maintenance are very significant. There are challenging factors, such as climate change which subsequently affect current changes in the sea and El-Nino storms. These factors should be considered as a threat to maintain the turtle population in the near future.

Environmental Services

The impact of good maintenance on ecosystem by stakeholders in Yakushima subsequently produced a sustainable of ecosystem, green forest, ecosystem, water supply and a conducive environment. Therefore, the sustainability of water supply from the forest to river and waterfall eventually could be used as "hydropower" for Electricity Company and for agricultural farm. There are several waterfalls, such as Senpiro-no-taki, Torohki-no-taki, and Ohko-no-taki and river in Anbo (Fig. 7.17). In this sense 'environmental service' derives from forest resources finally producing

Fig. 7.17 Ohko-no-taki waterfall near Kurio, Yakushima. *Source* Individual archive (February, 2013)

the water supply for "Yakushima Denko". The company produces 83,650 kW, and almost 90% is used for local supply energy for households, private companies, government office and public use in Yakushima. This company started operations under the original name '*Yakushima Denki Kogyo*' in June 1952. The shareholders of the company, for instance, 49.5% belongs to Taeheo Semento, Showa Denko 22.4%, Higashi-sho 5.9%, Mizuho Cop. Bank 4.9%, Nihon Rutsubo 4.4% and Nihonkoe 3.7%.

The total of hydropower produced 83,650 kW to turn on generator for electricity production (Table 7.4). The turbine generator engine uses Fuji Denki brand.

The Management System

The company privately uses a clean management and is usually audited on financially affairs by public accountant. It is necessary to implement the management as responsibility to all shareholders. The company is very healthy in operation and could get profits annually. It is obvious, that the company appeals in operation no economic 'incentive' or no giving of subsidies by government. It is based on

Table 7.4 The generator production (kW)

Type	Generator	Permitted power
Water	Tenpirodaki	1300
	Anbo river 1	23,200
	Anbo river 2	34,000
	Equal	58,500
Diesel	1	6200
	2	6200
	3	12,750
	Equal	25,150
	Total	83,650

Source Yakushima Denko (2011)

normal market. As a result, Yakushima Denko also distributed 'power' on electricity to consumers based on the rational market. In this context, the Company pays annually for 'environmental services' to maintain water supply about Y 85,000,000 to Kagoshima Prefecture Office (Interview, February 5, 2013).

This year 2012, The Yakushima Denko celebrates 60 years old. The campaign of the company to participate on keeping 'sustainability' of ecosystem Yakushima by issuing 1–2% from its profits for spending on *Corporate Social Responsibility* (CSR) for making donations and sport competitions to local communities and government office. The Company, for instance, donates one bus with medium size for 30 seats to Yakushima Town Office. They also hold some events for competition such as soccer, bicycle cycling to the mountain side and to do fishing to catch big fish in the sea. This event actively involved *NPO* (Non-Profit Organization), local communities and pupils in elementary, secondary and high schools. As a result of the events, friendship and mutual better understanding has developed between consumers and company. This condition is very necessary for nurturing healthy and accountable business in Yakushima.

There are some constraints to maintain water supply, at least twice annually happened. The rationale behind this temporary stop due to climate change, such as big El-Nino, storm (Taifu), Thunder, etc. If the supply of water is stopped and eventually affects to stop 'hydro power' for the electricity, the Company effort to use with "Diesel" engages to switch on electricity always runs.

Yakushima protected areas was designated as a World Heritage site in 1993. The water utilization is very strict to be monitor, because the public and NPOs (non-profit organizations) have access to participate in management control by reading information and empirically know when tourist go to for hiking and climbing mountain. The company utilizes hundred percentage water supplies from forest resources. In this case, the Company is actively involved in maintaining forest areas, especially protected forest and wilderness areas by making partnership with Forestry Agency and Office of MOE.

Local Farmer Responses to Plant 'Orange' Farm

The farmers own land that is outside the national forest Agency of Yakushima. Local farmers occupied the land which is on the border of national forest for a long time. According to the informant, their land originally comes from their grandparents (third generation). In Mugio district near Anbo live 156 households. Thirty percent are categorized as farmers. Thirty years ago, approximately in 1993, when Yakushima Island was determined to be World Heritage site, most the farmers cultivated orange farm and were very fruitful and productive. But recently, most of them are become an aging society (65–70 years old) and there is no regeneration. In contrast, most their young generation now works in Kagoshima City and other cities. A few households still continue working on orange farm. For instance, Michihiro, 60 years old and his wife Chiyomi 58 years old. Michihiro has two daughters (Fig. 7.18). One of his daughters has been married and now she and her husband would like to continue working as farmers. This family owns 4 ha of orange farm area.

Michihiro previously worked as shop seller in Miyanoura. He gets more experiences in agricultural local products, distribution and marketing of agricultural farm. Therefore he is familiar with trading and obtained a license certificate of agricultural management from institutional training in Kagoshima City, before he finally decided to work as an agricultural farmer. In operation of farming, this family is supported by volunteers from Tokyo Agricultural University. As an illustration, in 2012, the volunteers work temporarily in orange farm for 1–2 months with no salary, just accommodations. According to Daisuke, one of the informants (interview February 6, 2013), he prefers to be a volunteer to learn in the

Fig. 7.18 The owner of tankan orange brand in Mugio, near Anbo. *Source* Individual achieve (February, 2013)

field for practical farm, production, distribution and market. This valuable experience is necessary for his future job, as an entrepreneur.

The specific Tangkang Orange from Kagoshima has been originally developed in Okinawa, Ogasawara, and Miyazaki where most areas experienced lowered production. For example in Okinawa, previously suffered insect contamination and reduced production for a decade. In contrast, the orange products which have been planted with no fertilizer and chemical used are very fruitful. The orange appeals so called Tankang orange from an organic farm (Fig. 7.19). Based on the observation that Yakushima is suitable for the climate, weather and soil to plant and develop this orange. In its operation of the farm originally just use fertilizer from waste and animals. The brand of production of this eco farm is called "Hankan Tankang" (Fig. 7.20). The production in 20 years ago (1993) while the orange plant is young reached 30 tons per hectare. The most production occurs in January and February (winter) or mostly during rainy time. But, recently the production just reached

Fig. 7.19 Orange trees owns by Tankang brand in Mugio. *Source* Individual achieve (February, 2013)

Fig. 7.20 Hankang Tankang brand of orange company. *Source* Archive (February, 2013)

15–18 tons, because most plants are aging. The aging plants of some trees will be cut, because theoretically their plants will be renewing soon after 20–25 years old (Interview on 6 February, 2013).

The government of Kagoshima Prefecture throughout Yakushima Cho Office gives an economic 'incentive' to the farmers, especially for orange farming. The incentive has been a 'subsidy' to farmers; not direct money, but by making fences about 50% off its cost to protect from monkeys for maintaining farm. Besides, the government also establishes water supply pipe under the road to water farming area. The rationale is highlights on maintenance the brand image, that Yakushima Island is the largest orange production in Kagoshima Prefecture. Therefore 'Tankan orange' is very fruitful and fascinating for '*Omiyage*' (special gift) for tourists and displayed in various shops in Kagoshima City, particularly in Miyanoura and Anbo.

The philosophy of "Tankang" Orange should be sustainable business. Therefore to achieve the aim, the company shares knowledge with the younger generation, especially who wish to explore the voluntary work for orange farm. The company also develops on the principle teachings on mutual help, understanding, and friendship, to deliver the product over Japan market on time, entrepreneurship, and accountable management. The company has been prepared for regeneration and should be able to compete with other companies in the future. To implement this aim, the Tankan Company is open for any students to practice agricultural principles, production, and management. Because of this, the company has succeeded ib maintaining her existence as orange farm production and other processing industries from orange materials, such as juice orange, ice creams, chocolate and sweets. Processing industries economically is viable, and give additional values such as income generating, job creation and diversification business.

The constraints of orange farm in the field is facing a threat because of growing diseases such as the coming of birds, monkeys, heavy rain and sometimes occurs *Taifu (*storm) in the winter time.

Coming of Tourists and Hotels

Tourists visited Yakushima was considered low while the island still becomes National Park (Tables 7.5 and 7.6). While Yakushima was designated as World

Table 7.5 Condition of three largest tourists visited in the Yakushima islands

Island	Fiscal year 1970	Fiscal year 1971	Fiscal year 1972	Total
Nagashima	19,000	21,000	31,500	71,000
Tanegashima	26,000	51,000	71,000	148,000
Yakushima	97,000	122,000	147,000	366,000

Source Kagoshima Kikakubu Rito Shinkoka (Promotion of Developing Project of Tourism Kagoshima Prefecture), (1974)

Table 7.6 Availability of the capacity of accommodation for tourists (in 1972)

Island	Number of Minshuku and hotels	Capacity/number of people
Nagashima	6	169
Tanegashima	7	260
Yakushima	24	1420

Source Kagoshima Kikakubu Shinkoka, 1974

Heritage in 1993, the tourists visited Yakushima by plane and ships were rapidly developed in 1996–2010. And also the tourist agency is registered 94 units to guide the tourist. According to Yakushima Town officer the rationale of its rapid development highlights two factors. First, the establishment of natural recreation forests in *Jomon Sugi, Shiratani Unsuikyo, Yakusugi Land*, etc., are well properly managed and maintained by stakeholders. Second, the effective campaign launched by tourist agencies in media, TV and website for eco-tourism display has had positive impacts on domestic and foreign tourism as well (Interview, January 23, 2013). The increasing number of tourists after 1993 has a positive correlation with rapid development of infrastructure in terms of Minshuku,[9] Ryokan[10] and Hotels in Yakushima.

In case of transportation, tourist use of ships and planes also rapidly developed (Table 7.7). And visitors to Yakusugi Museum also increased in numbers (Table 7.8). It is an indication that eco-tourism to National Park and World Heritage sites still continue to be a favorite among tourist in Japan from 1990s until 2011.

The paper would like to appeal the profile of Yakushima Nature Activity Center (YKNA) as Tourist Agency and also the witness of three owners of Minshuku in Yakushima after they decided to establish Minshuku after Yakushima was designated to be World Heritage site in 1993.

[9]Minshuku: A Private house owned by an individual person providing bed and meals. The Minshuku is rapidly developed in number more than 41 units; the price reaches around Y 8000–Y 10,000 per night, after Yakushima was decided a World Heritage site in 1993. Meanwhile the Hotel totally accounted 11 units in 2012. For example the big hotels, such as Iwasaki, Sankara, JR Hotel, Yakushima Grand Hotel, Seaside Hotel. Etc. The price of Hotels registered more than Y 15,000 per night (See Practical Travel Guide, Yakushima Island 2012) (http:www.jnto.go.jp).

[10]Ryokan are Japanese Style Hotels. Actually there are few Ryokan in Yakushima, such as Shisuikan, Kamome-so, Tashiro Bekkan. The price of Ryokan reaches more than Y 8000–15,000 per night. But most aging society are prefer to select living in Ryokan than in Hotel. For the student, Youth Hostels, such as Portside and Yakushima Youth Hostel are provided and the price is below Y 8000. And also guest houses, such as Wa no Cottage Sen no ie and Chinryu-an are available.

Table 7.7 Tourist visit Yakushima (by Ship and Plane) after designated world heritage site

Year	Ship	Air plane	Total	Compare with previous year (%)
1996	195,880	56,958	252,838	98.5
1997	202,721	61,013	263,734	104.3
1998	211,288	68,447	279,735	106.1
1999	193,927	66,234	260,161	93.0
2000	191,570	71,507	263,077	101.1
2001	209,697	76,580	286,277	108.8
2002	204,531	85,004	289,535	101.1
2003	228,436	86,330	314,766	108.7
2004	203,271	90,561	293,832	93.3
2005	231,332	85,552	316,884	107.8
2006	251,239	81,985	333,224	105.2
2007	332,028	74,359	406,387	122.0
2008	310,531	75,456	385,987	95.0
2009	251,931	75,930	327,861	84.9
2010	258,062	75,157	333,219	101.6

Source *Tane-Yaku-Kanko-Renraku Kyogikai* (Yakushima Tourist Bureau), Published by Yakushima Town Office Statistic (2012)

Table 7.8 Yakusugi Shizenkan (Visitor Visit Yakusugi Museum) from fiscal year 1991–2011

Fiscal year	People	Fiscal year	People	Fiscal year	People	Fiscal year	People
1991–1992	25,049	1996–1997	47,275	2001–2002	45,088	2006–2007	39,991
1992–1993	31,889	1997–1998	38,776	2002–2003	51,028	2007–2008	44,202
1993–1994	32,258	1998–1999	45,256	2003–2004	58,271	2008–2009	50,379
1994–1995	38,354	1999–2000	42,253	2004–2005	54,219	2009–2010	41,523
1995–1996	40,975	2000–2001	40,688	2005–2006	46,716	2010–2011	41,915

Source *Yakusugi-Shizenkan* (Visitor in Yakusugi Museum), published by Yakushima Town Office Statistic (2012)

Tourist Guide Agency

Matsumoto Takashi acts as President of Yakushima Nature Activity Center (YKNA/Tour Agency). He previously stayed in Tokyo. Now his wife acts as 'editor' of the Journal 'Shemeno shima' (The Life in island). He likes to stay in Yakushima island, because of the environmental sphere, good education for his children and no pollution. He thought that tourist sector is rapidly developing and will be a challenging industry, and the excellent environment inherently appreciated by urban people (interview on February 6, 2013).

The company was established in 1993. He needed just 8 months to decide to establish company, after he eventually carried out field work and feasibility study to start the business. The *philosophy* of his company is to maintain the ecosystem

Fig. 7.21 The profile of owner Yakushima nature activity center with writer. *Source* Archive (February, 2013)

namely nature, forest and waters. Besides is how to actively involve tourists to appreciate the mindset of tourist to keep the nature well. In this sense, eco tours are the most favorite target for tourist going to 'Jomon Sugi'. Because the tourist can see the ecosystem of forest, landscape, geology, plants, animals, insect interaction, etc. Or tourists, by traveling to Jomon Sugi of about 5–6 h by walking can utilizes their five senses: eyes, nose, hearing, taste, and hand. Yakushima National Park and World Heritage could represent as complete environmental ecosystem such as forest, plants, waters and animals (Fig. 7.21).

The Contribution

The company has an obligation to keep and conserve nature programs by giving lectures to *Shogakko, Chuogakko and Kotogakko* (from elementary until high school) at least once annually. The subject of training and discussion with pupils in school consists of practical in the field on ecosystem interaction. Besides, the lessons on eco guide pupils' mind in order to appreciate the nature. Some high school pupils are interested in becoming guides for tourist and they can join with the company. This company is open for any guide who would like to practice and generate income. The rationale is Yakushima has many resources and is still interesting for many tourists to visit here.

The Manpower

The permanent staff of the company is 4 persons. But the guide profession is around 200 persons. They can join with the company especially in golden time (in May) and in summer time (June–August), when the peak amount of tourists visit Yakushima for hiking, climbing the mountain, playing Kayak in Anbo and visiting Nagata Inakahama for seeing turtle in the beach. In 2012, in July–September tourism reached 1300 visitors/per day. Ten days in September, tourist spent 2 and 3 days, reached 300–500 people. The flood of tourists in the summer has made the company very busy providing guide service. The guide fees for a one-day trip to Jomon sugi costs Y 15,000/per person.

The peak time for tourists to visit "Jomon Sugi" is summer and most spend one night; eventually discussions are considering a regulation to bring "Portable toilet" and to pay every tourist Y 1000 to enter the Jomon Sugi area to maintain the environment and nature. The stakeholders related Tourist Association includes the travel agency Hotels, Individual Hotel, NGOs on Environmental nature, Ministry of Environment, Branch Office, Forestry Agency have been invited by Yakushima Town Office to discuss the regulation to bring portable toilet and entrance fees this year 2013. According to the Yakushima *Cho* (town) Officer, this will be decided this year. So far, up to now the company is not paying tax to the Yakushima town office. The company just gives voluntary donations to the Yakushima office town.

Minshuku Owner

Toki Shoichi (51 years old) owns Minshuku Takenko. He originally comes from and was born in Anbo, Yakushima. He followed his parents and attended secondary and high school in Osaka. He is married his wife and has one son (8 years). His son now attends Elementary School for five years. He was interested to return home to Yakushima, after he thought that Yakushima would be a tourist industry after being designated as a World Heritage site in 1993. He opened his business of Minshuku in 2005 with his capital and experience after he worked in Osaka. His experience on management in a private company subsequently affect to his spirit to develop his business. The capacity of his Minshuku is just three rooms and can be used approximately for 12 persons. The price releases just Y 6200 include for meals dinner and breakfast per night (Fig. 7.22).

The philosophy of his business to provide best services to customers, keep in touch with visitors via e-mail and address, and extend networks. Additionally, to display his Minshuku in internet, media and the board in Anbo to Miyanoura destination. As a result of this strategy and campaign, some tourists spend 1–2 nights in his Minshuku. The peak season of tourist session comes in golden month (May) and summer (July–September) the visitors reached 10–12 persons/per day. Because most tourists come from outside Kagoshima Prefecture, they would like to go hiking and climbing the mountainsides. But, in winter session (December and February), the number of visitors to his Minshuku were decreased, just 2–4 persons per week.

He has ambitions to develop his business, if the Yakushima town office gives 'facility' to obtain 'credit' from the financial institution such as bank. So far, his

Fig. 7.22 Dinner meal provides by Minshuku Takenko in Anbo. *Source* Individual archive (January, 2013)

ambition is very limited because lack of capital and manpower to work to provide meals for visitors (Interview, January 23, 2013). The income from his business, he decided to save in the local bank. He would like to spend his savings to extend business, to continue his study for his son to university in Kagoshima and to maintain his health.

Asada is (58 years old). He is owner of Minshuku Rokmeikan near Yakushima Airport, Anbo. He was born in Osaka and previously worked in the Stock Security Exchange Bonds (*Soken Toritsu*) in Osaka. He decided to quit his job before he was retired. The company gives early retirement fees about Y 10,000,000. And he got also normal retirement Y 35,000,000. He married his wife in 1980s and has one daughter and one son. Her daughter married and lives with her husband in Chiba Prefecture. His daughter has one son.

His son is still not married yet and works in Osaka. Their family visited Yakushima at least twice in a New Year and summer vocation in August.

He previously visited Yakushima as tourist and he was interested in the fishing area and no parking fees for boat in the beachside. The Yakushima Island provides a fascinating fishing area and much fish production and no parking fees. So far while he stayed in Osaka, he paid for his boat ship annually Y 800,000 for parking fees. This background eventually interested his decision to remove in Yakushima Island. He bought land 600 Tsubo,[11] approximately 2244 m width from a private company. He paid more than Y 40,000,000 for the land. His land is far from the highway 200 m, but very strategic near the Yakushima Airport in Anbo. He finally established Minshuku in 2008. The Minshuku has good design and architecture,

[11]*Tsubo* is Japanese calculation style for buying land. For instance, 1 Tsubo is designated for 3.3 m in calculating. The tradition to buy land still be practiced in the urban and village areas until now.

Fig. 7.23 The interior design of Minshuku Rokmeikan, near Nagamine, Anbo. *Source* Archive (February, 2013)

owns three rooms and the capacity for 12 persons, and onsen[12] room for taking bath. Besides, the rest of his land provides for timber hand crafts making and products, such as ball point accessories, *Hazaire* (place for smoking ashes) and key holders are sold to visitors. But, he opens training to make cups, key holder, plate and statues from timber Yakusugi waste for visitors. Therefore for this training he provides some tools and machine equipment from cutting wood, design and crafting timber (Fig. 7.23).

The philosophy of his business is spirit; survive with a good management; diligent work, keep in touch with visitors and give excellent service to customers (Interview on February 5, 2013). Based on his experience for this strategy most visitors visit again for 2–3 times to his Minshuku. The visitors who spend a night in his Minshuku usually contact him via the internet and make reservation two months prior in peak seasons, especially for golden week holidays (in May) and for summer (July–September). The price reaches Y 7000 and includes for meal in dinner and breakfast per day. The owner of Japanese home style (Minshuku) in Yakushima is very important due to support tourist development and gives excellent service to customers. So far, his business not yet expands for the making branch in Yakushima, Island because limited capital and manpower who ready for temporary works.

Tsutomu and his wife Takama run business Minshuku Maedake near Haramaki. Tsutomu originally comes from Fukui and his wife Takama from Okinawa. Tsutomu worked in a Bank Office in Fukui and when he quit working, removed to Okinawa and opened a liquor shop. He met his wife in Okinawa. Tsutomu has three boys and the eldest one is a Sumo wrestler in Kagoshima City. The second son runs the second Minshuku branch in Ambo. He traveled to Yakushima, because he likes climbing the mountainside. Therefore, after he visited three times this area, he

[12]Onsen is hot water utilization in the country side. The traditional *Minshuku* in the country side, including most in Yakushima provide Onsen for taking baths.

Fig. 7.24 The dining room for visitors to have meals, Minshuku Maedake. *Source* Archive (March, 2013)

decided to move in 1990. He bought the land in 1992 and in 1993 when Yakushima was designated as a World Heritage site, the tourists boom started. The company location is very strategic: near a highway and it faces the views of Mae-dake Mountain with the height of 965 m. In the beginning, his business passed through hard times, because of lack of capital and manpower. According to Takama, she opened lunch box during one year and made marketing to school and private company worker. The family (husband and wife) have to wake up from 2 am to prepare meals for lunch box and to deliver the meals to the target (Interview on March 11, 2013) (Fig. 7.24).

The Maedake Minshuku subsequently opened in 1993 with hard working and experiences for making meals from lunch box, liquor shop, banking management and combined with a diligent work. The Minshuku has 6 main rooms and other buildings provide 7 rooms and also the 5 rooms in Anbo branch. The total owned 17 rooms provided. The Philosophy to develop her business is to give satisfaction to visitors based on her service in meals and bath facilities, such as onsen, clean rooms and internet. She provides also pamphlets, advertising in the media, website and board in the street. She has an obsession to expand the business to other Minshoku braches in Miyanoura and other related business with Minshuku. The campaign eventually produced results on the coming of visitors, especially in the peak session in golden week May and in summer holidays, a tourist groups comes for hiking and climbing the mountainside and visitors reached 75–80 persons per day. In the peak season, this company is assisted by temporary workers reaching 10–15 persons.

Besides the business of Minshuku, the company also develops “chicken” for breeding eggs and agricultural vegetable organic products, without utilization of fertilizer and chemicals. Poduction reaches 100–200 units of egg every day and sells to Anbo market and provides for visitors’ meals.

The company is now facing difficulties to continue the business, because Tsutomu’s (husband) suffered a stroke and his son is still not yet ready to handle the management. Therefore, to solve the problem, an option might be to hire the "professional" person to properly manage the company in the near future. But, so far, Takama (wife) still can manage this company for better performance by giving excellent service to customer.

Concluding Remarks

Yakushima National Park and World Heritage is categorized as very unique and a fascinating area for eco-tourism in Japan. On Yakushima Island, nature and humanity have worked together to create conducive environment. In this sense, the close relationship and coexistence of local community and environment that can be seen on Yakushima is being encouraged as community partnership model in Kagoshima prefecture. The impact of best quality on environment could invite urban people from domestic and international as well with a high income as tourists to return back again to the natural life. Because they are deeply aware that to recover their health, spirit and motivation for better performance in work, nothing else, just to visit the best quality environment and ecosystem such as forest, plants, animals, waters and seaside (beach). This requirement for better performance as an eco-tourism destination is provided by the quality of Yakushima environment. The boom in visiting tourists after Yakushima was designated a World Heritage site in 1993 has rapidly developed, from 147,000 in 1972 to become 252,838 in 1996 and 333,219 in 2010. The impact of tourist industry has eventually been affected the need of public transport (*airplane, ship, bus, and rental car*), restaurant, *hotel and Minshuku*, Tourist Guide Trvel Agency, Electricity supply, etc. have correlated with economic growth in Yakushima particularly and Kagoshima City at large.

Hence Yakushima as a National Park and World Heritage site should be properly managed by "partnership management" in some strategic issues, such as a landscape protection, protection of animals and plants, patrolling to control protected and special zone, to maintain the *Jomon Sugi* environment (tourist most destination) by bringing portable toilets and to conduct research in wilderness areas, which actively involve related stakeholders. For instance, the operation of 'partnership management' model, which has been practiced by the Ministry of Environment Office (MOE), Forestry Agency, Local Government (Yakushima Town Office) and Park management Organization to share the construction of Natural Recreation Forest in *Shiratani Unsuikyo, Jomon Sugi*, and *Yakusugi Land* in terms of public infrastructure (roads, bridge), feet step, signs of board, to control population of animals such as deer and monkeys, forest rehabilitation and to conduct research in wilderness area. Besides the role of other stakeholders, such as NGOs Office, Yakushima Electric Company (*Yakushima Denko*), Tourist Guide Agency Association, and Hotel, *Minshuku, Ryokan*, Youth Hotel and Guest House Association are also very significant. In this context, the protection and the socialization to local people and visitors of Turtle life on breeding eggs in Inaka-hama Beach, Nagata is subsequently very productive and fruitful. Because the turtle is categorized as endangered species based on *CITES* Agreement, and Japan one of member who made signature in the agreement.

One recommendation to Kagoshima Prefecture as the second prefecture rich of natural resources in terms of surrounding islands which stretches for 600 km from its northern to southern and 605 islands, including 28 inhabited ones, it should prepare construction infrastructure and manpower to maintain Yakushima Island as a Natural and World Heritage site. Besides to adopt and to nominee Amami-Oshima Islands as the next step for World Heritage area to keep its environment and to grow tourism industry.

Chapter 8
USA: Sustainable Forest Management

Introduction

The Ministers attending the 9th Session of the United Nations Forum on Forests (UNFF-9, 24 January-4 February 2011, New York) recognized that 1.6 billion people—nearly a quarter of the world's population—depend on forests for subsistence, livelihood, timber, fuel and income generation.[1] Inherently issued with the theme of 2011 as the International Year of Forests—Forests for People, delegates at UNFF-9 adopted a Ministerial Declaration and Omnibus Resolution on Forests for People, Livelihoods and Poverty Eradication. This note inspired many forest holders, such as Tom Tidwell, Chief, U.S. Forest Service, Robert K. Davies, Director, Division of Lands and Forest, New York State Department of Environmental Conservation (DEC) and Michael R. Blomberg, Mayor of New York. The statement of Chief, U.S. Forest Service in celebrating International Year of Forests on February 3, 2011, said: "In the United States, conservation groups, landowners, businesses and governments all levels are involved more than ever before in efforts to address environmental and economic challenges on many fronts, including in the development of a new Forest Service Planning Rule that will govern how we manage and conserve America's 193-million-acre National Forest System." In line with this statement, New York State Forester, said, "… planting trees, practicing sustainable forestry, and conserving our forest lands are among the most cost effective strategies we can take to keep New York green, while building our communities and contributing to the state's economic recovery."[2] Meanwhile, the Mayor of New York launched a program the so-called: "The New York Green Infrastructure Plan in 2002." This project cost more than $6 billion and was aimed

[1]Available at: http://www.un.org/esa/forests/pdf/session_documents/unff9 (accessed on October 9, 2014).
[2]"2010 Statewide Forest Resource Assessment & Strategy," in New York State Department of Environmental Conservation (accessed on September 8, 2014).

H. Hidayat, *Sustainable Plantation Forestry*,
https://doi.org/10.1007/978-981-10-7653-4_8

cleaning up New York's City's rivers, creeks and coastal waters, clean water and air, and protect forests as ecosystems.[3]

Observing those statements above and the UNFF-9th results, it indicates that forests resources are increasingly valued and significantly utilized in worldwide. Trade in forest products was estimated at $327 billion in 2004. Meanwhile, forests are home to 300 million people around the world.[4] So, it is necessary to understand that forests are more than trees—that they are ecosystems whose health is reflected in the quantity and quality of water, air, carbon sink, biodiversity and so forth—emerging from them. This paper would like to discuss the utilization timber products from forest resources, the USA as the second largest paper producer in the world. Its productions reached 75 million metric tons in 2011. Revenue of the USA's forest paper, and packaging (FPP) industry reached $106 billion, and create job absorption of 1.2 million in manpower.[5] In addition, per capita consumption of paper reached 231 kg. The USA was categorized as one of the biggest paper consumers in the world. As know that the USA's timber consumption reached 592 million cubic meters in 2005 and slightly decreased to became 490 million cubic meters in 2007 (Wang and Wang 2010). On the other hand, forests are regarded as ecosystem services for open space and outdoor recreation among the nation's urban and rural on recreational opportunities for 66 million Americans. This means nine in ten Americans (87%) participated in outdoor recreational activity in 2003. In the case of wildlife recreationists, they spent $108 billion in 2001. According to the survey in 1997, an estimated 1.1 million jobs was derived from wildlife-based recreation in the USA, and directly contributed more than $20 billion in income to the entire economies across the USA.[6] In addition, forests are viewed as ecosystems to preserve water catchment area and as sources of hydrology. From this point of view, this is why Adirondack and Catskill Forest Parks, which encompass nearly 3 million acres and function as forest preserves are considered very significant to maintain due to the supply of clean water and clean air to produce carbon sink for inhabitants in New York State. Showing this strategic synergy among stakeholders to promote "green infrastructure" has been endorsed by U.S. Environmental Protection Agency (EPA) and New York State Department of Environmental Conservation (DEC).[7]

---

[3]See, "New York City Green Infrastructure Plan: A Sustainable Strategy for Clean Ways," (accessed on September 12, 2014).

[4]See, UNFF-9 declaration "Forest for People," (http://www.un.org/forests) (accessed on October 13, 2014).

[5]See, "World's Paper Products by Countries," from Wikipedia Paper Industry, 2010 (accessed on October 7, 2014). The largest paper producer in the world was China in 2011. Its production reached 99 million metric tons. Japan as the third paper producer reached 27 million metric tons.

[6]"Rationale for a National Policy for Sustainable Forests," in Sustainable Forests Task Force (accessed on September 11, 2014).

[7]"New York City Green Infrastructure Plan: A Sustainable Strategy for Clean Ways," (accessed on September 12, 2014).

Obviously maintaining healthy forest resources and forests as ecosystem services eventually could positively contribute access on economic, social and ecologically, so, it is very significantly to pay attention the implementation of sustainable forest management among stakeholders in the USA, from federal, state, county, municipality, and city levels; industry timber companies; local communities; environmental groups; and even community at large.

The paper focuses discussion on the need to apply sustainable forest management in U.S., the implementation background from the perspective of external and internal factors as well as actors of forest management, whether the federal role, state government (New York) and private land ownership from the viewpoint of the transformation of timberland ownership to timberland investment (TIMOs and REITs). In addition, the paper also explores a case study of New York State in implementing sustainable forest management by examining working forests and conflicts of forest politics in the surrounding Adirondack and Catskills Park, to launch green infrastructure program plan, and strategy and issues to maintain future forests.

The Conceptual Approach

The Ministerial Declaration Preamble in UNFF-9 emphasized that forests were an "integral part of the global environment and human wellbeing". This declaration shows a strong commitment to improve the livelihoods of local communities, by a creating the conditions needed for sustainable forest management, including strengthening cooperation in finance, transfer of environmentally sound technologies, capacity building and governing, promoting secure land tenure, participatory decision making and benefit sharing, in accordance with national laws and priorities (Morgera 2011: 75). This declaration invites on stakeholders, especially the role of government as regulator and inspector to implement sustainable forest management, due local people and general people at large are mostly depend their livelihoods on healthy forests condition.

What does sustainable forest management mean? The concept is generally defined as a balance between the social, economic and environmental values associated with forest resources, with some consideration of these values for future generations.[8] Actually, the concept of sustainable forest management (SFM) has been relatively well defined since 1995. Substantial progress has been made in science and policy fields related to SFM internationally. On the other hand, Jeffrey Sayer, a conservationist, prefers to use the term "ecosystem approaches";

[8]For further information, "Evaluating sustainable forest management," available at www.sciencedirect.com. Elsevier Publisher (accessed on September 9, 2014). In addition, sustainability is defined as "development that meets the needs of the present without compromising the ability of future generations to meet their own needs. So sustainability used here, is a process and includes ecological, social and economic dimensions." See, Berkes and Folke (1998).

this phrase is now widely used to describe approaches relating to the management of the social-ecological systems to which our forests belong. He further says that ecosystem approaches require that we have an understanding of all the actors and interest groups that are influencing our forests. The rationale is highlighted because every forest is different and what constitutes good forestry in one place today will not be appropriate at another location. Therefore, this means that those who are professionally involved in forestry management face challenging times. The concept has been developed and endorsed by the Convention on Biological Diversity and the 180 countries that are signatories to that convention (Sayer 2005: 6).

In 2005, the Food and Agriculture Organization (FAO) appealed the question: "Is the world progressing towards sustainable forest management?" The result came up as the Global Forest Resources Assessment (FRA) 2005, which reported forest statistics for 229 countries and areas, aggregated into 12 regions. The report was based on six of the seven agreed thematic areas (criteria) of sustainable forest management (SFM), drawn from the nine ongoing regional/international processes on criteria and indicators (CI) for SFM, as follows:

1. Extent of forest resources.
2. Biological diversity.
3. Forest health and vitality.
4. Productive functions of forest resources.
5. Protective functions of forest resources.
6. Socio-economic functions
7. Legal, policy and institutional framework.[9]

It is shown that at the global level, the Forest Resources Assessment (FRA-2005) revealed that the world's forest resources performed well, with most major trends more positive and neutral than negative. However, these observations changed significantly when examined at the sub-regional level, where North America (US, Canada and Mexico) was clearly contributing to the positive trends and South and Southeast Asia and South America were contributing mainly to the negative trends. Nevertheless, all of the regions and sub-regions displayed a mix of data that made it very difficult to define the level of progress towards SFM, particularly when the varying spatial and temporal dimensions of sustainable ecosystem management are also considered.

Actually, sustainable forest management (SFM) is strongly interrelated with the economic dimension of decision making. Therefore, the globalization of the forest industry, which facilitates international trade in forest products (timber and non-timber), must also ensure best practices and promote the harmonization of both national and management level policies and management standards. In addition, Jeffrey Sayer et al. proposed significantly to utilize "technologies" for sustainable forest management. He further anticipated a significant shift in timber production

[9]See, "Evaluating sustainable forest management." Available at: www.sciencedirect.com (accessed on September 9, 2014), p. 110.

from natural forests to plantations. This phenomenon is affected by several driving factors, such as the reliability of supply, uniformity of raw material and competitive pricing. In this case, technology promises many advances in processing in terms of the best quality seedling, fertilizing, planting, inspecting, pests of wood, production, recycling and so forth that may have a big impact on forest management (Sayer et al. 2005: 264–265).

The FRA, 2005 report also revealed that development of regular and comparable monitoring and information reporting programs, especially at the local, sub-national, and international levels continue to be paramount objectives of SFM. As an illustration, at the international level, this report notes that Criteria (C) and Indicator (I) frameworks have emerged as a powerful tool in implementing SFM internationally, with around 150 countries now using them in their forest management and reporting processes. Meanwhile, at the national level, the report notes a highly successful approach to developing a national C&I framework for Australia, based on a modified set of the Montreal Process C&I. This process significantly improved Australia's ability to report comprehensively and consistently on economic, environmental and social forest values. In addition, at the sub-national (regional) level of forest management, Hickey and Innes' findings based on the field study was designed to identify 'common ground' between the various SFM indicator lists that have been generated through different process-based initiatives.[10] There are two main driving factors affecting sustainable forest management (SFM) in the USA. Firstly, from external perspective is certification of sustainable management; Secondly, the primary internal factor is the increasing demand for timber.

The Background to Implement SFM

External Factor: Certification

The intensive logging and illegal logging in 1980s and 1990s, mostly occurred in developing countries, who own tropical forests and treated forest resources as commodity to obtain foreign exchange earnings. In line with this was the weakness of government as regulator and law enforcement practice among timber traders who carried out illegal logging to supply the demand of timber to forestry industries. Therefore, to counter further loss of tropical forests and to implement sustainable forest management, the United Nations held Conference on Environment and Development (known as the Earth Summit) in Rio de Janeiro, 1992. One of the most important agendas was to promote the idea of setting standards of sustainability in order to "certify" the wood harvested from those forests were from a sustainable source.

[10]"Evaluating Sustainable Forest Management," in www.elsevier.com (accessed on September 9, 2014), p. 112.

When environmental groups returned home after attending the United Nations Conference on Environment and Development in Rio de Janeiro, they formed a movement in Europe. The movement was concerned about the loss of tropical forests and hoped to certify forests products more sustainable productivity. Therefore, environmental groups promoted the idea of setting standards of sustainability in order to certify that wood harvested from those forests was from a sustainable source. Its development idea of "green certification" quickly caught on among environmentally conscious consumers and the retail outlets they patronized and rapidly spread to North America, including the USA. It was not so easy to define sustainable forest management, and a debate raged in forestry circles as to what factors should be considered (Journal of Forestry 1998; Lober and Eisen 1995). By 2001, there was already registered 60 forest certification systems around the world, presented by various groups with different ideas about what should be included in the standards to be met to earn certification. As an illustration, in Europe the forest owner associations of the European Union came together to create their own certification process—the Pan European Forest Certification (PEFC) system. On the other hand, in the USA, the American Timber Forest Service (ATFS) was already in existence, and for decades had been inspecting and certifying that its member tree farms were properly planned and managed (Simpson 2001).

The World Wide Fund (WWF) for nature based in Switzerland, decided in 1993 to help resolve confusion and differences among certification systems. It subsequently brought together representatives of the forest industry and environmental organizations to form the Forest Stewardship Council (FSC). The FSC adopted a set of principles and criteria to apply to the management of tropical, temperate, and boreal forests worldwide, and established a process for third-party verification of a criterion (Sample 2000). In the case of Indonesia and other developing countries, logging concession holders who wanted certification followed FSC, in order that their timber products could be profitably sold in the domestic and international market as well.[11] Thus, following a successful third-party FSC assessment and verification, a producer would be entitled to affix the FSC label directly to a product to inform consumers that it was produced from a forest managed in accordance with the FSC principles and criteria.

In its development, the USA eventually established the *Sustainable Forestry Initiative* (SFI) in 1995 to develop rigorous standards for sustainable forest management, and required their landowner member companies to follow the SFI's comprehensive system of sustainable environmental and forestry principles, practices, and performance measures as a condition of membership in AFPA. Today, SFI, Inc. is a fully independent charitable organization that promotes sustainable forest management with certification standards, including measures to protect water

[11]Hidayat, Herman, "From Deforestation to Sustainable Forest Management: A case Study of Dwima Jaya Utama Logging Company in Katingan District, Central Kalimantan," Research Report for LIPI, August, 2014.

quality, biodiversity, wildlife habitat, species at risk, and forests with exceptional conservation. In this matter, SFI works through a network of grassroots implementation committees to provide third-party verification and is widely used across North America and around the globe.

All three leading third-party verification systems for certifying sustainable forest management, such as ATFS, FSC, and SFI, are widely applied and credible. In addition, in 2008, the National Association of State Foresters passed a resolution that all three systems include credible elements, make positive contributions to sustainability, and no program can claim to be the best approach. The final result is a general hope that through the existence of green certification, forest management may perform better in responding to society and consumer demands for greater accountability and sustainability of forests.

What is the Response from Timber Industry on Certification?

George Pacific (GP) Company is among the three largest paper industries that includes International Paper, George Pacific and Weyerhaeuser in the USA. The Company takes an active role in organizations to support sustainable forestry at the national and state level, including Sustainable Forestry Initiative (SFI) and Implementation Committees (SICs). George Pacific, since 1995, has made a commitment to support the sustainable forestry initiative (SFI), by sending 140,000 logging professionals that have been trained through SFI-approved training programs. Besides, the Company (GP) supports credible forest certification programs such as SFI, Forest Stewardship Council (FSC), and Program for Endorsement of Forest Certification (PEFC) that are currently being utilized at numerous George Pacific facilities. Accordingly, George Pacific is committed to sustainable practices and makes the following commitments:

(1) The Company will continue regularly third-party certification of its wood and fiber procurement practices across all of its operating areas;
(2) All things being equal, George-Pacific will give preference to wood certified by established and recognized certification systems;
(3) George-Pacific will continue to require its timber suppliers to be trained in sustainable forestry practices;
(4) George-Pacific will continue to participate in the industry effort to recover 70% of paper consumed in the United States by 2020.[12]

[12] "George Pacific's Commitment to Sustainable Forestry and Certification," www.gp.com (access on September 8, 2014).

Internal Factor: (The Timber Demand)

It must be highlighted that overall there is a worldwide 'shortage' of timber. Several studies have addressed global supply and demand for wood. Sedjo and Lyon (1996) indicated that average annual demand for industrial roundwood would increase from 1700 million m^3 in 1995 to about 2300 million m^3 in 2045. In this case, consumption of pulpwood for wood-based panels and paper is expected to approximately double to 1330 million m^3 by 2045 (Evans and Turnbull 2004: 13).

Timber consumption, which has become a global concern, includes fuel wood, energy, (biomass), fiber, furniture manufacturing, paper making and house construction. Such consumption has continued to evolve in developed nations, increasing and leading to stricter utilization standards and innovative products. Demand for timber and its products, driven by expanding populations and consumer oriented lifestyles, continues to grow, but at the same time, appreciation of forest resources has grown. Some stakeholders, such as NGOs (like WWF, Greenpeace, Walhi, and The Nature Conservation) advocate for environmental conservation and biodiversity protection. In the case of the USA, according to statistics from forest service (James 2008), the population of the USA increased by 52.70%, from 194 million to 296 million between 1965 and 2005. Meanwhile, the consumption of timber increased by 58.89%, from 440 million m^3 in 1977 to 590 million m^3 in 1987, and 594 million m^3 in 2005. It decreased to 538 million m^3 in 2007 (Fig. 8.1). It is predicted that by 2050 the USA's population will eventually reach 500 million. In addition, during that same time period, the country went from being primarily rural to being nearly 80% urban and suburban dwellers, with the areas of greatest growth in the West and the costal South.[13] As a result, the nearly tripled population will need to convert forest land at a rate of 1 million acres per year for activities such as agriculture, estate plantation, industry, settlement, and so forth. For comparison, the example of timber consumption in U.S. and China as illustrative. In a global context, the USA and China are both among the biggest timber consumers. Known as the largest paper producer in the world, the USA's paper industries produced 89 million m^3 ton and China was the second largest producer reaching 58 million m^3 ton in 2006.[14] Timber consumption in the USA per capita is 2.01 m^3 on average, almost 3.2 times the average level in the world (State Forestry Administration 2008). Because of the large population, though the per capita level in China is relatively low about 0.24 m^3, China consumes lots of timber. Historical trends indicate that China's timber consumption is rapidly increasing; for instance, in 2002 it reached 100 million m^3, 200 million m^3 in 2003, and it rapidly increased to 300 million m^3 in 2004 and 350 million m^3 in 2005 (Fig. 8.2). While in the USA, the

[13]See, "Rationale for a National Policy for Sustainable Forests," in Sustainable Forests Task Force. (accessed on September 11, 2014).

[14]*World Pulp and Paper Statistic* in 2007.

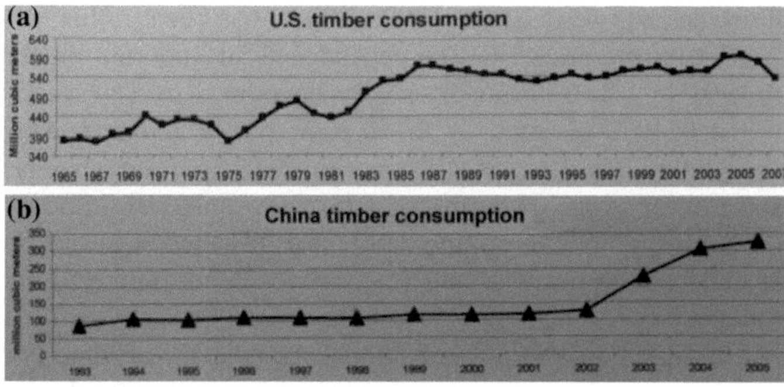

Fig. 8.1 **a** U.S. timber consumption, **b** China timber consumption. *Source* Wang and Wang (2010)

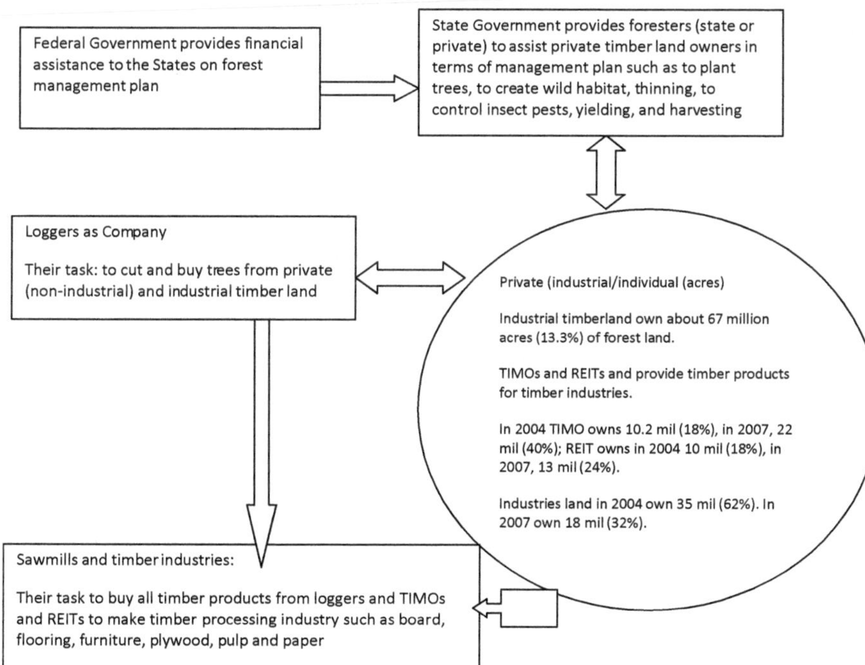

Fig. 8.2 The scheme of timber industry and sawmill find out supply of timber. *Source* Interview with Dr. Peter Smallidge, on August 27, 2014

consumption in the last two decades (1990s–2000s) has been in flux, that has been small compared with the past.[15] USA's dependence on imported timber increased from 2% in 1991 to 9% in 1996; it rapidly developed to 16% in 2002 and is predicted to reach 19% in 2050, if the current policy remains (State Forestry Administration 2008). The USA is still a net timber import country, because production costs are very high in terms of labor, raw material (timber cost), and efficiency of paper companies in the USA; for instance, George Pacific, The International Paper, Weyerhaeuser and so forth could not compete with overseas paper companies.[16]

In addition, a Task Force from the National Association of State Foresters and the Society of American Foresters was established in April 2006. In realization of the program, the Association cooperates with the government Agency. The aim of this association is to create a renewed national commitment and social contract to understand, enhance, and protect the health, productivity, and sustainability of America's forests. So the rationale is a compelling national interest in sustainable forests for the USA, as follows:

(1) To provide a clean and healthy environment for the nation's urban and rural citizens:

It is known that forests produce water resources. Therefore, it is very vital to provide clean water for drinking water. It is estimated that about two-thirds of the nation's scarce water resources originates on forests and forested watersheds provide a source of drinking water for over 180 million people.

(2) To provide employment and economic opportunities:

The U.S. is the second largest paper producer in the world. Production reached 75 million metric tons in 2011.[17] U.S. paper and wood manufacturing generates 1.2 million jobs and $230 billion dollars in annual sales. The forest industry is the number one industry in 10 of 13 states in the Southeast, providing the raw material for a multitude of wood products, including lumber, paper, laminates, engineered structural wood, and composite materials. In addition, forest products also require less energy to produce than other building materials and are highly recyclable.

---

[15]See, Wang and Wang (2010).

[16]Interview with Steven A. Wolf, Associate Professor in Department of Natural Resources, College of Agriculture and Life Sciences, Cornell University, September 16, 2014. He further predicts if the U.S. Department of Agriculture (USDA-Federal government) and New York State cannot issue more strategic policies in terms of limited 'regulation' and 'incentives' for better forest land management who belong to individual private land, which reach 38% of total national forests land to be more productive in terms of quality of timber, production, and competitive on prices (reasonable), the US is going to become a net importer timber from overseas (e.g., Latin America, Canada, Australia, Southeast Asia, etc.).

[17]See, "List of 10 main countries Paper producers in the world," from Paper Products in Wikipedia, 2010 (accessed October 7. 214). The global paper production approximately reaches 400 million metric tons. More than half production (200 million metric tons) predominantly occupied by three countries China, the USA and Japan.

(3) Open space and outdoor recreation:

Forest services such as recreation and hunting provide many jobs for the people. For instance, wildlife recreationists spent a total $108 billion in 2001. It is estimated 1.1 million jobs were derived from forest-based recreation in the U.S. in 1997, contributing more than $20 billion in income to economies across the U.S.

(4) America's cultural and traditional heritage:

The frontier history, such as conservation accomplishments and private family forests are partly integrated into the U.S.'s cultural and traditional heritage that define who we are. In this case, many of the best examples of forest management can be found on tribal lands, where Native American values have influenced and encouraged long-term on sustainable forests.

(5) Energy self-sufficiency:

The benefits of forests as ecosystems are enormous for human beings and ecology. Forests can provide to reduce energy use and greenhouse gas emissions. Forests also can produce bio-energy as a new industry, with the potential to provide substantial amounts of energy using woody biomass as fuel. The U.S. could feasibly derive 10% of its energy from wood (currently 3.1%). In this case, this could help to reduce utilizing fossil fuel and contributing to forest health by reducing fuels.[18]

Actors of Forest Management

Forests are valuable for their timber products, the global ecosystem and humanity; much of this value is not marketable and tends to be ignored in economic considerations. Obviously, those goods and services that have a monetary market value are usually considered (e.g. timber), and often these alone do not compete favorably with other land uses, such as agriculture. In the case of the USA, forests have been a long standing concern of the states, and forest management is a significant activity due to the economic importance of forestry in many states and in the nation. According to the forest facts (USDA 2009), the total forestland in the USA is approximately 751 million acres (303.93 million ha)—about one-third of the Nation's total land area. The forest area coverage is about 33%, while the forest stock is 26,393 million m^3. There are many actors. for instance, federal government (central government) in terms of forest regulation which are represented by the United States Department of Agriculture (USDA), state government, and individual/private forestland owners (interview, 29 August 2014). These actors are actively involved in forest resource management in the USA.

---

[18]See, "Rationale for a National Policy for Sustainable Forests," in Sustainable Forest Task Force (accessed on September 11, 2014).

Federal Government

Forest Governance in Federal System

The government structure of the USA is divided into federal (central government), state government (province), municipalities, counties, and cities. In case of the federal system, the central government usually has responsibilities to manage natural resources, including forest, activities and events that eventually affect more than one state. The federal system can be simultaneously decentralized in some respects and centralized in others. In the USA, for example, some 31 federal entities interact directly with the Forest Service in planning and managing federal forest-lands and many others have a more indirect linkage. In addition, federal ownership is substantial, including in the USA, where the federal government owns about 35 per cent of all forests, the state government just owns about 5% and the private sector owns the majority at about 60%. Therefore, federal and state governments have established regulation and programs to enhance and regulate private enterprises.[19] As an illustration, while the federal and state government raised forestland taxes for the aim of government income and efficiency of timber business, most industrial timberland owners sold to timberland investment such as TIMOs and REITs (Interview, 16 September, 2014).

Participation of Civil Society and Private Sector

The role of civil society such as non-governmental organization (NGOs) and private sector on forest governance is very significant. As previously explained, the private sector owns the majority of the forestland in the USA, so the role of state government is necessary for regulating the private sector, with the objective on forest planning, production, environmental impacts and fair business practices of commercial forest activities. Meanwhile, federal government is actively involved with the private sector through incentives and fiscal programs.

The United States Department of Agriculture (USDA)

The US Department of Agriculture (USDA) as representative of federal government consists of three principal agencies that manage renewable resources: (1) the Forest Service; (2) National Institute of Food and Agriculture (NIFA); and (3) Natural Resources Conservations Service (NRCS). The NIFA and NRCS reflect a recent organization in the USDA to consolidate natural resource program responsibilities

[19]See, Gregersen et al. (2005).

into fewer agencies. The tasks of the three agencies are to function as the principal sources of technical, educational, and financial assistance to improve renewable resources management on private lands (Handee et al. 2012: 368).

The task of the *Forest Service*[20] *Agency* is the management of the 193 million acres[21] (78.2 million ha) in the National Forest System, which includes 155 national forests and 20 national grasslands. National forests keep those resources coming in abundance by using them wisely: at present, national forests produce the most valuable kinds of timber, wood and forage. These areas are located in 44 states, Puerto Rico, and the Virgin Islands. National forests are found from Maine to California and from Alaska to Texas and Florida. In some of this vast area there is annual fire danger, and multiple risks and hazards threaten forest resources and local communities. In addition to the control of wildfires, there are other aspects of fire, such as removal of forest debris as a preventive measure.

The task of NIFA (*National Institute of Food and Agriculture*) based on Farm Bill 2008 replaces the Cooperative State Research, Education, and Extension Service. NIFA's mission is to advance knowledge for agriculture, the environment, human health and well being, and communities by supporting research, education, and extension programs in the land-grant university system and other partner organizations. Besides, this Agency (NIFA) helps fund such activities at the state at local levels, and provides national program leadership in these subject areas.

Meanwhile the Natural Resource Conservation Service (NRCS) previously called the Soil Conservation Service celebrated its seventy-fifth anniversary in 2010 and its federal commitment to conservation programs on private lands. Because in the USA 60% of all USA's land is registered as privately owned, the proper stewardship of these lands is essential to the health of the USA's environment. The NRSC assists conservations districts in preparing conservation agreements with landowners and implementing planned conservation work on individual lands. Those holdings contain forestland, wildlife habitat, erosion problems, or any land that would benefit from planting trees.

The Federal Role in American Forests

Federal budgets today show a long-term trend of spending in federally and private owned forest lands across the full range, such as recreation, wilderness, wildlife, water and timber. The federal government is actively involved in the USA's forest

---

[20]The task of United States Forest Service includes: (1) advocating a conservation ethic in promoting the health, productivity, diversity, and beauty of forests and associated lands; (2) protecting and managing national forests and grasslands so the best demonstrate the sustainable multiple-use management concept (Ibid).

[21]One ha consists of 2.47 acres. For instance, 50.5 acres equals 20.43 ha. In the USA, acres are used to measure land, rather than hectares.

sector in both extensive and multi-dimensional ways. The specific roles of the federal government in forest sector today is as follows:

(1) Manage the national forests, national parks and other federal lands;
(2) Provide financial assistance to the States for delivery to forest landowners;
(3) Partially finance and set overall quality standards for national program designed to assist in protecting forests from fire, insects and diseases;
(4) Directly carry out forest research, as well as help finance research at state educational institutions;
(5) Set policy/standards for air and water quality, pesticide use, protection of endangered species and wetlands management on both public and private lands;
(6) Directly enforce federal wildlife laws and regulations (dealing primarily with migratory birds and endangered species);
(7) Establish and enforce worker safety rules for industrial facilities and forest operation;
(8) Assess federal taxes on income derived from forests and federal inheritance taxes on the estates of decreased forest landowners.[22]

Federal responsibilities which are shared with state and local responsibilities for policy making and governance range from: (1) federal policy regulatory oversight, with state and local enforcement on private lands (e.g., Clean Water Act, pesticide labeling and enforcement, etc.); (2) indirect federal control of standards through cost sharing leverage (e.g., wildfire protection standards, the federal cost sharing of private reforestation, etc.); and (3) state and local control (e.g., regulation of private forest lands, except for listed endangered species).

National policy can clarify and enhance the roles of federal, state, and local governments, promoting regional collaboration, joint planning and coordinated action.

A National Policy for sustainable forests would say that:

(1) The management and conservation of forest resources in the United States should be guided by a mandate to meet the forest related needs of the present generation without compromising the ability of future generations to meet their needs.
(2) Doing so requires that economic, social and environmental values from forests be provided within a framework where these values are mutually supportive.
(3) Laws and programs that promote this vision of sustainable forests and the inter-connectedness of environmental, social and economic values are acceptable expression of federal policy.[23]

[22]"Rationale for a National Policy for Sustainable Forests," in Sustainable Forests Task Force (accessed on September 11, 2014).

[23]See, "Rationale for a National Policy for Sustainable Forests," (access on September 11, 2014).

State Government

The USA consists of the federal government, fifty (50) states. Each state is divided into counties. There was limited forest regulatory activity in some states in the 1860s and 1870s. In 1885, several states moved for the first time to create a state forestry agency. Thus, the state forestry programs were formed between 1885 (New York, Ohio, California) and 1966 (Arizona). All 50 states now have official forestry programs, regardless of whether they own state lands or forest management or just provide programs.

After 1900, the federal government assumed a leadership role and in the ensuing decades shaped the development of the forestry program nationwide. From this point, the unique relationship between federal and state governments in forestry matters makes the history of US forest regulations important to the evolution of federal state relationship in general.

State Forestry Departments

Forestlands nearly reach two-thirds (59%) of the total forest land in the USA owned privately. Federal forestlands are 31% of the total, and state and local governments hold 10% of it. But the states' 10% amounts to 49 million acres (19.8 million ha), one-third more acreage than is owned by the forest industry. According to a 2008 survey of state forestry organizations sponsored by the National Association of State Foresters (2010) emphasis "state forestry agencies are also involved in overseeing and helping improve the management of private forestlands, they play a critical role for our nation's forests" (Handee et al. 2012: 368).

Ten states are responsible for directly managing forestland in the U.S. (Table 8.1). Some states have some forestland and some have only a few acres to

Table 8.1 The largest width of forest land area in ten states

Number	States	Acres (million)	Hectares (million)
1.	Alaska	24.7	10
2.	Minnesota	4.5	1.8
3.	Michigan	3.9	1.6
4.	Pennsylvania and Washington	2	0.85
5.	Florida	1	0.4
6.	Idaho	0.985	0.399
7.	New York	0.977	0.396
8.	Wisconsin	0.924	0.374
9.	New Jersey	0.809	0.328

Source Hendee et al. (2012, p. 389)

directly manage. Timber harvests from state lands totally reached 2.3 billion board feet of timber in 2008, providing significant revenues to their respective states.

The state government is responsible, for instance, for compliance, monitoring, and policy development. But the largest state forest responsibilities are in providing services and outreach to landowners to help improve the protection, management and sustainability of the state's private, nonindustrial forests. Among all states collectively, 5% of their budget is spent on forest practices. State forestry agencies employed a total of 26,000 workers in 2008, 29% of them were foresters and 37% of the totals were seasonal. The majority of state forestry outreach expenditure (64%) was related to fire prevention, protection, control and management as all firefighting costs have been increasing. Other important programs administered by states include protecting working forests threatened by development or encroachment, forest recreation, urban and community forestry, water quality protection, and field monitoring of best forest management practices (BMPs). To sum up, many states have forest practice regulations that are administrated by state forestry organizations. Regulations of forestry activity by local jurisdiction have increased rapidly (Ibid, 403).

Private Lands Ownership

In the USA, private forestlands, in both corporate and non-corporate ownerships, comprise 56% of all USA's forest land and 70% of timberland (Table 8.2) and, in general, occupy the better timber growing sites. The major reason for this is that early federal land disposal laws encouraged the sale and settlement of the better and more accessible timbered lands (Ibid, 406). Private forestlands in the USA included 423 million acres (171.3 million ha) owned by close to 11 million forestland owners. The composition of owners with parcels of 10,000 acres (4050 ha) or more account for 22% of all forestland and this includes many of the corporate owners (TIMOs and REITs)[24]; the rest, 61% of forest landowners have between only one and nine acres of forest land. This group, previously referred to as NIPFs, are now generally referred to as family forest owners (i.e., individuals, couples. estates,

[24]TIMO is the abbreviation of Timber Investment Management Organization. This Agency is very strategic in providing timber supply to the pulp and paper industry, sawmills, plywood factories and so forth. TIMO has bought about 30 million acres during two decades (the end of the 1990s to the 2000s) from private corporations (paper industries). TIMO also changes terminology from using the categories of forest industry and nonindustrial private forest (NIPF) landowners to the categories of private corporate and private non-corporate owners (including Native American Lands).

REITs is Real Estate Investment Trusts. Currently REITs cooperates with TIMO in providing fundraising from pension funds, private capital (private equities) to buy forest land from companies, especially paper industries such as George Pacific, International Paper and so on. By a new regulation issued by the Federal government, paper industries must concentrate on paper making, for providing raw material (timber supply) is provided by TIMO.

Table 8.2 Ownership of forestland and timberland in the USA

Ownership	Forestland acres (Million)	Forestland area (%)	Timberland acres (million)[a]	Timberland area (%)
Private, noncorporate	285 (115 ha)	38	250 (101 ha)	49
Private, corporate	138 (56 ha)	18	106 (43 ha)	21
Public (national forest)	147 (60 ha)	20	99 (40 ha)	19
Other Public (federal, state, and local governments)	181 (73 ha)	24	59 (24 ha)	11
Total	751 (304 ha)	100	514 (208 ha)	100

[a]Timberlands are lands that grow 20 cubic feet of wood per acre per year
Source Handee et al. (2012)

trusts, and groups of unincorporated individuals) and they comprise 92% of private forest owners and own 62% (264 million acres; 106.9 million ha) of all private forestland. The majority of family forest ownership is for reasons other than management for timber production and includes motives such as aesthetic, family legacy, privacy, recreation, and others.[25] Although family forestlands contain an important component of the nation's timberland, only 19% of their owners report having a written forest plan and only 58% of their owners report commercial harvesting of trees from their lands (Butler 2009).

Private forests are generally separated into two categories: industrial and non-industrial. This distinction can be confusing, since industrial in this context means a company that owns both forestland and mills to process forest products (Best and Wayburn 2001: 6–7). Thus, industrial forestland may be held in million acre tracts by Multinational Corporation, or it may be a 160-acre-parcel owned by a small local sawmill. A large oil company or institutional pension fund that owns forestland but does not own a wood processing facility will be found in the non-industrial category. The nonindustrial category encompasses land that may be held by an individual with 5 acres as well as hundreds of thousands of acres held by Alaska Native corporations.

So private, corporate forestlands owned by the forest industry include lands held by both small companies and large corporations that grow and process wood into products, as well as TIMOs and REITs. Private, noncorporate forestlands include small family forest lands held by persons from all walks of life, for which forestry is at best an adjunct to their major vocation. This is the largest ownership category and includes 49% of the nation's timberland.

[25]One respondent said that he owns more than 48,000 acres of individual forestland in Georgia State, which he has planted with trees, for future secondary income. He would like to use the income from timber harvesting for preparation of his daughter's wedding (Interview, 8 September 2014). Another respondent said that he owns about 50,000 acres (20,430 ha) of forestland in Essex town, part of New York State. This individual forest land was bought in the 2000s from an individual landowner and is used for planting trees, hunting and recreation with the family (interview, 29 August 2014).

Table 8.3 Forest owners by type of entity

Ownership type	Ownerships	Ownership (%)	Acreage (000s)	Acreage (%)	Average acreage
Forest industry	13,300	0.13	79,715	20.26	5994
Farm	2,431,300	24.55	111,450	28.33	46
Industrial					
Business	30,700	0.31	9031	2.30	294
Real estate	236,200	2.39	15,948	4.05	68
Other business	74,600	0.75	3986	1.01	53
Recreation club	116,200	1.17	7768	1.97	67
Public utility	800	0.01	2248	0.57	2810
Individuals	6,765,000	68.32	133,521	33.94	20
Other	233,600	2.36	29,722	7.56	127
Total	9,901,700	100.00	393,389	100.00	40

Source Best and Wahburn (2001)

In Table 8.3, which describes ownership types compiled from the 1994 Birch study, provides some illumination on the kinds of entities that own forestland. From this table, we can begin to appreciate the diversity of nonindustrial owners. While individuals and farms are the largest ownership types overall, the forest industry clearly has the most concentrated control.

The Transformation of Timberland Ownership

In the history of the USA, the existence of timber industries, for instance the International Paper, George Pacific, Weyerhaeuser, Champion International and so forth are very significant, because they have been among the largest private land owners in the USA. Throughout the 20th century, industrial timber companies continued to add to their timberland stocks and by the end 1970s, these holdings peaked at approximately 68 million acres[26] (Table 8.4). But these holdings were primarily owned by 40 companies in terms of approximately 80% of private industrial timberlands, which enabled them to dominate the forest products industry and gave them a considerable advantage over smaller companies that often relied on public timber to supply their mills. The rationale behind this condition is that timber industries considered that substantial timberland holdings were vital for the firm's success. Timberlands were owned for a variety of reasons in terms of holding timberland; for instance, to provide security, for meeting present and future raw

[26]Gunnoe and Gellert (2010) and also read Yin et al. (1998).

Table 8.4 Private US timberland ownership 1953–2007, in 1000 s of acres

Year	Total US Forest lands	Total public forest lands	Total private forest lands	Forest industry (private corporate %)	Non-industrial private
2007	514,213	157,704	356,488	106,131 (20.6%)	250,357
2002	503,540	147,276	356,264	65,596 (13.0%)	290,667
1997	503,664	145,967	357,698	66,858 (13.3%)	290,840
1987	486,317	132,406	353,911	70,347 (14.5%)	283,564
1977	492,355	138,169	354,186	68,937 (14.0%)	285,249
1953	508,854	145,436	363,419	58,979 (11.6%)	304,440

Source Smith et al. (2004), Tables 10 and 11

Note The ownership classifications changed with the 2007 data to 'private corporate' and 'private non-corporate'. In the private years, the ownership was classified as forest industry or non-industrial private

In 1997 and 2002, Native American lands were included in non-industrial private, whereas previously they 'may be included' in other public owner

material requirements were the primary justifications offered in support of forest ownership. But the most important was that these lands provided reliable resources for the industry's capital intensive production facilities.[27] Other reasons, according to showed the insurance that in-house timberland provided against short-term price fluctuations and the fact that timberland was a low-risk investment that provided asset liquidity and a hedge against inflation.

As seen in Table 8.4, almost three-quarters 356,488 million acres (71%) in 2007 of these timberlands are privately owned, and this private area accounted for 92% of the harvest (growing stock removals). However, as Table 8.4 shows, only 13.3% of timberland, or 67 million acres, are industrial private forests (Smith et al. 2004). It is obvious these less fragmented timberlands that have become the primary targets for conversion of their management to TIMOs and REITs and other financial institutions.

The transformation actually happened in the early 1980s, when deindustrialization was sweeping through America's rush belt; what some referred to as the 'invisible forest crisis' began to develop as industrial timber companies started to sell off millions of acres of timberland to a host of newly established investment organizations. Two forms of institutional ownership, timberland investment management organizations (TIMOs) and real estate investment trusts (REITs) have been at the center of this transformation of timberland ownership.[28] TIMO's role is very

[27]As with the deepening extraction of timber in commodity frontiers in the past, access to and control of land is crucial.

[28]Regarding the TIMOs and REITs role of a new timberland ownership has been Interviewed Dr. Steven Wolf on September 16, 2014 and Dr. Peter Smallidge on August 27, 2014. Both lecturers emphasized the role of timber investment institution. They are lecturers in Department of Natural Resources, College of Graduate School of Agricultural and Life Sciences, Cornell University, Ithaca, NY.

significant as institutional investors in timberlands. This company got capital from institutional investors, such as pension funds, endowments, foundations, and universities. To date, TIMOs manage more than 22.4 million acres of timberland worth an estimated value of more 18 billion dollars, which is a significant increase compared to even three years earlier. Generally, the TIMOs' investors are looking for 'long-term' investment in timberlands usually ranging from 10 to 20 years. (Harris 2007). Timberland became a viable investment class due to strong historical risk adjusted returns and low correlation with other types of investment classes (Zinkhan et al. 1992).

The next institutional investors are REITs,[29] tax designations for corporations that invest in timberland real estate that reduce or eliminate corporate income taxes. These companies now owns more than 13.5 million acres of timberland worth an estimated value of more than $11 billion dollars (Harris 2007). Fewer timber companies have taken this route, but Mendell (2007) cites four: Longview Fiber, Plum Creek Timber, Potlatch and Rayonier. REITS are corporations with special tax designations that allow them to invest in timberland real estate while vastly reducing their corporate income taxes. As a result of this scheme, REITs are required to distribute 90% of their income back to investors. In recent years, timber companies have accelerated their divestiture of timberlands. International Paper has been one of the largest sellers in recent years, selling more than 9 million acres. Other industry leaders, such as Mead Westvaco, Boise Cascade, and Bowater have sold off the majority of their timberland holdings (Andrew and Paul 2010: 269). In fact, more than a dozen of the largest and best known timber companies no longer own major timberlands. Other timber industries have been reluctant to succumb to the pressure of institutional investors. Weyerhaeuser, for example, has been slower to sell off its timberlands, despite increasing pressure from stakeholders to follow industry trends.[30] According to industrial analysts, expect these trends to continue and the old model of vertically integrated timber companies holding large tracts of timberland will come to an end (Clutter et al. 2005). As a consequence of this scheme (Fig. 8.2), most timber industries and sawmills source raw material (timber supply) domestically from private (non-corporate/individual), TIMO and REITs with competitive price and internationally as well as from timber imports from overseas, such as Canada, South America, New Zealand and so forth.[31] In line with this statement, according to Steven Wolf, why do industrial timber companies sell their timberlands? There are 4 rationales to argue this issue: (1) the timber industries are facing the rise of

[29]'The modern idea of a REITs originated in 1960 with the establishment of the real estate investment tax provision, which reestablishes special tax considerations for distributed income'. It was not until the 1980s that REITs were used for timberlands investment.

[30]Weyerhaeuser decided to convert its timberland holdings to a REIT. See, http://crosscut.com/2010/01/04real-estate/19483/Behind-Weyerhaeuser-s-moveto-REIT-hood/.

[31]Interview with Dr. Peter Smallidge, on August 27, 2014.

finance in the global economy in terms of products efficiency, labor wages, and market. Financialization provides the overarching macroeconomic context to explain the processes of deregulation and institutional reconfiguration in the transformation of timberland ownership; (2) many timber industries in USA were involved in deficit and debt in 1980s. Therefore, they needed cash money to sell their timberlands to pay their debts; (3) most of timber high officers argue that they must return to making paper with efficient management as core business; and (4) Federal government issued regulations on a new land tax system (raising taxes on land prices) with a consequence that any companies should concentrate their core business including TIMOs and REITs for land investment management.[32]

In 2007, TIMOs and REITs owned and managed more than 35 million acres (64%, more than half of timber land) that previously were owned by industrial timber companies (67 million acres, in 1997). Meanwhile, the rest owned by industrial timber land just reached 32 million acres in 2007. From this point, the transformation means look at net gains and net losses in industrial timberland. It shows that from 2004 to the year 2007, the timber industry has sold around 35 million acres. In 2004, it was registered that International Paper Company was the largest seller, selling more than 9 million acres, followed by Mead Westvaco and Boise Cascade, and Weyerhaeuser. TIMOs have bought about 22 (40%) million of those acres, and the rest, REITs have acquired 13 million acres (24%) (Fig. 8.2). By the end of 2007, almost all of those timberland properties will be under TIMOs and REITs management. Actually, the timberland holdings of the TIMOs and REITs are geographically dispersed across the USA, but the most significant concentrations are found in the pine plantations of the Southeast, conifer plantations in the Pacific Northwest and the mixed softwood and hardwood stands of the Northeast. As a result, since institutional investors scan the globe for the investment outlets, it is not surprising to find evidence that most industrial companies are looking overseas, to South America in particular, for future timberland investments (Mendell et al. 2006). Interestingly, TIMOs are institutional investors, so managers acting as representative on behalf of their financially-oriented institutional clients, have expressed several interrelated reasons why they have sought out timberland as an investment. In interviews conducted by the USDA Service, the managers' point to: (1) the strong historical risk and adjusted returns; (2) the low correlation, or counter-cyclicity on timber lands with other traditional asset classes such as stocks and bonds and (3) timberland's ability to provide a 'hedge' against inflation (Clutter et al. 2005). From this point of view, therefore, timberlands offered one avenue for over-accumulated capital in a financialized economy and an ideal investment class for large institutional investors seeking to diversify large portfolios.

[32]Interview with Steven Wolf, on September 16, 2014.

A Case Study of New York State

It is necessary to implement sustainable forest management as a source of clean air and water supply for the New York State. This rationale background is worth to quote two critical statements. Firstly is come from Tom Tidwell as the Chief of the U.S. Forest Service, an Agency of Department of Agriculture. He said in a welcoming speech on International Years of Forests in February 2011 that "the USDA Forest Service is committed to investing in the future of our public lands through our working with partners at home and abroad. The value of our forests cannot be underestimated. In addition to providing clean air and water and wildlife habitat, the livelihoods of 1.6 billion people rely of forests worldwide. At home, one in five Americans depends upon clean water originating in national forests." Secondly derives from Robert K. Devis, New York State Forester, Director Division of Lands and Forests, New York State Department of Environmental Conservation. He further told: "… at a time of great environmental and economic challenge, planting trees, practicing sustainable forestry, and conserving out forest lands are among the most cost effective strategies we can take to keep New York green, while building our communities and contributing to the state's economic recovery."[33]

Looking at the two valuable statements above, they begin with an emphasis on clean air and water that is closely related with the mega project of Mayor New York, Michael Bloomberg who launched the city's investment for a "Green Infrastructure Plan" worth more than $6 billion in water quality management in 2002. The aim of this priority project is improving water quality, but also assisting the city achieve cleaner air and greener streets for 8.4 million inhabitants.[34] The green infrastructure component is planned to respond to public needs and government supports that will be necessary to make additional water quality investment. Therefore, to achieve this goal, Mayor of New York emphasizes "collaborative management" among government officers, such as city government officers with New York State's Department of Environmental Conservation and the United States Environmental Protection Agency (EPA) to make the plan a reality.

Actually, this collaborative approach for this project is not new. Based on the fact the city and its planners have a long track record of successfully meeting water quality standards with natural solutions that have substantial and quantifiable co-benefits. In addition, stakeholders such as New York City, New York State, Community, and Environmental groups agreed that preserving forested areas and natural buffers was a better way to keep drinking water clean. New York has a long history of maintaining important forest resources, from the Forest Preservation Act in 1885, to the 2009 revision of the State's Open Space Conservation Plan (OSP).

---

[33]See, "A Letter from New York's State Forester," in New York State *Forest Resource Assessment & Strategy* (Summary). New York State Department of Environmental Conservation (an access on September 8, 2014).

[34]See, "New York City Green Infrastructure Plan: A Sustainable Strategy for Clean Waterways" (accessed on September 12, 2014).

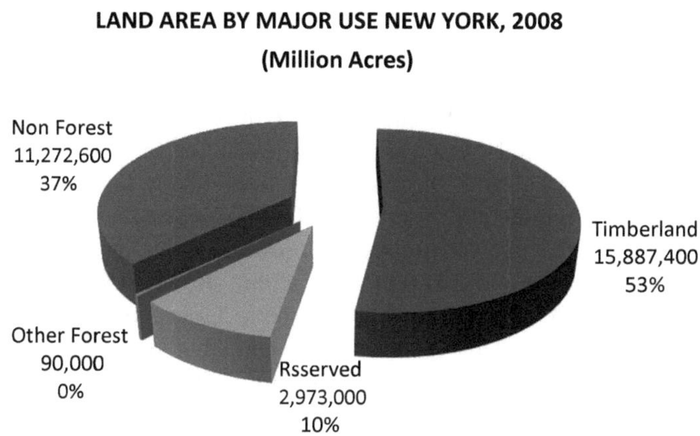

Fig. 8.3 Land area by major use New York, 2008. *Source* 2010 Statewide Forest Resource Assessment and Strategy. New York State Department of Environmental Conservation, 2011, p. 5

It is also concerned with the impacts of global climate change; it is increasingly clear that healthy forests are very significant to the future. Therefore, by improving sustainable forest management we can keep New York's forests as forests, working for our future.

New York seriously faces the challenges of changing climate change that could have far greater impacts than the 1930s drought. Forests, including urban and rural forests, provide front-line defenses against the impacts of global warming, such as climate change, drought, forest fires, flood, green gases, storms (*Catharine, El Nino*, etc.). It is known that forest areas act as sponges during storms; they absorb rainfall and reduce flooding. Trees work as filters to clean the air we breathe; they catch and remove airborne particulate matter that cause respiratory irritation and illness. Trees use carbon dioxide (a greenhouse gas) and give off oxygen, an element essential for animal life.[35]

New York land area reaches 30.2 million acres. New York State is blessed with forests that cover nearly 63% of the state, about 18.9 million acres (equivalent 9.4 million ha) (Fig. 8.3). The entire population of New York State is 19 million. The ownership composition of forest land namely is privately owned: 14.4 million acres (76% of forest land area) are owned by more than 687,000 private landowners. The composition of land includes the forest industry and non-industrial owners such as individuals, hunting clubs, and forest industry corporations. Meanwhile publicly owned forest land reaches 4.5 million acres. On the other hand, New York's urban/community land area encompasses 3.42 million acres and tree coverage 1.32 million acres (40.4%). This condition shows that New York State is categorized as one of

[35]See, "Introduction." 2011. in New York State Forest Resource Assessment and Strategy (an accessed on September 8, 2014).

the most heavily forested states in the country with more than 60% of her land covered with forests, according to Robert K. Davies, New York State Forester.[36]

Forest and open spaces, such as conservation, preservation and stewardship in New York has a long history. They trace back to the end of 1800 s: New York State has functioned forest land by a variety of means, for instance, forest preserve, public recreation, parks, and so forth. The total forest land managed by the Department of Environmental Conservation (DEC) reaches 4.5 million acres; nearly 3 million acres, or 63%, are categorized as forest Preserves. It contains 2.6 million acres in the Adirondack and 291,000 acres within the Catskills. In this case, the Adirondack and Catskill parks have some of the largest block of unfragmented forest land in the state. This area is very significant as a catchment area for hydrology and watershed in upstream. Therefore, DEC officers have to protect some wildlife species, such as Lynx and fish, and some birds and other biodiversity. The benefit of preserving forests in the Adirondack and Catskill forests are conservatively estimated to provide clean water and clean air for the people and also to reduce 3 million tons of carbon dioxide annually. In addition, there are 786,000 acres of "state forests" across New York and managed by Department of Environmental Conservation (DEC) that are utilized for open space, watershed protection, timber production and recreation. For timber products, 771,000 acres are categorized "green certified", including sustainable forest management standard established by the Forest Stewardship Council (FSC) and the Sustainable Forestry Initiative (SFI).[37]

Working Forests

Working forests are defined as forests that are capable of producing crops of timber or wood products. The objective of working forest is to "maintain forests" as forests and "preserve open space" to promote sustainable forestry.[38] New York State has 889,000 acres of new working forest mostly located in the Adirondack Park. There are many groups, such as public agencies, private conservation groups, financial investors and local communities that have cooperated in terms of integrated land protection with timber harvesting and economic development through "working forest" land transactions. In this case, many working forest transactions occurred through negotiation among stakeholders, such as industrial timber company, Finch

[36]A Letter from New York's State Forester in New York State Forest Resource Assessment and Strategy (Accessed on September 8, 2014).

[37]"2010 Statewide Forest Resource Assessment & Strategy," New York State Department of Environmental Conservation (DEC) (accessed on September, 8, 2014), p. 15.

[38]Ibid, p. 17.

Paper mill and other companies to sell thousands of acres or hectares to environmental groups and the state that are worth millions of dollars in investment for instance:

(1) In 1998, the Conservation Fund bought more than 121,000 ha of land in New York, Vermont, and New Hampshire from Champion International Timber Company worth $76.2 million. The aim of this purchase was to establish public protected areas and private timber management lands for conservation and recreation;
(2) In 2006, famous NGOs which were known the Nature Conservation (TNC) and its colleague the Conservation Fund bought 114,000 ha of lands in 11 states from International Paper Timber Company worth $375 million. The aim of this land transaction was improving ecological restoration, recreation and timber harvest (Neugarten et al. 2012: 1206–1207).

Observing these working forest land transactions above, to address ecological, economic, and social demands on forests reflects on a commitment to rural multifunctionality. In addition, some of these easements offer public forest-related recreational facilities. On the other hand, if the buyer comes from investment institution such as TIMOs and REITs, the purchase could be utilized for land investment and timber products which range 15–20 years.

Conflict in Forest Politics

The question rises why surrounding industrial private land of Adirondack is so significant? There are two rationales beyond this strategic land. First, based on expert analysis, 2.6 million acres in the Adirondack and 291,000 acres in the Catskills are categorized as forest preserve. So New York state and the federal government, as well, would like to maintain as catchment area for upstream watershed to provide clean water and air and also to protect biodiversity for recreation for the people. Second, environmental groups and environmental funds would like to buy thousands of private timber and individual land to protect ecological restoration, recreation and timber harvest.

From this rationale above, conflict around forest land use is rooted in two tensions: First is how land is used; second is whose interests are represented (who has access to decision making processes)? The first tension of land conflict happened between conservation, extraction of trees, and real estate development groups. The interest of environmental groups is to expand land with the aim to provide ‘use values’ for environmental protection and ecosystem services regarding clean water, air, and recreation. In contrast, extraction groups and investment institutions, such as TIMOs and REITs emphasize forest land as “exchange values” (marketable commodities and for investment). Therefore, this conflict obviously has divided environmental and community interests by portraying as advocates for

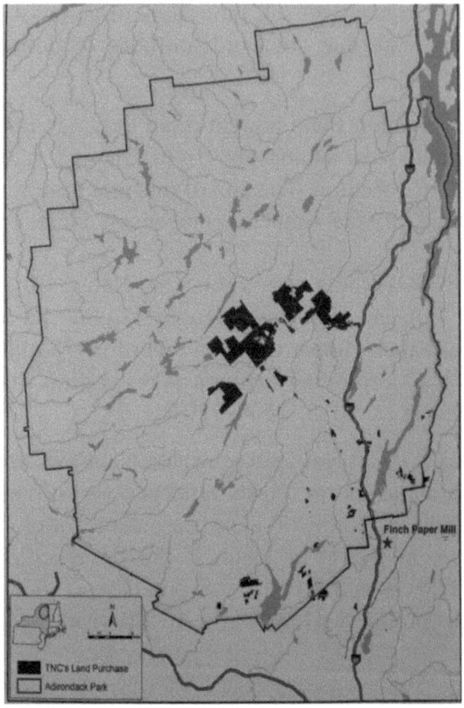

Fig. 8.4 The former Finch Pruyn Lands (with black boxes), bought by The Nature Conservancy (TNC) in 2007. Map Craig Chessmen, The Nature Conservancy; Neugarten et al. (2012, p. 1209)

ecological protection rather than complementary or synergistic. The second tension applies to conflicts between actors with different levels of influence over land use decisions. The dominant actors highlight timber industry companies and investment institutions versus local community, large conservation NGOs, financial resources and the political influence of their membership (Neugarten et al. 2012: 1207–1208). Meanwhile, the role of the state carries dual functions of responsibilities for fostering economic development by enabling access to natural resources, while another mission is safeguarding those resources from exploitation and supporting environmental groups for protection ecological restoration.

According to findings study carried out by a team (Neugarten et al. 2012) focused on the Finch Pruyn working forest (FPWF) in the Adirondack region, the former Finch Pruyn lands are located and around the 2.4 million ha Adirondack Park (Fig. 8.4). This park is a combination of state-owned forest preserve lands and private lands, including rural residential areas and villages. The question of the investigation study on surrounding Adirondack region, highlighted two critical questions: (1) how will the land be used in terms of distribution of property rights on the forest? And (2) whose interests are represented in decision making?

When the timber industry declined in the Adirondack region, that eventually lead to forest land conflict among stakeholders. This conflict happened among the state, timberland investment and private Timber Land Company regarding the acquisition of land surrounding of park land. As an illustration, from 2005 to 2007 the state bought conservation easements over 104,000 ha of land formerly owned by International Paper. On the other hand, Finch, Pruyn Co was the large industrial forest landowner within the park, holding about 65,000 ha. The buyer of Finch's timber land was owned by an investor group led by Atlas Holdings and Blue Wolf Capital Management in 2007. In its development the owners directly sold the forest land again for $110 million to the Nature Conservancy (TNC). TNC bought this forestland for the mission of selling ecologically valuable parcels to New York State to become part of the forest preserve land. The forest land contains many important ecological features, including large tracts of intact temperate forest and wetland, and species such as Bicknell's thrush. In addition,, the land also provides for private recreational use by hunt clubs, scenic areas, for instance, Hudson River Gorge, and other public recreation. According to the plan, TNC will allocate about 26,000 ha to New York State for becoming part of forest preserve. On the other hand, TNC agreed to sell 400 ha of land to municipalities for community development projects, such as housing and recreation facilities. (Neugarten et al. 2012: 1210).

Another important decision in terms of land acquisition is that The Nature Conservancy (TNC) sold 37,000 ha in 2009 to ATP, a Danish pension fund, worth $32.8 million. This land was eventually managed by RMK Timberland Group and TIMOs. In 2010, TNC sold a conservation easement to New York State for $30 million. In line with this selling, the easement requires that ATP, a Danish pension Fund maintains FSC certification for carrying out sustainable forest management. The products of timber from ATP are distributed to the Finch Paper mill for supplying raw material, because it is based on a fiber supply agreement. Unfortunately, the real condition of mill factory has been declining for years, since forestry jobs begin to disappear in the region, based on informant recognized (Neugarten et al. 2012: 1210–1211).

Strategy and Issues to Maintain the Future Forests

How New York State Maintains Forests?

There are two strategic ways for maintaining the future of New York's forests: Firstly, New York State launches a policy to move forward in "partnership" with other stakeholders whether who own public forest land from individual and industry private land. These partnerships related with other stakeholders such as public agencies, individuals/industry, environmental groups and organizations that aware to care for the future forest. Secondly, New York State Forestry Officer launches 10 strategic issues to be implemented by making 'partnerships' with other stakeholders. The 10 strategic issues as follows: (1) keeping forests as forest; (2) sustaining working landscapes; (3) sustainable forestry practices; (4) sustainable markets for sustainable timber products; (5) water quality and supply; (6) biodiversity; (7) forest health; (8) climate change; (9) urban tree canopy and green infrastructure; and (10) connection between people and outdoors.[39]

This discussion would like to focus on just five issues: as keeping forests as forests, sustainable forestry practices, water quality and supply, climate change, and urban tree canopy and green infrastructure, among the 10 strategic issues. The rationale of choice, it depends on the subjective of writer regarding the length of description on the explanation.

Strategic Issue 1: "Keeping Forests as Forests."

Background: How to invite awareness among individual/family forest land owners as the majority? This question is very significant, because the New York State forest owned by representing 76% of individual and family owner, approximately over 11 million acres from total 14 million acres. These private forests areas have public benefits, such as clean air and water, wildlife habitat, carbon dioxide reduction, scenic beauty, job absorption, and local economic growth. What is the rationale of individuals or families who struggle to maintain private forests? The strong argument based on economics, related costs of buying, holding and managing forest land, property values, cost of management, taxes, and future investment are all significant drivers. In additional, local, national and global markets factors are also affect the returns from direct investments in forest lands.

To implement strategy, action is needed to retain forests across New York State. First, to reduce threat of subdivision and land use change on private forestland by:

(a) State prefers to buy working forest for the purpose of conservation easements from willing sellers;

Second, to actively involve of local governments in conserving their forests through:

(a) To launch promotion, education & technical assistance with local land use planning for forest conservation and biodiversity protection.
(b) To launch policy on application of local open space protection measures for instance community preservation acts and community forest programs.

[39]See, 2011. 2010 Statewide Forest Resource Assessment & Strategy. New York State Department of Environmental Conservation (accessed on September 8, 2014), pp. 19–36.

Strategic Issue 2: "Sustainable Forestry Practices."

Background: External factors, such as Sustainable Forestry Initiative (SFI) and Forest Stewardship Council (FSC) are driving causes for State Forest managed by the Department of Environmental Conservation to practice model "Sustainable Forest Management" for other landowners. In this case, forestland owned by individual, family and industry/company must practice their forestland in sustainability to provide a broad array of environmental, social and economic benefits, to meet the needs of present and future generations.

To implement first strategy in action by cultivating a long-term Forest Stewardship Ethic through:

(a) To educate private forest landowners, the general public and elected official about the necessary of forest for conservation and sustainable forest management.

To implement second strategy in action by developing a systematic agenda with other stakeholders, such as State Extension Forester, local cooperative extension associations, The Nature Conservancy (TNC), State Environmental Facilities Corporation for maximizing efficiency through:

(a) To disseminate information to small family forest landowners about the need for sustainable forestry certification programs;
(b) To continue assisting landowners with developing Forest Stewardship management plans;
(c) To explore emerging areas of sustainable forestry and natural resource science;
(d) To support "peer-to-peer" private forest landowner networks.

Strategic Issue 3: "Water Quality and Supply."

Background: One of the function of forests is to provide catchment areas for hydrology to supply water for many aims, including clean drinking water, and to supply dams for water irrigation to paddy rice fields. In this case, the New York City Department of Environmental Protection estimated the cost of installing water filtration equipment costs nearly $7 billion, with over $300 million for operating annually costs. Therefore, to reduce the cost of the project, New York City eventually decided to sustain forests and improve the quality of land management in its "source watershed" surrounding the Catskills and lower Hudson Valley region. As a result, New York City finds water quality with more efficient cost for watering her inhabitants.

To implement this strategy, an action to protect high quality watershed and shorelines, through:

(a) To launch protection of high quality areas based on watersheds, amount of forest land, population served and threats;

(b) To increase forest lands along riparian corridors and shoreline;
(c) To promote conservation easements and acquisition, including community forests o protect watersheds, water quality and control flows;
(d) To review studies and reports of critical ecosystem management projects, especially for public drinking supply watersheds.

Strategic Issue Four: "Climate Change."

Background: Climate change is currently become the critical issue among countries, including U.S. The connection between forests and climate change was discussed at the Ministerial Conference on Atmospheric Pollution and Climate Change held in Noordwijk, the Netherland, in 1989.[40] The role of forest to tackle greenhouse effect of global warming as the effect of climate change is very significant. Forests also acts as climate buffers, moderating temperature extremes and creating local microclimates. Trees are about 50% carbon and represent the most dynamic component of the forest ecosystem carbon pool. Therefore, reforestation and rehabilitation program in the state forest including national parks are very necessary to expand tree plantation.

To implement strategy in action is to recognize the role of forests to mitigate and adapt to climate change, through:

(a) To promote the use of sustainably produced bio-energy to replace fossil fuel use. Because fossil fuel will rapidly decreased;
(b) To promote economic return to landowners from carbon reduction;
(c) To increase the practice and recognition of carbon reduction through sustainable forest management;
(d) To understand trends in climate change and its effects on wildlife occurrence and potential.

Strategic Issue 5: "Urban Tree Canopy and Green Infrastructure"

Background:The threat of climate change in line with the increasing of carbon dioxide, flood, storms, and so forth—eventually threatens human being safety, health and agricultural products. To prevent this critical condition, city planners in big cities, such as New York City should launch policy "green infrastructure." The green infrastructure aims to promote natural resources, healthy functioning systems like forests, trees planting, wetlands, riparian areas, and river corridors in order to prevent water pollution, flood and soil erosion.

Tree canopy cover averages 40.4% in urban and community areas. it is also could reduce urban heat and substantially reduce energy demands and related

[40]Haug and Gupta (2013).

greenhouse gas emissions. Therefore, it is important to disseminate information regarding tree plantation in urban areas. In addition, city planners should establish "green infrastructure" and to realize this program to be in action.

To implement strategy in action is to educate communities on the significance of urban forestry and green infrastructure, through:

(a) To expand assistance programs to increase tree canopy in local communities, by stressing health and societal benefits of trees;
(b) To educate planning, zoning boards and encourage green space and tree planting;
(c) To continue making 'partnerships' with the International Society of Arboriculture (ISA), State Nursery Landscape Association, and the New York State (NYS) Urban and Community Forestry Council.

Concluding Remarks

The implementation of sustainable forest management is very significant due to external and internal factors. Regarding the external factors, is the necessity of certification of timber products. Meanwhile, in terms of internal factors, the demand for timber supply is huge for the U.S. timber consumption. The potential of sustainable forest management (SFM) to be efficient and reasonable depends upon the representativeness of stakeholders, such as the significant role of federal, state government, timberland industry, environmental groups, timberland investment agencies (TIMOs and REITs), and communities.

If we take a closer look at net gains and net losses in industrial timberland, the findings become clearer that from 2004 till 2007, the timber industry owners sold millions of acres of their timberlands to timberland investments, such as TIMOs and REITs. As a result, TIMOs and REITs emerged as major land holdings and become large players in the timberland investment community. By the end of 2007, almost all of those properties will be under TIMO management. In contrast, industrial timberland owners will look for timberland areas overseas, which are considered cheaper, such as South America, New Zealand, Australia, Canada, and Southeast Asia to provide timber raw material. The rationale is the huge timber consumption of USA which reached 590 million cubic meters in 2005 that subsequently affects timber import from overseas. In addition, the paper companies of the USA's lack of competitiveness due to inefficiency, cost of labor, and rising of timber price. Interestingly, the condition since one decade ago suffered industrial timber companies eventually indicated the necessity on sustainable forest management practices and managerial performance should be reform.

Interestingly, the emerging of environmental groups versus industrial timberland in terms of competition on land acquisition and purchasing timberland area. Environmental groups, for instance, The Nature Conservancy (TNC) supported by environmental funds appears as a winner in purchasing timberland from industry owners with the objective for ecological restoration, conservation and recreation in surrounding Adirondack and the Catskills. In addition, New York State also provides funds for timberland purchasing for the aim of conservation and ecological restoration in the upstream. This policy is closely related with the match of launching "Green Infrastructure Plan" in 2002 to provide clean water, air and carbon sink for rural and urban inhabitants in downstream.

In order to maintain and protect healthy forests and environment, New York State launched a policy for ten strategic issues to be implemented, such as keeping forests as forests, sustainable forest practice, to prevent climate change, to provide water and air quality and so forth. In addition, New York State moves forward in "partnership" on forest protection with other stakeholders, such as New York Infrastructure construction, Department of Environmental Conservation Agency, public forest land Agency and environmental groups.

Chapter 9
Tanjung Puting National Park Central Kalimantan-Indonesia

Introduction

Forests are one of the renewable natural resources that provides elements for human beings to produce and consume. Yet, forest has regeneration potential and limited assimilation, if during its exploitation under the limited assimilation, forest resources can be utilized in a sustainable manner. In contrast, if it is exceeded, forest resources could be degraded and forest resources' function as production and consumption factors will be threatened (Soemarwoto 2001: 59). Therefore, as potential resources, forests are important, not just for the production of the timber, but also for many social and ecological functions, such as conservation of biodiversity, the supply of water, and the prevention of global warming (Inoue and Isozaki 2003: xi). On the other hand, forest resources often become capital that are utilized for national development in any country, as done for timber industries (*plywood, sawn wood, pulp and paper*, etc.) in the New Order (Soeharto regime) period and became the second largest foreign exchange earnings after of oil boom in 1980s–1990s.

National parks in Indonesia, which in (2007) accounted more than fifty units (at least 23.5 million hectares) are one of the last resorts for *forest conservation* in order to prevent forest degradation and to maintain sustainability of natural forests, with their specific ecosystems and biodiversity of flora and fauna. The management authority is under the Ministry of Forestry, Directorate General of Forest Protection and Nature Conservation. But, in realization in the field, the government made an effort to make 'partnership' management with other stakeholders related to interest groups, such as local governments, academics, NGOs and local people. At the Earth Summit in June, 1992 in Rio de Janeiro 'forest conservation' was one of the key issues in this Summit meeting. Although countries adopted Agenda 21, which called for actions to prevent 'deforestation', and the Forest Principles, the Earth Summit failed to conclude with the creation of a Forest Convention.

H. Hidayat, *Sustainable Plantation Forestry*,
https://doi.org/10.1007/978-981-10-7653-4_9

After the Earth Summit, a number of international initiatives emerged, such as the Intergovernmental Panel on Forest (**IPF**), the World Commission of Forests and Sustainable development (**WCFSD**), and others, in order to find ways to halt worldwide deforestation and degradation of all types of forestlands. At the Special Session of the General Assembly of the United Nations to Review and Appraise the Implementation of Agenda 21, in June 1997, it was agreed that work should be continued in order to reach an international consensus on forest conservation.

Obviously, in the past, most debates regarding various aspects of forests tended to focus on the forest sector and the direct causes of deforestation and forest degradation and not on the cross-sectoral aspects of the underlying causes linked to them, such as the connection between forests and societies. After the UN Special Session, non-governmental organizations took the initiative on one of the most pressing agendas and started researching the underlying causes of deforestation and forest degradation.

In recent years, the world's forests have been affected by over exploitation, over harvesting, over grazing, pests and diseases, climate change, global warming, floods, soil erosion, droughts, storms, air pollution, forest fires, as well as economic crises in Asia and other regions—all leading to an overall decrease in world's forest cover. Forestry in Asia, particularly in Southeast Asia, which traditionally was a timber supplier to forest industries in Japan and others, has been strongly impacted. A number of initiatives have suggested forest policy reforms, reforestation programs, and sanctions against illegal logging actors, and the need for the sustainable management of forest has been widely recognized and encouraged. But because implementation of reforms at the local level has been insufficient, it is imperative that local people begin to effectively participate in forest planning and management as well as in protected-area management.

This paper discusses the role of Tanjung Puting National Park in Central Kalimantan from the political ecology perspective, which emphasizes the role of stakeholders/actors in collaborative management and its implication on various ecological issues. This paper also will analyze conflicts of interest between central and local government on the policies of natural forest resources, particularly on national park management.

Theoretical Review

This study uses "political ecology" as an analytical framework which emphasizes the stakeholders' movement (Fig. 9.1).[1] Now we must clarify, what political ecology means. Many scientists (Peterson 2000; Bryant 1992; Blaikie and Brookfield 1987; Abe Ken-ichi 2003) define it differently (Peterson 2000) notes

[1]See Bryant and Bailey (1997). Further the implementation for political Ecology concept, see Seki (2001).

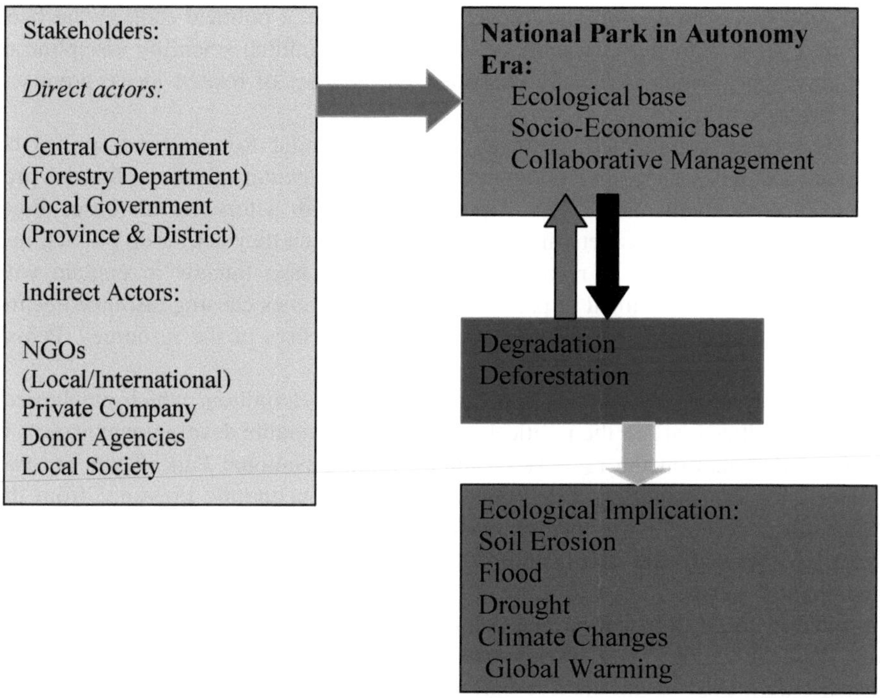

Fig. 9.1 Stakeholders on national park management and its implication. *Source* Improvement from Bryant and Bailey's Concept of Political Ecology (1997)

that, "political ecology as an approach that combines the concerns of ecology and political economy to represent an ever-changing dynamic tension between ecological and human change, and between diverse groups within society at scales from the local individual to transnational as a whole." Other scientists define it as, "political ecology", a framework to understand the complex interrelations between local people, national and global political economies, and ecosystems (Blaikie and Brookfield 1987). The concept has been adapted in a variety of ways, such as Third-World political ecology, where (Bryant 1992) notes that: "political ecology may be defined as the attempt to understand the political sources, conditions and ramifications of environmental change." Most current political ecology tends to overlook ecological dynamics and focus upon the structure of human systems (Rocheleau et al. 1996). Abe Ken-ichi (2003) defines political ecology, as "a collective name for all intellectual efforts to critically analyze the problems of natural resource appropriation and political economic origins of resource degradation, be they for the purpose of academic study or practical applications".[2] In other words, political ecology is concerned with the political dimensions of natural resource use

[2]See Abe et al. (2003).

and subtleties of those politics. Apparently, the scope of political ecology has been referred to as 'a method of analysis', rather than a unified scientific discipline or sub-discipline, which is usually characterized by a set of related ideas, premises, and theories.

Meanwhile, commented: political ecology is similar to a method applied by human ecologists analyzing policy-relevant environmental questions that is 'progressive contextualization'. This approach starts with actors, in this case, direct resource users, and considers the contexts within which they act, or do not act, in a particular way towards a resource. This approach also intends to explain why people use the environment in particular ways, sometimes causing resource decline or degradation detrimental to their own and others' uses of the resources (Peluso 1992).

From the above definitions, apparently, Bryant's definition, which emphasizes 'putting politics first' on the political ecology of sustainable development aspects is more operational on the reviewed Tanjung Putting National Park. There are two reasons for this condition: First, that 'political and economic pressure' from the Soeharto government was predominantly placed on forest management for three decades. Second, the implication of political and economic pressure upon 'ecological' perspectives was ignored by forestry bureaucrats, which subsequently resulted in forest degradation and deforestation.

'Political ecology' is a framework to approach the subjects mentioned. It is a generic term used for the field research connecting two types of study by bringing the point of view of politics into the study of environmental disruption. It includes a small-scale study centered on a local society (e.g., cultural anthropology, applied anthropology) and a large-scale study from national and worldwide standpoint (e.g., political economy).

Description of Tanjung Puting Area

Administratively, Tanjung Puting National Park was initially established and consisted of two conservation areas, namely Kotawaringin areas which encompassed 100,000 ha, based on Zelfbestuur van Kotawaringin number 24 June 13, 1936, and Sampit conservation areas of 205,000 ha, based on letter decision by Governor General of Dutch colonial number 39 August 18, 1937. Both conservation areas in Kotawaringin and Sampit eventually merged to become Tanjung Puting, which covered 305,000 ha.[3] This national park is very rich in flora and fauna, etc. It categorized fauna such as twelve various birds and 38 mammalians. Among the famous mammalian are Orangutans (Pongo Pygmaeus), and Owa-Owa

[3]See Book two on *Planning and Tanjung Puting Management: 1999–2024*, published by Department of Forestry and Plantation, Directorate General of Protection and Forest Conservation, Tanjung Puting National Park Institute, p. 5–10.

(Hylobates Agilis), Beruang Madu (Helarctos Malayanus), and Bekantan (Nasalis Larvatus). In the case of birds, Sindanglawe (Storm Stork, Ciconia Stormil), etc. From the vegetation perspective, Tanjung Puting areas is the center for biodiversity storage. For example, the Dipteocarpus forest type which covers 50–60% and wet forest (10%) from all areas. It contains valuable trees, such as Meranti (Shorea spp), Gaharu (Aquilaria Malaccensis), Kayu Ulin (Eusideroxylon Zwageri)), Ramin (Gonystylus), Rattan Damar Batu, etc.

In 1978, through the decision from Ministry of Agriculture number 698/Kpts/Um/II/1978 on November 13, 1978, 30,000 ha areas between River Serimbang and Segintung were excluded. Then, the total width of Tanjung Puting National Park was reduced to become 270,040 ha. In 1981, Tanjung Puting National Park was declared a 'world heritage' site for Biosphere Conservation by UNESCO. This means that from the management perspective, the responsibility of Tanjung Puting maintenance lies not just with the Indonesian government through Department of Forestry (Directorate General of Forest Protection and Natural Conservation) but also the international community actively involved in NGOs formation, such as Orangutan Foundation International (OFI) and World Education (WE) that is sponsored by UNESCO and Illegal Logging Response Centre (ILCR) which is sponsored by European Community (EC).[4]

Ecotourism Potential

Besides, as a site of flora and fauna, National Park Tanjung Puting is well-known as a recreation resort for domestic and foreign tourists. There are three driving factors Tanjung Puting is an appointed destination by visitors: (1) the status of Tanjung Puting National Park as a 'world heritage' site for the biggest Orangutan habitat in Kalimantan since the mid-1990s; (2) A documentary film about the Orangutans' condition that told of the bad condition of Orangutans because of factors, such as forest fires, deforestation, hunters that caught and sold them by traders to third parties (black market). This film was used as a campaign tool to change international views on Orangutan as being an endangered animal in Kalimantan; (3) the positive support by local and central government because it has provided good access to the location, infrastructure, security, safety and social and political stability in the district level.

Some eco-tourism areas which are most visited by visitors:

(1) Tanjung Harapan area which has been designated for utilization and research and is well equipped with a resort home for tourists, location for Orangutan Rehabilitation Center, tracking and Orangutan Meal Exhibition at certain times every morning and in the afternoon.

---

[4]See Hidayat (2006).

Table 9.1 Visitors to Tanjung Puting national park in 2004

No	Activities	Foreign/man	Domestic/man	Foreign/days	Domestic/days
1.	Visitors	580	444	4380	1479
2.	Research	3	17	148	366
3.	Film shooting	11	–	–	75
4.	Volunteer	24	–	–	432
5.	Official	–	–	–	
	Total	618	461	5528	2352

Source Tanjung Puting national park office's statistic (2004)

(2) Tanggui Camp is location used for specific utilization for tracking in night visiting and Orangutan rehabilitation activities and might to see to feed up at 8.00 o'clock. This location is the habitat for deer, forest pig, various bird such as Rangkong, Paruh Bangau, and others.
(3) Leaky Camp was designated as a specific utilization zone since 1970 for research and protection of Orangutans. At this camp, one could find out about wild Orangutan (Pongo Pygmaeus) rehabilitation and Owa-Owa ((Hylobates Agilis). In the upstream of Sekonyer river in Leaky camp tourists could learn about Buaya Muara (Crocodilus Porosusu) and Buaya Senyulong/Sapit (Tomistoma Schlegelli) which is well-known to be very wild.

There are visitors as well domestic and foreign in 2004 (Table 9.1). For example, visitors based on their nationality derived from: USA, 115, United Kingdom, 101; Australia, 51, Germany, 50, and Japan, 12. Mostly foreign tourists visited in September and October and domestic tourists in July and September. From the incomes perspective, in 2004, Rp. 226,922,000 and slowly reduced in 2005 to become Rp. 178,827,000. The factors of income reduction related to the national security in terms of the terrorism issue and boom in Indonesia in 2000s.

After the realization of local autonomy, since January 2001, the income from visitors' fees US$5/one day for foreign visitors and domestic Rp. 5000/day/person.[5] The distribution of income is divided between 80% for local government and Tanjung Puting National Park Office received 20%.

Stakeholder's Perception of National Park

Central Government

National Parks (such as Tanjung Puting) have the significant role and strategic function of conservation and protection for biodiversity, flora and fauna. To achieve

[5]The decision of ticket fee to Tanjung Puting National Park based on *PERDA* (Local Government Regulation) number 11, 2002 and Head of District (Bupati) Kotawaringin Barat, July 2002.

this role, central government and other stakeholders have tasked strategic functions as maintenance for catchments area, hydrology resources, watershed, to produce O_2 (carbon sink) and micro climate, and as educational and research facilities, eco-tourism and ecological services, etc.[6] There are three main purposes of national park management: (1) the protection of ecological process in order to guarantee as sustainable its function and role as ensuring systemic life; (2) to preserve various natural resources and the ecosystem in order to maintain genetic preservations; (3) and to generate sustainable benefits for improvement of social welfare for society who live in and around national park particularly, and society at large, in general.[7] Obviously, these three purposes of national park are appropriate with the law number 5, 1990 about "Natural Conservation and its Ecosystem" which highlighted that the central government has the responsibility to manage the national park. In line with these purposes, national park management in any districts in Indonesia, the National Park Officers (as representative of Directorate General of Forest Protection and Natural Conservation) must always pay attention to three dimensions: namely ecological, economic and social. Hopefully this dimension has positive implications for other stakeholders' especially local people. Therefore, to realize this dimension, officers should make strategic steps by establishing internal zones (core zones, forest zones and research zones) and external zones, so called "buffer zones" as the border line with society's land.

In fact, to prevent national park management from criminals among society who enter national park and conduct *illegal logging* and *mining* activities, the active role of local government (province and district) in the autonomy era in establishing "Buffer zones" (zona penyanggah) is very necessary. The function of this zone that is used by Agro-forestry programs can be facilitated by local government and National Park Officer. The role of Agro-forestry, which could plant leading local species trees and plantation (such as Potatoes, Kayu Manis, Albazia, Durian, Rambutan, Kemiri, Jengkol, etc.) could have positive implications for income generation of the local people. But, the reality until now is that none of the 'buffer zones' established by local government cooperated with national park officers.

Local Government

Although local government has no real 'authority' to manage national parks, based on the law (number 5/1990), co-management with central government is very necessary. As told by the head of forest agency in local government (district) the function of Tanjung Puting national park is very strategic and significant for hydrology resources, watershed, catchments area for the purposes of agricultural

---

[6]See Wiratno et al. (2004).

[7]Interview with Ady Susmianto, Director of Conservation area, Directorate General of Forest Protection and Natural Conservation (Department of Forestry), May 27, 2005, in Jakarta.

water supply. Therefore, in order to maintain this national park, local government has a moral obligation to cooperate with National Park Office. For example, for increasing local people's income generating, local government (districts) had established four (4) resort homes for tourist in Sekonyer village. The management of resort homes was organized by local people. The charge for an average room is Rp. 120,000/per day, plus breakfast for foreign and Rp. 80,000 per day for domestic tourists. While this research was carried out in May 2006, 4 tourists came from Belgium and spent for two days. The distribution of income is divided, namely 80% for local people and 20% for local government.

On the other hand, in crushing illegal logging activities, such as Balak Telabang Operation One, which was held on December 5–24, 2000 and continuously launched with Balak Telabang Operation Two from January 22 until February 10, 2001. The synergistic operation between local government, National Park Officer sand local people captured 50 men in jail and more 1500 M3 mix forest trees and 1176 M3 Ramin trees. Six driving factors caused deforestation in Tanjung Puting areas: (1) the economic crisis that occurred from 1998 to 2001; (2) the change of political order from centralization of power to decentralization (local autonomy) since January 2001; (3) the weak coordination between law officers and the function of court in central and local as well; (4) the KKN practices (*corruption, collusion and nepotism*) that happened between government officers and private business); (5) the weakness of the forest security system and the inspection of forest products; (6) and the price of illegal logging is cheaper than the formal wood from HPH (Logging Forest Concession) holders.[8]

Collaborative Management

There have been paradigm shifts from government based management to become *collaborative management,* which involve other stakeholders.[9] It means there is an indication to realize effective management on protected areas, social justice and to democratize natural resources management. Moreover, because Indonesia is categorized as a member that has ratified the Biological Diversity Convention. Therefore, Indonesia must adhere to this convention as relate to biodiversity conservation. NGOs' perception on national parks is a very strategic function for conservation areas, and it should be maintained on the principles of sustainable forest management.

Tanjung Puting National Park has carried out collaborative management with other NGOs, such as OFI (*Orangutan Foundation Indonesia*), FNPF (*Friends of*

[8]Interview with Tanjung Puting National Park Officer was held on May 4, 2006. And Forester, staff of Forest Agency in District, West Kotawaringin, May 11, 2006.

[9]The practice of collaborative management among stakeholders on national park management program especially on forest rehabilitation and empowering socio-economic program of local people was familiarly carried out in 1990s.

National Park Foundation), WE (*World Education*), Yayorin (*Yayasan Orangutan Indonesia*) and ILRC (*Illegal Logging Response Center*) on establishing a journal about program and national park activities.

Friends of National Park Foundation (FNPF)

It entered into cooperation with Tanjung Puting Office in 1997. The fundraising came from national and international agencies, such as the Gibbon foundation and local private companies who have responsibility for forest and environmental conservation and empowering social economic conditions of local people. Actually, the FNPF in Kotawaringin has a branch office and the head office is located in Denpasar, Bali. In realizing the program, FNPF was supported by other volunteer staff and local workers. As previously mentioned, Tanjung Puting areas had 65% forest degradation because of illegal logging practices and forest fires in 1990s and the early 2000s. FNPF was seriously involved in conducting rehabilitation and land conservation programs in many areas provided by self-preparation of local trees seeding, such as Meranti (Shorea sp), Gaharu, Ulin (Eusideroxylon Zwageri), Keruing (Dipterocarpus sp), Jatimas, Jelutung, Tengkawang, Ramin (Gonystylus), etc. In 2004, it carried out rehabilitation or reforestation program about 24 ha width and slowly reduced to become 16 ha in 2005, and in 2006 increased to become 29 ha in Camp Pesalat and Beguruh. According to Basuki, alumni of Mulawarman University, Samarinda, Head of Rehabilitation program, he is very optimistic about the acceleration of reforestation program provided by local trees seed and actively supported by other volunteers from high school pupils and local people, etc. Most of the pupils actively engage in reforestation program campaign were previously taught conservation subjects as extra curriculum in High School (SMA 2) in West Kotawaringin. But, a critical problem exists regarding the limited funds from other parties (international and local agencies) to carry out reforestation program in large areas.

Meanwhile, FNPF cooperated with local people to carry out empowering socio-economic programs. This NGO bought 6 female cows and two goats for a breeding program in 2005–2006. The target of these breeding activities after several years will be fruitful. FNPF also established *Demplot* (agricultural and fishery practices demonstration) area. This training of Demplot invited local people to develop their capacity and knowledge in order be to increase fishery and agricultural products using technical and mechanical tools, and excellent seeds. As told by FNPF field officer, agricultural was a potential sector from the viewpoint of width area, quality of land, irrigation from Sekonyer River and manpower. On the other hand, on the improvement of skills, FNPF officers gave facilities to 2 persons from Sekonyer village to attend training on carving statues and *Batik* printing for about one month (2003) in Bali. Hoping that after they returned home, they would be actively involved in teaching statue carving and Batik printing technical knowledge to local people. Based on the village report (2006), they eventually obtained about

20 persons who practiced as statue and Batik trainees. Most of their products, such as Orangutan statues, garment products (kaos/sport clothes) and kitchen tools-were made by local people as souvenirs and sold to tourists. The income generated from these activities, such as souvenirs could become best sellers for domestic and foreign tourists as well and improve their social welfare.

World Education (WE)

The *WE* is categorized as an international NGO, because it has programs on conservation and empowering local socio-economic in many countries. The head office of WE is located in Boston, USA. The *WE* has a *MOU* (Memorandum of Understanding) with Doctorate General of Forest Protection and Natural Conservation (PHKA) and is represented by Balai Taman Nasional (National Park Office).

There are two principles aims for developing WE on the promotion of *ICDP* (Integrated Conservation and Development Project) namely: (1) to improve the process of planning and decision making from stakeholders' movements in terms of forest conservation, reforestation programs in Tanjung Puting National Park; and (2) to launch empowering socioeconomic programs through improving agricultural and poultry products.

In general, the achievements target in period October 2003 and February 2006 were:

(1) The establishment of WE operational office in Pangkalan Bun for Tanjung Puting Integrated Conservation and Development Project (TPICDP);
(2) Intensively cooperate with Tanjung Puting National Park Office and other stakeholders on synergistic constructing program;
(3) The food endurance and empowering economic program through implementing of paddy rice, poultry, fishery, agro-forestry products;
(4) And improvement of Group organization.

We could realize about fifty up sixty percent the above program. For example, the endurance of food in twelve villages to give service about 1000 poor farmers, namely 5 villages located in East Kotawaringin (Seruyan) and 7 villages located in Pangkalan Bun (Fig. 9.2). As told by the WE symbol: "because we learn together and produce excellent products." WE develop leading program such as fishery (fish pool, breeding), agricultural products (paddy rice, fruits, vegetables, etc.), poultry (hen, a leading chicken/ayam buras) in some villages in Kumai sub-district.[10] For example, in terms of agricultural products, the intensification of mechanisms through using various leading seeds (punggur, mendawa, martapura, etc.), technical instructions, participation of local people eventually produce good results. In 2005, the production of rice increased to 4.5 ton/per ha, which previously produced average 2 ton/per ha in 2004. This condition has positive impacts to improve social benefits for local people, particularly to provide their food sustainability stock in the future. The response from local people is also positive, as told by informant

[10]See Quarterly report of WE program Period October 2003 to February 2006, Pangkalan Bun.

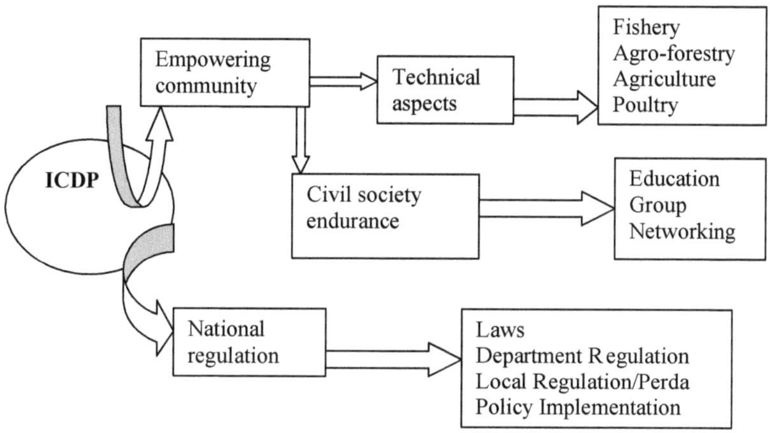

Fig. 9.2 Grand operational program of world education. *Source* Booklet of WE (World Education), (2005)

(Yd) that cooperation between NGOs (such as WE, FNPF, etc.) and National Park Officers and local people to promote socio-economic and ecological environments could impact on increasing benefits for income generation.

Conflict of Interest Central and Local Government

The spirit of reformation (1998–2005) in autonomy era placed a strong demand on positive values such as *democratization, accountability and transparency*. As we know that 'democratization' is haracterized by 'conflicts of interest among stakeholders.' Therefore, freedoms of expression eventually cause freedom to criticize other parties who have different ideas.[11] In context, the relationship between central and local governments has caused 'conflicts of interest' particularly on forest resources and natural resources in general. Centralization, which had been carried out by the central government in the Soeharto regime has been shifted into a new paradigm of 'decentralization'. It is because local autonomy is one of the formations of the real practice of democracy to guarantee individual, group, community rights and freedom in society. In line with central and local government relations, democracy demands freedom for local government to manage their affairs in many aspects. In this case, autonomy era hoped by local government to develop self-reliance and independence and to promote many things to be more progress and subsequently drive local people to be more welfare and prosperity. Related with "national park management" happened conflict of interest.

[11]See Rauf (2002).

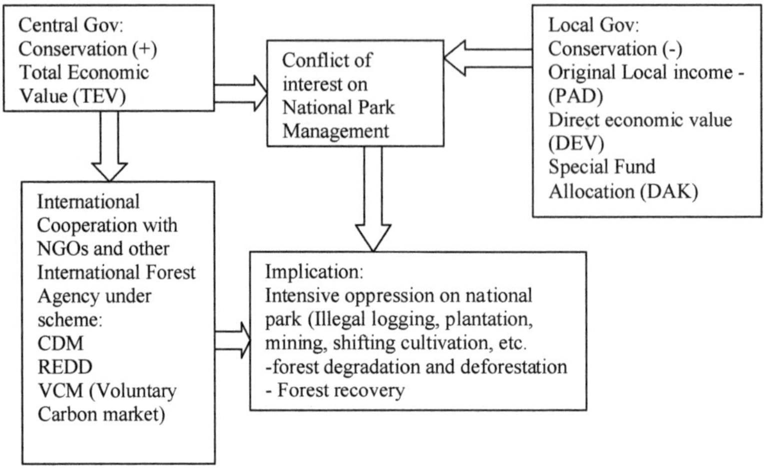

Fig. 9.3 Conflict of interest on national park and other cooperations. *Source* Data improvement (2006)

Local government sees the national park as having potential resources, minus conservation, economic orientation that could be exploited from timber and non-timber exploitation in order to expand their *PAD* (original income of local people). But so far, this planning just reaches into intellectual discourse among political elites and is still not yet published *Perda* (local government regulation) as regulation from the association decision of executive (head of district) and legislative institutes (local parliament). This condition is based on interpretation of law No. 32/2004 about "Local Autonomy". Even if local government's land occupied more than half, it is worth asking for "Special Compensation Fees" (DAK) from central government. This DAK can be used to establish infrastructure and to empower socio-economic conditions of the local community who live in and around the border of national park (Fig. 9.3). From this point, actually there is a plan to manage the national park by authorities. But, because of the real conditions suffered by local government, namely there is limited manpower and allocation of local budget beyond their capacities. As a result, local government (Kotawaringin Barat) just want to cooperate with the central government (through National Park Officer) and other stakeholders for this areas, based on consideration as follows: (1) strategic function of Tanjung Puting National Park as an asset for local, national and international stakeholders and its function for bio-diversity conservation, protection of fauna and flora especially Orangutan and other animals; (2) As an eco-tourism area for domestic and foreign tourists; and (3) to improve socio-economic programs for local people.

Central government's perspective highlighted that conservation in national park areas is very significant.[12] Because this area is categorized by its strategic function as catchments area, hydrology resources, bio-diversity storage, flora and fauna conservation, eco-tourism, etc. Therefore, any illegal activities, such as illegal logging and mining, land encroachment, land cultivation by local people and others is intolerable and these actors must be captured and jailed. For this purpose, the authority rights to manage still belong to 'central government', based on Law No. 5/1999 about "Natural resources conservation and Its Ecosystem").

The increasing of global warming which eventually affect into climate changes as indication of fail the 'Kyoto Protocol' agreement in 1997. There are many effects of global warming: (1) from 1974–2004 the effects of *green house* (*rumah kaca*) CO2 increased 70% (from 28.7 billion ton to 49 billion ton CO_2); (2) the sea level increases of about 18 cm until the end of 21 century; (3) the availability of clean water decreased in Asia; (4) in Africa, until 2020, about 75–250 million of people will suffer from the lack of water; (5) and the intensity of heat waves as the effects of climate change have been felt in South Europe and other parts of Europe (*Kompas,* November 19, 2007). Underlying factors of the failure of the Kyoto Protocol, are because of three critical rationales: First, the USA and Australia, which are considered significant contributors to gas emissions did not agree to make signature. Second, the concept of CDM (*Clean Development Mechanism*) which highlighted reforestation programs among developing countries, who especially own 'tropical forest', and are reluctant to perform this program. The concept of CDM is unrealistic because it emphasizes that critical forest land before 1960s must be carried out for reforestation program. Even the practice of logging industries (HPH) among developing countries, including Indonesia began at the end of 1960s. The impact of logging concession (HPH) which was ignored on the implementation of sustainable forest management (SFM) principles eventually affect on forest degradation and deforestation mostly occurred in 1980s, which do not properly perform a sustainable forest management. Third, the emerging of China, Brazil and India as industrialized countries, which greatly contribute to carbon dioxide must ratify the 'Kyoto Protocol' agreement on reduction of gas emission.

Then, what is the role of national parks in Indonesia which encompass about 23.5 million ha? Its role is very significant for production of carbon sink to protect global warming. Actually, the real challenge is how central government, which is represented by Department of Forestry encourages international foundation agencies, whether among advanced countries, *OPEC* (*Organization of Petroleum Exporting Countries*), international NGOs (Greenpeace, World Education, WWF, etc.), United Nations institute on climate changes, etc., to cooperate for maintaining national park management. In contrast, advanced countries (include G-8) which master science, technology and wealth must cooperate with developing countries,

---

[12]Interview with Bappeda staff (Local Planning Agency) in Kotawaringin Barat, on May 4, 2007. See, Hidayat, Herman (ed.) (2006). *Op cit.* p. 77.

especially those who own tropical forests, for realizing reforestation programs and empowering of socio-economic local people.

Vice President, M. Jusuf Kalla launched the idea of *oil for education* and *oil for forest* in the Summit OPEC meeting on November 17–18, 2007 in Riyadh. Oil for education is considered to become a significant, issue, because oil consumer countries (include Indonesia) really suffered heavy burdens as oil's price went up on their budgets. Meanwhile, oil for forest is related to environmental issues, because oil is the greater part to produce (fossil elements) that eventually cause global warming. The Conference subsequently agreed to the idea and finally established 'fund rising' to take US$50 cent/per barrel from OPEC members. This fund intends to contribute to sustainable forest development and to overcome global climate change and educational programs in developing countries (*Kompas*, November 19, 2007). And even King Abdullah bin Abdul Aziz, at the closing of summit OPEC meeting, gave donations about US$300 million for carrying out an initial energy research, environment and climate change issue to developing countries (*Media Indonesia*, November 19, 2007).

Currently, the Department of Forestry launched the concept of REDD[13] (*Reduction Emission on Forest Degradation and Deforestation*) on November 6, 2007, when it a held national workshop on reduction of 'gas emission' in Jakarta. This concept highlights maintaining a sustainable forest that could continuously contribute its benefits to local people and other parties. But, global responsibility must be performed by advanced country's contributions for giving 'incentives' as fund raising schemes to this program. In fact, this concept was adopted in *UNFCCC* (on climate change) in Bali on December 3–14, 2007. But in case of Indonesia, REDD concept schema has examined three areas. First, it has done on 'Kawasan Hutan Harapan' areas 101,000 ha (*Hutan Harapan*) as forest production restoration project that located in the border of Jambi Province and South Sumatra in 2006. Actually, allocation of concession from Department Forestry for 100 years to restore the forest areas done by *Restorasi Ekosistem Indonesia* Company (*Kompas*, November 7, 2007). According to Sukianto Rusli, Director of *Wild Bird Conservation* that "forest hope (Hutan Harapan) is tropical forest in low land which a final rest areas about 500,000 ha and previously existed 3 million hectares. This forest land contains rich mammalian species in Sumatra." Second, Malino district in East Kalimantan's efforts to categorize its protected forest areas of about 325,0416 ha to voluntary carbon market (*Kompas,* November 9, 2007). The cooperation of three parties initiates by *Borneo Tropical Forest Rain Forest Foundation* (BTRF) with Malino district and *Global Eco Rescue* (*Ecological Service Company*) from Nassau, Bahamas Caribbean Archipelago. This cooperation for forest utilization under schema of voluntary carbon market (VCM) to protect forest areas and maximize its benefits to improve social welfare of people.

---

[13]See Wahyudi Wardojo, "about Forest and Climate Change Anticipation", in *Tempo*, Edisi Khusus 3 Tahun SBY-JK. 29 Oktober-4 November 2007. Wahyudi Wardojo, "about Forest and Climate Change Anticipation", pp. 137.

The VCM schema is one of the carbon trade schemes out of CDM (*Clean Development of Mechanism*). According to John Alexander Embiricos, CEO of Global Eco Rescue (GER), economic values are considered uncertain from this project cooperation for two years. But the GER will give 1 Euro/per ha of forest land which covers of its agreement. That amount of money will be utilized for the identification process of carbon sink production from this forest absorption. On the other hand, Marthin Billa, head of Malino district said that carbon trade schemes could not obtain benefits anymore. But at least, he is very optimistic that this project could elevate social welfare of his people and maintain of forest areas. Third, to prevent global climate change until 2025, Emmy Hafild, Executive Director of Greenpeace Southeast Asia cooperates with Local government Riau to protect *peat swamp forest* (Hutan Rawa Gambut). Riau owns 4 million ha of 22 million ha of peat swamp forest; Indonesia could produce 14.6 billion tons of carbon dioxide which is currently saved in 4 million ha. This cooperation between both parties (local government of Riau and Greenpeace), if Riau could manage 4 million peat swamp forest ha from forest fires and conversion forest land into plantation will be paid US$5 until 20 per ton/carbon dioxide.

Another example of ‘carbon emission reduction’ scheme trade between Japan and Indonesia Power and Fajar Futura Company supported by *BPPT* Agency (Technology Assessment and Application Agency) highlights five projects of micro hydro in Indonesia (Cilencak, West Java with 1megawatt/MW, Siteki and Blumbungan in Banjarnegara, Middle Java 1.2 MW and 1.6 MW, Ketenger in Purwokerto 0.5 MW and Rante Bala, South Sulawesi 2.4 MW) (*Kompas*, November 10, 2007). This cooperation to reduce carbon emission bought by Japan about 30,000 tons of carbon dioxide per year under the scheme of CDM (*Clean Development* of Mechanism) under the Protocol Kyoto mandatory.[14] According to Irhan Febijanto, Coordinator of CDM Team in BPPT, “electric power engine by using micro-hydro energy is well known ecological friendship and low investment compared with other renewable energy resources”. But this scheme is actively engaging local people on managing electric distribution and benefits of energy and encouraging local people to maintain ecological conservation by protecting rain water in catchment’s areas.

Its Implication on Ecological Damage

There are two ecological impacts that will eventually affect Tanjung Puting areas in the near future. First, illegal logging practice by local people and big wood traders who own companies for wood trading in local and export to overseas. Second, illegal mining which affect water contamination.

[14]The estimation of carbon emission reduction market currently cost US$5–20 per ton.

Illegal logging means exploitation of forest products (timber) from forest products, conservation and protected areas, through illegal log cutting and its wood process networks.[15] The intensive illegal logging occurred in 1998 and the early 2000s during the reformation era. When the economic crisis happened in Indonesia, it seriously resulted in difficulties of life among the local people, causing instability and, loss of security and law sanctions. These conditions were driving factors in illegal logging in Tanjung Puting National Park. The protected trees, such as Ramin (Gonystylus bancanus), Ulin (Eusideraxylon Zwageri), Meranti (Shorea sp), Keruing (Dipterocarpus sp), etc., were massively cut down by local people and newcomers. The big wood trader from Tanjung Lingga Group, which is led by Abdul Rasyid, was actively involved as an actor of illegal logging. The illegal logging products, after shipping from Seknyer River to Kumai port, were subsequently exported to Singapore, Malaysia, China, Hongkong, and Taiwan.[16] Obviously, illegal logging practice in the field, could be identified actively by six main actors: (1) wood traders (cukong), capital holders, and bureaucrats and military); (2) local people and newcomers; (3) factory holders (plywood, sawmill, molding, pulp and paper, etc.); (4) HPH holders or IPKH as thief and timber collectors; (5) government officers from the forest agency; and (6) and foreign businessmen.

The six actors above were eventually supported in their operations by various parties, such as Indonesian bureaucrats, soft sanctions and regulations, and collusion between bureaucrats and businessmen. The ecological impact from illegal logging eventually caused forest fires in 1998–1999, floods in Sekonyer River in the rainy season, drought in summer season, soil erosion, loss of biodiversity, extinction of flora and fauna, such as Orangutans and other animals.

On the other hand, it caused 'water pollution' in the Sekonyer River because of the Indo Turba factory company's the processing of CPO from oil palm. This factory initially processed CPO from an oil palm plantation, which is located in upper Sekonyer River and eventually got damage on water treatment. It caused 'water pollution that seriously affected animals in river, such as Crocodiles, turtles; various fishes in Sekonyer River died. The water pollution also affected local people who previously used water for their needs, such as agricultural, taking baths, kitchen water utilization, clothes washing, etc. The effect of mining activities which produce Pasir Zirkon (Zirko sand) and Pasir Puya' (Puya sand) as raw material for ceramics and asbes. According to the Bappeda staff in Kotawaringin, there was no legal 'permission' for mining exploration. Obviously, this mining exploration, which was carried out by migrant workers from Java, Banjar, Bugis, Madura and some local people seriously affected local government in terms of income loss, ecological damage: soil erosion, forest degradation, water pollution, etc. This condition resulted in ecological and tourist costs because of water pollution

[15]See Suarga (2005).

[16]*See Illegal Logging in Tanjung Putting National Park: And Update on the Final Cut Report,* by Telapak and Ela NGOs (ELA: Environmental Investigation Agency), 2000, p. 13–19.

experienced by local people from their primary and secondary subsistence, supply water for their households' affairs (taking bath, clothes washing, etc.), which finally obliged local people to buy water for their needs. The tourist sector's income loss happened in the resort places surrounding in Kumai and Sekonyer village. The domestic and foreign tourists that usually visited in July and September annually decreased to 865 persons for foreign and 393 for domestic in 2002, as compared with tourist visitors in 2001 that reached 2380 foreign visitors and 506 domestic visitors.

Concluding Remarks

The impact of human activities on utilizing fossils (oil and gas) species and land changes (forest land conservation into plantation areas) have been causing 'global warming'. This phenomenon currently manifests as 'climate change' which eventually has an impact of on reducing food products, water distribution constraints, floods, droughts, sea water tides, and plantation diseases, etc. According to a *Stem Review* report, deforestation in developing countries, such as Indonesia, Brazil, India, etc.' contributed gas emission (CO_2) of about 20% from global gas emission. Meanwhile, carbon sink, which is currently saved in forest ecosystems is considered to produce greater amounts compared with that saved in the atmosphere. From this perspective, in order to establish conducive spheres, international support is very significant to protect existed forest areas.

Actually, Indonesia owns 23.5 million hectares of forest conservation (national parks, Suaka Alam/natural forests for flora and fauna, etc.), from total Indonesian forest areas of 123.4 million hectares. This condition actually contributes to produce carbon sink (O_2) to protect 'global warming'. Therefore, implementation of gas emission reduction through the concept of *REDD* (reduction on forest degradation and deforestation) could eventually give positive 'incentive' to developing countries who own 'tropical forest'. This incentive from advanced countries by paying *REDD* (per ha/1 Euro) could be intensively utilized for forest conservation through huge reforestation programs and empowering socio-economic programs for local people. As told by Susilo Bambang Yudoyono, Indonesian President in the seminar on the 'Climate Change Convention' in New York, in September 2007, he launched positive ideas, which are supported by other developing countries 'on the necessary to cooperate between developing and advanced countries on managing environmental issues'. Therefore, this cooperation will be established on forest sustainable management and even to formulate developing countries' position who own tropical forest to realize affirmative action on forest conservation. Hopefully the 'financial mechanism scheme on *RADD*' eventually agreed to by participants in the COP-13 in the international seminar on UNFCCC (Convention Work Scheme United Nations on Climate Change) in Bali December 3–14, 2007.

Currently forest degradation and deforestation which is caused by illegal logging, forest land conservation to plantation and mining activities, etc., has occurred in many districts in Indonesia based on the report from Department Forestry officer that deforestation reached 1.5 million ha annually. Certainly, the real phenomenon of floods, drought, soil erosion, climate changes, global warming, etc., provide us with serious impacts of ecological disaster from forest degradation and deforestation. Therefore, to overcome this serious threat, the 'political ecology' concept that actively engages stakeholders' movements from central and local government and the private sector as direct actors and NGOs, whether local and international as well as academicians, and local people as indirect actors, are very significant to be involved as integrated solutions for national park management.

'Collaborative management' which is currently promoted in the reformation era as an alternative concept to be implemented on national park management is very fascinating. It occurred in conflicts of interest of national park management between central and local government. This concept highlights stakeholders' roles in promoting rehabilitation or reforestation programs and empowering socio-economics of local people, apparently looked at as 'positive solution' to recover forest degradation and deforestation in the near future. Therefore, the commitment and consistency to perform 'affirmative programs' among stakeholders, such as the central government encouraging international donation agencies' cooperation, and the National Park Office and NGOs, which invited local people and the serious attention to establish "buffer zone" for *agro-forestry program* by local government as means of 'incomes generating' for local people is very necessary to be realized in autonomy era.

Bibliography

A. Books

Abe, K. et al. (2003). *The political ecology of tropical forests in Southeast Asia: Historical perspective*. Japan: Kyoto University Press.

Akira, S. A. (1989). *Capital accumulation in Thailand 1855–1985*. Japan: The Centre for East Asian Cultural Studies (CEACS).

APP. (2007). *Growing a Sustainable Future: Environmental and Social Sustainability Report for Indonesia*. Annual Report, published by APP.

Arbhabhirama, A., Phantumvanit, D., Elkington, J., & Ingkasuwan, P. (1998). *Thailand natural resources profile* (p. 168). London: Oxford University Press.

Aruan, A. L. P. (2005). The future role of plantation forests and forest-based industry. January 29, 2009 from http://www.fao.org.

Berkes, F., & Folke, C. (Eds.). (1998). *Linking social and ecological system* (p. 4). UK: Cambridge University Press.

Best, C., & Wahburn, L. A. (2001). *America's private forests: Status and stewardship* (p. 7). Washington: Island Press.

Blaike, P., & Brookfield, H. (1987). *Land degradation and society*. London: Methuen Press.

Boado, E. L. (1988). Incentive policies and forest use in the Philippines. In R. Repetto & M. Gillis (Eds.), *Public policies and the misuse of forest resources* (pp. 165–204). New York: Cambridge University Press.

Broad, R., & Cavanagh, J. (1993). *Plundering paradise: The struggle for the environment in the Philippines*. Manila: Anvil Publishing, Inc.

Bryant, R., & Bailey, S. (1997). *Third world political ecology*. London: Routledge Press.

Butler, B. J. (2009). Forest Ownership. In B. Smith et al. (Eds.), *Forest Resources*. Washington D.C: U.S. Forest Service.

Carrere, R., & Lohmann, L. (1996). *Pulping the South: Industrial tree plantations* (p. 57). London: Zed Books Ltd.

Carrere, R., & Lohmann, L. (1999). *Pulping the South: Industrial tree plantations and the world paper economy* (pp. 229–230). London and New Jersey: Zed Books Ltd.

Carter, A. (1993). Towards a green political theory. In A. Dobson & P. Lucardie (Eds.), *The politics of nature: Explorations in green political theory* (pp. 43–44). London: Routledge Press.

Chuntanaparb, L., & Wood, H. I. (1986). *Management of degraded forest land in Thailand*. Bangkok: Northeast Thailand Upland Social Forestry Project. Kasetsart University.

H. Hidayat, *Sustainable Plantation Forestry*,
https://doi.org/10.1007/978-981-10-7653-4

Coronel, S. S., & Severino, H. G. (1996). *Patrimony: Six case studies on local politics and the environment in the Philippines*. Manila: The Philippine Center for Investigative Journalism.

Craig, T. (Trans.). (1994). *The autobiography of Shibusawa Eiichi: From peasant to entrepreneur* (pp. ix–x). Japan: University of Tokyo Press.

Dempster, P. (1969). *Japan Advances: A Geographical Study*. London: Metheun Press.

Dern. (1998). *Development and management of forest plantations: A guide book*. Department of Environment and Natural Resources Ecosystems Research and Development Bureau, College 4031: Laguna, Philippines.

De Jong, W., & Van Hung, T. (2006). *Forest rehabilitation in Vietnam: Histories, realities and future*. Jakarta: CIFOR (Center for International Forestry Research).

Dickens, P. (1992). *Society and nature: Towards a green social theory* (pp. 1–16). Philadelphia: Temple University Press.

Directory Indonesian Pulp and Paper. (1999).

Hoshino, Y. (1992). Japan's post-second world war environmental problems. In J. Ui (Ed.), *Industrial pollution in Japan* (p. 65). Tokyo: The United Nations University Press.

Evans, J., & Turnbull, J. (Eds.). (2004). *Plantation forestry in the tropics*. London: Oxford University Press.

Forest Area and Forest Land. (2008). Hanoi: Agricultural Publishing House.

Forestry Agency. (2012). Forestry Statistic of Japan.

Forsyth, T. (2003). *Critical political ecology: The politics of environmental science*. London: Routledge Press.

Fujisaki, S. (1995). Rethinking economic growth and its implication for resources and the environment. In R. Kojima (Ed.), *Development and the environment: The experiences of Japan and industrializing Asia* (p. 1). Tokyo: Institute of Developing Economies.

Garrity, D. P., Kummer, D. M., & Guiang, E. S. (1993). *The upland ecosystem in the Philippines: Alternatives for sustainable farming and forestry*. D.C, Washington: National Academy Press.

Gregersen, H. M., Contreras-Hermosilla, A., White, A., & Phillips, L. (2005). Forest governance in federal systems: An overview of experiences and implications for decentralization. In C. J. P. Colfer & D. Capistrano (Eds.), *The politics of decentralization: Forests, power and people* (pp. 15–27). USA: Earthscan Publisher.

GSO (General Statistic Organization). (2003). *Statistical year book*. Hanoi: Statistic Publishing House.

Guizol, P. H., & Aruan, A. L. P. (2006). Impact of incentives on the development of forest plantation resources in Indonesia. http://www.fao.org/pdf.

Harootunian, H. D. (1970). *Toward restoration: The growth of political consciousness in Tokugawa Japan* (pp 321–327). Los Angeles: University of California Press.

Haug, C., & Gupta, J. (2013). The emergence of REDD on the global policy agenda, In J. Gupta, N. van der Grijp, & O. Kuik (Eds.), *Climate change, forests and REDD*. UK: Routledge Press.

Hendee, J. C., Dawson, C. P., & Sharpe, W. F. (2012). *Introduction to forests and renewable resources* (p. 406). Illinois: Waveland Press.

Hidayat, H. (2004). *Dynamism of Forest Policy in Indonesia: Focusin on the Movement and Logic of Stakeholders under the Soeharto Government and reformation Era*, Ph.D Dissertation in Department of Forest Science, Graduate School of Agricultural and Life Sciences, The University of Tokyo.

Hidayat, H. (2005). *Forest Policy in Indonesia: Focusing Stakeholders Movement in the Soeharto and Reformation Era* (Ph.D. Dissertation), in Department of Forest Science, Graduate School of Agricultural and Life Sciences, The University of Tokyo.

Hidayat, H. (Ed.). (2006). *Conflict potential between central and local government on national park management in local autonomy (A Case Study of Tanjung Puting and Kutai)* (p. 44). Jakarta: LIPI Press.

Hurst, P (1990). *Rainforest Politics: Ecological Destruction in Southeast Asia*. London: Zed Book Press.

Iwai, Y., & Yakutake, K. (2002). Japan's Wood Trade. In Y. Iwai (Ed.), *Forestry and the forest industry in Japan* (pp. 245–246). Canada: UBC Press.

Jansen, M. B. (1968). Tokugawa and modern Japan. In J. W. Hall & M. B. Jansen (Eds.), *Studies in the institutional history of early modern Japan* (p. 318). New Jersey: Princeton University Press.

JOFF (Japan Overseas Plantation Center for Pulpwood). (2005). A guide book for pulpwood products. In S. Kant & A. R. Berry (Eds.), *Economics, sustainability, and natural resources: Economics of sustainable forest management*. The Netherlands: Springer Press.

Kartodihardjo, H. (2006). *Ekonomi dan Institusi Pengelolaan Hutan: Telaah Lanjut Analisis Kebijakan Usaha Kehutanan*. Bogor: Penerbit IDEALS.

Kiat, I. (2008). *Sustainable growth: Committed to sustainability in all operations*. Riau: Annual Report of Indah Kiat Pulp and Paper Company.

Kummer, D. (1992). *Deforestation in the post-war Philippines*. Quezon City, Philippines: Ateneo de Manila University Press.

Kuwahara, Sueo et al. (2010). *The Monograph of Live in Kosogidani and Ishizuka Village in Yakushima Island*. Japan: Kagoshima University.

Kuwahara. (2013). Culture and Society in the Islands of Kagoshima. In K. Kawai (Ed.), *The Islands of Kagoshima: Culture, Society, Industry and Nature*. Kagoshima University Research Centre for the Pacific Islands.

Lye, T. P., Jong, W. D., & Abe, K. I. (2003). *The political ecology of tropical forests in Southeast Asia: Historical perspectives* (pp. 3–4). Japan: Kyoto University Press.

Magno, F. (2003). Forest Devolution and Social Capital: State Civil Society Relations in the Philippines. In Contreras (Ed.), *Creating space for Local Forest Management in the Philippines*, (pp. 17–35). Manila: De La Salle Institute of Governance.

Muhtaman, D. R., Siregar, C. A., & Hopmans, P. (2000). *Criteria and indicators for sustainable plantation forestry in Indonesia* (pp. 1–2). Bogor: CIFOR and ACIAR.

Nakayama, I. (1975). *Industrialization and Labor Management Relations in Japan*. Tokyo: The Japan Institute of Labor.

Narita, K. (1954). *A life of Ts'ai Lung and Japanese paper making* (pp. 2–3). Tokyo: Paper Museum Publisher.

Noda, H. (2002). The Japanese pulp and paper industry and its wood use. In Y. Iwai (Ed.), *Forestry and the forest industry in Japan* (pp. 225–226). Canada: UBC Press.

Nonberg, J., & Cunming, G. (Eds.). (2008). *Complexity theory for a sustainable future*. New York: Columbia University Press.

Ogushi, T. et al. (1998). *Water pollution in Fuji district and contaminated rotten mud of Tagonoura Harbor*. Paper in Institutes of Social Science Journal, Faculty of Engineering, Chuo University, p. 131.

Ogura, T. B. (1995). *A step towards the shift of agriculture, forestry and fisheries to ecological industries*. Tokyo: Food and Agriculture Policy Research Center.

Peluso, N. L. (1992). *Rich forests, poor people: Resource control and resistance in Java*. Berkeley, USA: University of California Press.

Phongpaichit, P., & Baker, C. (1996). *Thailand's boom*. Chiang Mai: Silkworms Books.

Pye, O. (2005). *Khor Jor Kor: Forest Politics in Thailand*. Bangkok: White Lotus Co Press.

Rauf, M. (2002). Local government and horizontal conflict. In S. Haris (Ed.), *Decentralization, democratization and accountability: Local government* (p. 145). Jakarta: AIPI and Partnership for Government Reform in Indonesia.

Reimann, K. D. (2010). *The rise of Japanese NGOs: Activism from above*. USA and Canada: Routledge.

Rocheleau, D. et al. (1996). *Feminist political ecology: Global issues and local experiences*. London: Routledge Press.

Sam, D. D. & Trung, L. Q. (2003). Forest policy trends in Vietnam. In M. Inoue & Isozaki, H. (Eds.), *People and forest-policy and local reality in Southeast Asia, the Russian far East, and Japan* (p. 159). The Netherlands: Kluwer.

Sato, Y. (Ed.). (1998). *Changing industrial structures and business strategies in Indonesia.* Tokyo: Institute of Developing Economies.
Sayer, J. (Eds.). (2005). *Forestry and Development in U.S and U.K.* Earth Scan Publisher.
Shiraishi, T. (2005). Introduction: States, markets and societies after the Asian crisis. In T. Shiraishi & P. N. Abinales (Eds.), *After the crisis: Hegemony, technocracy and governance in Southeast Asia* (pp. 1). Japan: Kyoto University Press.
Soemarwoto, O. (2001). *Ekologi, Lingkungan Hidup dan Pembangunan.* Jakarta: Djambatan Press.
Sonnenfield, D. A. (1996). *Greening the tiger? Social movements' influence on adoption of environmental technologies in the pulp and paper industries of Australia, Indonesia, and Thailand.* Santa Cruz: University of California Press.
Taylor, P. (Ed.). (2004). *Social inequality in Vietnam and the challenges to reform.* Singapore: Institute of Southeast Asian Studies (ISEAS).
Suarga, R. (2005). *Pemberantasan illegal logging: Optimisme di Tengah Praktek Premanisme global* (pp. 6–7). Tangerang: Wana Akasara Press.
The Japanese Pulp and Paper Industry in Charts and Figures. (2005). Published by Japan Pulp and Paper Co., Ltd.
The Ministry of Environment (MOE). (2010, 2011) on National Park of Japan. Tokyo.
Tingsabadh, C. (1989). *Employment effects of reforestation programs.* Bangkok: TDRI.
Vietnam Economy. (1998). *It was published by Central Institute of Economic Management.* Hanoi: Education Publishing House.
Vitug, M. (1993). *The Politics of Logging: Power from Forest.* Manila: Philippines Centre for Investigative Journalism Press.
Wang, L., & Wang, H. (2010). Comparison of timber consumption in U.S. and China. In *Global perspective on sustainable forest management journal* (pp. 290–291). Accessed September 3, 2014.
Wakabayashi, K. (2000). *Tokyo Wan no Kankyo Mondai Shi, (History of environmental problem of the Tokyo Bay)* (pp. 303–312). Tokyo: Yuhikaku Press.
Walker, B., & Salt, D. (2006). *Resilience thinking: Sustaining ecosystems and people in a changing world.* Washington: Island Press.
Weidner, H. (1989). Preface. In S. Tsuru & H. Weidner (Eds.), *Environmental policy in Japan* (p. 9), Berlin: Sigma Bohn.
Williams, J. (2007). *50 facts that should change the world.* UK: Icon Books Ltd.
Wiratno, I., & Ahmad, S. A. (2004). *Berkaca di Cermin Retak: Refleksi Konservasi dan Implikasi bagi Pengelolaan Taman Nasional.* Dephut dan Gibbon Foundation, hal. 200–202.
Yamasaki, K., & Yamazaki, T. (2000). *Sustainable development of tokyo bay: Collaborations of anti-reclamation movement and urban fishery* (pp. 1–4).
Yoshihara, K. (1999). *The nation and economic growth: Korea and Thailand.* Japan: Kyoto Press.
Zinkhan et al. (1992). *Timberland Investments.* Portland: Timber Press.

B. Paper and Journals

Achieving the ITTO Objective 2000 and sustainable forest management in Thailand. (2006). Report Submitted to ITT Council by Diagnostic Mission.
Andrew, G., & Paul, G. (2010). Financialization, Shareholder Value, and Transformation of Timber Ownership in the U.S. *Critical Sociology, 37*(3), 265–284. Sage Publication.
APP pulp mills & sinar mas group forest companies: Environmental compliance & wood supply audit: Audit Executive Summary. Reviewed by AMEC Simons Forest Industry Consulting. November 21, 2001.
Bai Bang Paper: The Vietnam Sweden friendship and cooperation project. (2000). http://www.vica.vnn.vn/uni/hop_tac/engl/baibang/evn_td.htm.

Bantayan, N. C., & Rosario, B. (2007). *Strategic Plan for the Forestry Cluster: 2005–2020 Pulp and Paper Industry Sub-Cluster* (Research Paper on Team Group). Department of Forest Products and Paper Science. College of Forestry and Natural Resources. Laguna: University of the Philippines Los Banos College.

Barney, K. (2005). *At the supply edge: Thailand's forest policies, plantation sector, and commodity export links with china, forest trends, center for international forestry research.* The York Center for Asia Research. http://www.forest-trend.org/documents/publications/Thailand.

Blower, N., et al. (1999). *Paper, prices and politics: An evaluation of the Swedish support to the Bai Bang Project in Vietnam.* Stockholm: Centre for International Economics. Sida Evaluation 99/3.

Bryant, R. (1992). Political ecology: An emerging research agenda in Third-World studies. *Political Geography, 11*(1), 12–36.

Chokkalingam, U. (2006). *One century of forest rehabilitation in the Philippines: Approaches, outcomes and lessons*. Bogor: CIFOR.

Chuntanaparb, L., & Wood, H. I. (1986). *Management of degraded forest land in Thailand.* Bangkok: Northeast Thailand Upland Social Forestry Project. Kasetsart University.

Clutter, M. (2008). Strategic Factors Driving Timberland Ownership Changes in the U.S South. *Ecological Indicators, 8*(2008), 109–114. Elsevier Publication.

Colchester, M. (1986). Banking on disaster: International support for transmigration. *Ecologist, 16*(2/3), 61–67.

Colchester, M., & Lohmann, L. (Eds.). (1995). *The struggle for land and the fate of the forests.* London: World Rainforest Movement, Penang, Malaysia (branch) Press.

Eyes of the Forest (EoF). (2009). *Riau forest rescue network for NGOs.*

FAO. (1998). *Global Forest products consumption, production, trade and prices: global forest products model projections to 2010.* Working Paper GFPOS/WP/01 Food and Agriculture Organization of the United Nations, Rome.

FAO. (1999). *State of the world's forests 1999.* Rome: Food and Agriculture Organization of the United Nations.

FAO. (2001). *Global Forest resources Assessment 2000.* Main Report. FAO Forestry Paper 140. Food and Agriculture Organization of the United Nations, Rome.

FAO. (2004). *Corporate document repository.* Advisory Committee on Paper and Wood Products, 38th Session, Proceedings. See: https://www.fao.org/docrep.

Fearnside, P. M. (1998). Plantation forestry in Brazil: Projection 2050. *Biomass and Bioenergy, 115*(6), Amsterdam: Elsevier Science Ltd.

FKKM Riau (Forum Komunikasi Kehutanan Masyarakat Riau). Number 3. Edition December 2006–March 2007.

Forestry Statistics of Thailand. 2550 or 2007.

Greenberg, J., & Park, T. (1994). Political ecology. *Journal of Political Ecology, 1,* 1–12.

Guiang, E. S., Borlagdan, S. B., & Pulhin, J. M. (2001). *Preliminary assessment of community-based forest management in the Philippines.* Project Report, Institute of Philippine Culture, Ateneo de Manila University and Department of Social Forestry and Forest Governance, College of Forestry and Natural Resources, University of the Philippines Los Banos. Quezon City.

Guizoi, P. H., & Aruan, A. L. P. (2008). Impact of incentives on the development of forest plantation resources in indonesia with emphasis on industrial timber plantations in the outer Islands. *Growth,* http://www.fao.org. December 4, 2008.

Gunnoe, A., & Gellert, P. K. (2010). Financialization, shareholder value, and the transformation of timberland ownership in the US. *Journal of Critical Sociology, 37*(3), 267.

Hamilton, H. (1989). Vietnamese pulp project helps protect ecosystem. *World Wood, 30*(3), June.

Harrison, S., Emtage, N. F., & Nasayao, E. E. (2005). Past and present forestry support programs in the Philippines, and Lessons for the future. *Small-scale Forest Economic Management and Policy, 3*(3), 303–317.

Heyd, H., & Neef, A. (2006). Public participation in water management in northern Thai Highlands. *Water Policy, 8,* 396–398.

Hie, P. S. (2004). The changing administration and Role of Forestry in the Economy of Vietnam. *Small-Scale Forest Economics, Management and Policy, 3*(1).

Jerve, A. M. (1999). *A leap of faith: A story of Swedish aid and paper production in Vietnam-the Bai Bang Project*, 1968–1996. Stockholm: Christian Michelsen Institute. Sida Evaluation 99/4.

Jiang, Z., & Zhang, S. Y. (2003). *China's plantation forests for sustainable wood supply and development.* http://www.fao.org.

Journal of Forest Management. (2008). *2*(3). January–June. Faculty of Forestry. Kasetsart University. Thailand: Bangkok.

Journal of Forestry. (1998). Sustainable Forest and Sustainable Communities, in Special Isuue, *96*(3).

Kono, Y., Sijapati, S., & Takeda, S. (1994). Dynamic of upland utilization and forest land management: A case study in yasothon province, Northeast Thailand. *Journal of Southeast Asian Studies, 32*(1), 4.

Korten, F. (1994). Questioning the call for environmental loans: Critical examination of forestry lending in the Philippines. *World Development, 22*(7), 971–981.

Kuaycharoen, P. (2004). Commercial tree plantations in Thailand: Flawed science, dubious politics and vested interests, *Watershed, 9*(3), March–June 2004.

Laemsak, N. (2008). The wood-based industries in Thailand. *Journal of Forest Management, 2*(3). Faculty of Forestry, Kasetsart University. January–June (2008).

Lang, C. (2000). Globalization of the pulp and paper industry: The Vietnam deforestation, reforestation and industrial plantations. http://www.wrm.org.uy.

Lang, C. (2002). *The pulp invasion: The international pulp and paper industry in the mekong region.* NOVIB (The Netherlands) and TERRA (Towards Ecological Recovery and Regional Alliance-Thailand).

Lang, C. (2003). Thailand: Eucalyptus, encroachment, deforestation and pollution linked to pulp and paper company, *World Rainforest Movement Bulletin, 70.* https://chrislang.org/2003/05/10/thailand-eucalyptus-encroachment-deforestation-and-pollution-linked-to-pulp-and-paper-company.

Lang, C. (2005). Thailand: The Fast-Growing Pulp and Paper Industry. *World Rainforest Movement.*http://www.wrm.org.uy/countries/Asia/thailand.html.

Lang, C. (2008a). *Plantations, poverty and power: Europe's role in the expansion of the Pulp Industry in the South.* World Rainforest Movement. Swedish Society for Nature Conservation.

Lang, C. (2008b). Vietnam: Paper shortages, price increases, new mills and more plantations. Published in WRM Bulletin 137, December 2008. http://chrislang.org/2008/12/22/Vietnam-papershortages.

Lechi, A. (1995). Vietnam. *Tappi Journal, 78*(6), 57–58.

Liu, D. S., Iverson, L. R., & Brown, S. (1993). Rates and patterns of deforestation in the Philippines: Application of geographic information system analysis. *Forest Ecology and Management, 57,* 1–16.

Lober, D., & Ersen, M. (1995). Retackling, Certification and the Home Improvement Industry. *Journal of Forestry, 93*(4), 39–41.

Lowe, P., & Rudig, W. (1986). Political ecology and the social sciences: The state of the art. *British Journal of Political Science, 16,* 513–550.

Mangaoang, E. O. (2002). A forester's perspective of the socio-economic information requirements for forestry in Leyte. In *Socio-economic Research Methods in Forestry: A Training Manual Cooperative Research Center for Tropical Rainforest Ecology and Management (Rainforest CRC) Cairns*, Australia.

Master Plan for Forestry Development. (1990). DERN: FMB Forest Management Bureau.

Miyakawa, H., et al. (2006). *Toward the brighter Future of CBFM (A field review on 23 CBFM sites).* Quezon City, Philippines: DERN-JICA Project Enhancement of Community-Based Forest Management Program.

Morgera, E. (2011). Sustainable Forest Management. *Environmental Policy and Law Journal, 41*(2), 74–76.

National Standard: For Sustainable Forest Management (Vietnam-FSC Standard). (2007). It was published by Research Institute for Sustainable Forest Management and Forest Certification (SFMI), Hanoi.

Neugarten et al. (2012). Forest Fights and Forest Rights. *Society and Natural Resources Journal,* 1205–1220. Routledge, November 25.

Nong Nghiep and Phat Trien Nong Thon (Journal of Agriculture and Rural Development). (2009). 50 The of Forestry in Vietnam. November 28.

Ohlsson, B., Sandewall, M., Sandewall, R. K., & Phon, N. H. (2005). Government plans and farmers intentions: A study on forest land use planning in Vietnam. *Ambio, 34*. http://www.ambio.kva.sc.

Orden, T. (1960). Brief history of Philippine reforestation. *Philippine Lumberman, 7*(1): 76–80. Paper and Wood Products-FAO, Vancouver, 31 May. http://www.fao.forestry/media.

Paper Industry in Vietnam. (2001). *News from Embassy* of the Socialist Republic of Vietnam in the USA, February 4, 2001.

Paperloop. (2000). *Country spotlight: Thailand economic growth fuels rise in demand.* Web-site: http://www.paperloop.com/newsinfo/regional/asia_australia/thailand_spotlight.shtml. Accessed January 12, 2001.

Peluso, N. L. (1992). The political ecology of extraction and extractive reserves in East Kalimantan-Indonesia. *Development and Change, 23*(4).

Pesonen, K. (1995, March). *Indo-China Pulp and Paper Industry-its future prospects, Paper Asia*. 12–18.

Peterson, G. (2000). Political ecology and ecological resilience: An integration of human and ecological dynamics. *Ecological Economics, 35*(3), 323–336. December.

Porter, G., & Ganapin, D. J. (1988). *Resources, populations and the Philippines' future: A case study.* World Resources Paper No. 4. Washington, DC: World Resources Institute.

Pulhin, J. M. (2006). Where goes thou CBFM? Reflections on the forum theme. A paper presented during the Ten Year review of CBFM in the Philippines: A forum for reflection and dialogue, 20–22 April, *International Institute for Rural Reconstruction, Philippines, Cavite..*

Pulhin, J. M., Inoue, M., & Enters, T. (2007). Three decade of community-based forest management in the Philippines: Emerging lessons for sustainable and equitable forest management. *International Forestry Review, 9*(4).

Pulp Industry in the South. (2004). World Rainforest Movement. Swedish Society for Nature Conservation.

Pulp and Paper Statistics. (2006). Japan Paper Association. Tokyo.

Sam, D. S. (2006). *Impact of Eucalyptus & Acacia mangium in Soil Fertility in Vietnam,* (special paper).

Sample, A. (2000). Spring: Forest Management Certification. *The Pichot Letter*, 8–13.

Sasrisang, A., Hoamuangkaew, W., Suksard, S., & Jarusombuti, S. (2008). Market structure of pulp industry in Thailand. *Journal of Forest Management*, 2(3), Faculty of Forestry, Kasetsart University.

Sayer, J. A. (1991). The Vinh Phu Pulp and Paper Mill. In N. Collin (Eds.), *The conservation atlas of tropical forests Asia and the Pacific*. London: IUCN/Macmilan.

Sedjo, R., & Lyon, K. (1996). *Timber Supply Model 96: A Global timber supply model with a pulpwood component.* Discussion Paper 96–15. Washington, DC: Resources for the Future.

Seki, Y. (2001). The political ecology of the Philippine restoration program: ODA, government, and local people. *Philippine Political Science Journal, 22*(45), 79–93.

Simpson, B. (2011). Your Forests are Certified. *Tree Farmer*. No. 5.

Sohngen, B., Mendelsohn, R., Sedjo, R., & Lyon, K. (1997). *An analysis of global timber markets.* Discussion Paper 97–37. Washington, DC: Resources for the Future.

Sonnenfeld, D. (1998). Logging and recycling: Problems in the industrial ecology of pulp manufacturing in South-East Asia. *Greener Management International*, *22*, 108–122.

Subhadhira, S., Apichatvullop, Y., Kunurat, P., & Hafner, J. (1987). Case studies of human-forest interaction in Northeast Thailand final report Bangkok. *Northeast Thailand Upland Social Forestry Project*, Thailand: Kasetsart University.

Sugihara, K. (2008). *Multiple paths of economic development in global history*. The Paper presented in the Symposium in commemoration of the Executive Committee Meeting of the International Economic History Association. November 8–9, 2008 in Inamori Foundation Memorial Hall. Japan: Kyoto University.

Small-Medium Business Opportunities on Seed Production, Collection and Trade for the Tree Plantation Programs in Eastern Mindanao. 2008. DERN Regions 10, 11, 13.

Thai Journal of Forestry. (2003). *22*, (29–35).

Thaiutsa, B., et al. (2003). *Complete report of site potentials for growing eucalyptus (Executive Summary)*. Thailand: Faculty of Forestry, Kasetsart University, Forest Research Center.

The Philippines Recommends for Reforestation, Tree Farming, and Plantation Development. (2008). PCARRD philippines recommends series no. 94. Los Banos, Laguna: Department of Science and Technology (DOST).

The Financial Times, October, 13 1981. and the Financial Times, June 20, 1997.

Thiep, X. T. (2005). Eucalyptus plantation in Vietnam: Their history and development process, Forest Inventory and Planning Institute, Ministry of Forestry. http://www.fao.org/docrep.

Tjiwi Kimia. (2008). *Sustainable Performance: Committed to Sustainability in all Operations*. Annual Report of Tjiwi Kimia Paper Company.

Tolentino, E. L. (2008). Restoration of Philippine native forest by smallholder tree farmers. In D. J. Snelder & R. D. Lasco (Eds.), *Smallholder tree growing for rural development and environmental services*. Berlin: Springer Science.

Tolentino, E. L. (2007). *The future of tree plantations in the Philippines: Phenotypic characterization of current seed sources*. Paper presented in University System Professorial Chair Lecture. Philippines: UP Los Banos.

Ubukata, F. (2009). *Getting villagers involved in the system: The politics, economics and ecology of production relations in the thai pulp industry*. (Monograph Paper) Kyoto Working Papers on Area Studies No. 40 (G-COE Series 38).

Utting, P. (2000). An overview of the potential and pitfall of participatory conservations. In P. Utting (Ed.), *Forest policy and politics in the Philippines: The dynamics of participatory conservation* (pp. 171–215). Quezon City and Manila: Ateneo de Manila University Press and United Nations Research Institute for Social Development.

Vietnam Paper Industry: Flying under the radar. https://www.glgroup.com/News/Vietnam-Paper-Industry.

Vietnam Paper Industry. (2009). Country Report. Vietnam Pulp and Paper Association.

Vietnam: What happening in the Pulp and Paper Sector? http://www.wrm.org.uy/bulletin/115/Vietnam.html.

Warta FKKM Riau (Hutan Bagi Kesejahteraan Masyarakat). Triwulan Pertama (2009).

Watershed. (2004). Permpongsacharoen (editor). Vol. 9, No. 3, March–June. *People's Forum on Ecology*. Published by *TERRA* (Toward Ecological Recovery and Regional Alliance. Bangkok.

World Growth Media released, December 2, 2008.

WRI (World Resources Institute). (2008). *Climate analysis indicators tool* (CAIT) *Version 6.0*. Washington, DC. http://cait.wri.org.

Yin, R., Caulfield, J. P., Aronow, M. E., & Harris, T. G., Jr. (1998). Industrial timberland: Current situation, holding rationale, and future development. *Forest Products Journal, 48*(10), 43–48.

C. Newspapers and Magazines

AA (www 1) in Advance Agro website: http://www.advanceagro.com Accessed July 8, 2000.

AA (www 2) Advance Agro Shareholders and Lenders. Website: http://www.advanceagro.com/company/shareholders.html Accessed July 8, 2000.

Barnes, W. (1996, January). International company news: Siam Pulp and paper buys 30 percent of Indonesian venture. *Financial Times*, 4.
Fernholz, K. *TIMOs and REITTs: What, Why and How*, in http://www.dovetailinc.org/report.
Gatra, 14–20 October, 2010.
Harris, T. (2007). *Industry TIMOs and REITs: The Changing Face of Forestry in the New Private Forest Landowners*, in http://www.afoa.org.
Kompas, 22 September, 2010.
Kompas, 6 June, 2010.
Paper projects kick off to cool down shortage. (2008). *Vietnam News*. 11 September.
Paper shortage leads to import tax cut. 2008. *Vietnam News*, 5 September.
Sam, D. L. (2006). *Impact of eucalyptus & acacia mangium in soil fertility in Vietnam* (published Scientific Report, Forest Science Journal).
Saigon Times, August 29, 2001.
Saigon Today, June 19, 2001.
Smith, et al. (2004). Forest Resources of the United States. U.S. Department of Agriculture. Forest Service. Centre Research Station.
Theparat, C. (2000, December). Surplus likely for 10 years. *Bangkok Post*, 18.
Tempo, 18–22 October 2010.
The Jakarta Post, August 16, 2010.
The Japan Times, October 21, 2010.
The Japan Times, August 17, 2010.
Tribun Pekanbaru, 15 December 2007. mill plans. *Financial Times*, June 20, 1997.
Umesh, P. (1999, February). *Core Pulp and Paper Unit seeks Further Acquisitions. Bangkok Post,* 16.
Uryu, et al. (2008). *Deforestation, Forest degradation, biodiversity loss and* Co_2*emissions in riau, Sumatra, Indonesia* (WWF Indonesia Technical Report). Jakarta, Indonesia. Published at: https://www.worldwildlife.org/wildplaces/borneo/updates/disappearingforest.cfm.
Woranuj, M. (1998). Six Stalled Projects top Supachai's talks in China. *Bangkok Post*, April 11.
Yakusugi Island. (2010). in Yakushima World Heritage abd Management Plan. Yakushima. Japan.

Index

H. Hidayat, *Sustainable Plantation Forestry*,
https://doi.org/10.1007/978-981-10-7653-4

MIX
Papier aus verantwortungsvollen Quellen
Paper from responsible sources
FSC® C105338

If you have any concerns about our products,
you can contact us on
ProductSafety@springernature.com

In case Publisher is established outside the EU,
the EU authorized representative is:
Springer Nature Customer Service Center GmbH
Europaplatz 3, 69115 Heidelberg, Germany

Printed by Libri Plureos GmbH
in Hamburg, Germany